PHTHALOCYANINES AND SOME CURRENT APPLICATIONS

Edited by **Yusuf Yilmaz**

Phthalocyanines and Some Current Applications
http://dx.doi.org/10.5772/65623
Edited by Yusuf Yilmaz

Contributors

Rodica-Mariana Ion, Xiaoquan Lu, Samrat Devaramani, Kazutaka Hirakawa, María Elena Sanchez, Jose Ramón Alvarez Bada, Angelina Romo Ubeda, Enrique Garibay Ochoa, Makhmut Yakubov, Dmitry Milordov, Svetlana Yakubova, Dmitry Borisov, Nickolay Mironov, Guzalia Abilova, Kirill Sinyashin, Elvira Tazeeva, Yulia Borisova, Pavel Gryaznov, Zheng-Hong Huang, Yingzhi Chen, Lu-Ning Wang, Marcela Socol, Nicoleta Preda, Anca Stanculescu, Florin Stanculescu, Gabriel Socol, Biplab Sanyal, Heike Herper, Barbara Brena, Sumanta Bhandary, Iseli Nantes, Juliana Araujo-Chaves, Jeverson Arantes, Carolina Santos, Érica Miranda, William Vallejo, Carlos Diaz, Karen Navarro, Mileidy Lerma, Catalina Fajardo, Cesar Quiñones

Published by InTech

Janeza Trdine 9, 51000 Rijeka, Croatia

Publishing Process Manager Mirena Calmic
Technical Editor SPi Global
Cover InTech Design team

First published June, 2017
Printed in Croatia
Legal deposit, Croatia: National and University Library in Zagreb

Additional hard copies can be obtained from orders@intechopen.com

Phthalocyanines and Some Current Applications, Edited by Yusuf Yilmaz
p. cm.
Print ISBN 978-953-51-3255-4
Online ISBN 978-953-51-3256-1

For EU product safety concerns:
IN TECH d.o.o., Prolaz Marije Krucifikse Kozulić 3, 51000 Rijeka, Croatia,
info@intechopen.com or visit our website at intechopen.com.

Meet the editor

Assistant Professor Yusuf Yılmaz was born in Tokat, Turkey. After completing his primary and secondary education in Tokat, he graduated from the chemistry department of Kafkas University in 2008. He started his master's degree education at Yildiz Technical University in 2008 and conducted researches on phthalocyanines during his master's degree education. He completed his master's degree in 2010 and started his PhD at Gaziantep University in the same year. In 2011, he was appointed as a research assistant to the chemistry department of Gaziantep University. In 2013, he worked as a research assistant at Rhodes University. In 2014, he received the title Doctor of Chemistry, and he was appointed as an assistant professor in Gaziantep University in the same year. Dr. Yusuf Yilmaz has worked on three nationally supported research projects until this time. He has eight international SCI papers, eight international reports, and six national reports about phthalocyanines in total.

Contents

Preface

This book contains synthesis, characterization, and current applications of some phthalocyanines. Since their initial accidental synthesis and characterization in Scotland in the late 1920s, there has been a strong research focus on the use of phthalocyanines (Pcs) as dyes and pigments. In recent years, active research fields have included their use in electrophotography, photovoltaic and solar cells, molecular electronics, Langmuir-Blodgett films, photosensitizers, electrochromic display devices, gas sensors, liquid crystals, low-dimensional conductors, and optical disks. Phthalocyanines possess interesting biological, electronic, optical, catalytic, and structural properties. The chapters in this book have been prepared by expert scientists and presented to the public.

Assistant Professor Yusuf Yilmaz
Naci Topçuoğlu Vocational School, Gaziantep University
Turkey

Chapter 1

Free-Base and Metal Complexes of 5,10,15,20-Tetrakis(N-Methyl Pyridinium L)Porphyrin: Catalytic and Therapeutic Properties

Juliana Casares Araujo Chaves,
Carolina Gregorutti dos Santos,
Érica Gislaine Aparecida de Miranda,
Jeverson Teodoro Arantes Junior and
Iseli Lourenço Nantes

Additional information is available at the end of the chapter

http://dx.doi.org/10.5772/intechopen.68225

Abstract

Porphyrins are tetrapyrrole macrocycles that can coordinate transition metal ions such as iron, cobalt and magnesium and are able to perform a diversity of functions and applications. In biological systems, these molecules are associated with proteins involved in photosynthesis, cell respiration, cell death, antioxidant defence, among others. The stability and versatile applications of porphyrins inspired the synthesis of derivatives including 5,10,15,20-tetrakis(N-methyl pyridinium-4-yl)porphyrin (TMPyP) that is the object of the present chapter. In synthetic porphyrins such as TMPyP, the catalytic and photochemical properties can be achieved by the coordination with a diversity of central metal ions. In photodynamic therapy (PDT), TMPyP and other porphyrins act as photosensitizers. The photochemical properties of TMPyP and other porphyrins are also useful for the fabrication of solar cells. The catalytic properties require the presence of a central metal. The MnTMPyP have antioxidant activity that is influenced the capacity of membrane binding, substituents, and meso substituents. Manipulation of the interfacial confinement properties is one of the newest application areas of porphyrins. The association of porphyrins with different surfaces modulates the electronic and physicochemical properties of these molecules. All of these properties are the object of experimental and theoretical studies discussed in the present chapter.

Keywords: porphyrins, TMPyP, antioxidant activity, photodynamic therapy

1. Introduction

Porphyrins constitute a group of aromatic organic molecules, composed of four pyrrole rings linked by methene (=CH–) bridges (5, 10, 15 and 20), that are the *meso*-carbon atoms/positions [1]. Free base porphyrins are able to complex with metal ions such as iron, zinc, copper and others at themacrocycle center to form metalloporphyrins. Therefore, the properties of a porphyrin can be modulated by the inserting or changing the central metal and appending different substituents at the *peripheral* (β-positions (2, 3, 7, 8, 12, 13,17 and 18)) and *meso* positions (**Figure 1**). Furthermore, the activity of a metalloporphyrin frequently involves redox cycling of the central metal. When peripheral and meso substituents are exclusively hydrogen atoms, and two of the four macrocycle nitrogen atoms are protonated, this molecule is known as a free-base porphine. When different organic groups are appended at the *peripheral* or *meso* positions, these compounds are known as porphyrins [2]. The manipulation of different substituents and central metal provides a wide diversity of biochemical functions for porphyrins.

In biological systems, the porphyrins are associated with proteins involved in important cellular processes such as photosynthesis, molecular oxygen transport, cell respiration, cell death, the combat of the oxidative stress, biological synthesis, fat acid oxidation and others [1, 3–5]. The iron protoporphyrin IX (known as heme group) is the biological metalloporphyrin present in almost all biological processes. Heme is the prosthetic group of myoglobin, hemoglobin and a diversity of enzymes such as peroxidases, cytochromes, $NO^{\bullet}$ synthase and others. Besides iron ion, other metals are found in biological porphyrins, the magnesium ion in chlorophyll, and the cobalt ion in vitamin B 12 [6]. Biological and synthetic porphyrins and metalloporphyrins have been extensively investigated and applied in medicine, chemistry, sensing and other technological devices due to their catalytic, photochemical and photophysical properties [6, 7]. In biological systems, free-base porphyrins are largely used as photosensitizer (PS) in photodynamic therapy (PDT) [2, 5, 8, 9]. Otherwise, metalloporphyrins have been used for mimicking the function of hemeproteins such as cytochrome P-450 in oxidative catalysis and superoxide dismutase SOD against oxidative stress. Porphyrins are also used as building blocks and in transport chains of molecular devices [4, 9–11].

Porphyrins are versatile catalytic and therapeutic agents. The properties of porphyrins can be modulated by changing the central metal, substituents at the *peripheral* and *meso* positions and the microenvironment. Different microenvironments respond for the diversity of functions of heme group in the hemeproteins: oxygen transport, electron transport, hydroxylation, peroxide cleavage and others. The versatility of functions can also be achieved for synthetic porphyrins by manipulating their structures and microenvironments. One example of interchangeable functions of porphyrins is the substitution of the central metal in TMPyP (5,10,15,20-tetrakis(N-methyl pyridinium L)porphyrin). MnTMPyP exhibits antioxidant function, and it has been attributed to the superoxide dismutase (SOD)-like and

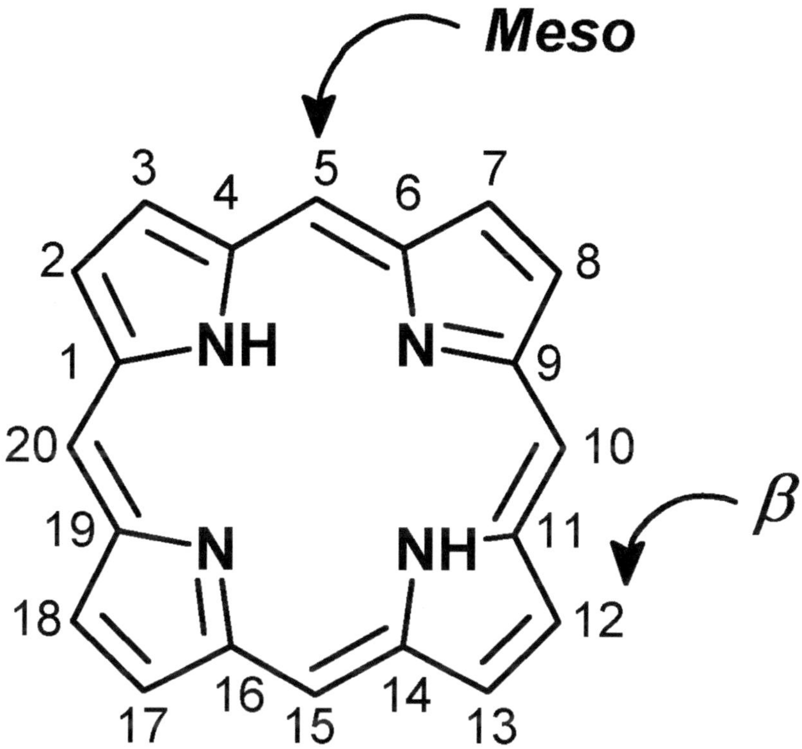

Figure 1. Free-base porphine with peripheral and meso positions.

glutathione peroxidase (GPx)-mimetic capacities [12, 13], while FeTMPyP exhibits pro-oxidant activity that responds to the toxicological effects of these compounds [14]. The pro-oxidant activity of FeTMPyP has been attributed to the generation of free radicals due to the homolytic cleavage of peroxides. The introduction and modification of substituents in a metalloporphyrin changes the redox potential and the solubility. In this regard, TMPyP and TPPS4 are examples of synthetic porphyrins made water soluble by the *meso* substitution of pyridine and sulfonate groups, respectively. Depending on the *meso* substituent, there is the possibility of a refined modulation of the porphyrin activity by isomerization. Previous studies comparing SOD activity of *ortho, meta* and *para* isomers of MnTMPyP (**Figure 2**) showed that the former exhibits the most effective SOD-like activity due to an appropriate combination of redox potential and electrostatic facilitation [15–18]. *Para* MnTMPyP exhibits a lower redox potential value that disfavors SOD activity [19]. However, the association of *para* MnTMPyP to negatively charged membranes (phosphatidylcholine (PC)/phosphatidylserine (PS)) modulates its redox potential toward a more efficient SOD activity [20]. Thus, the study of Araujo-Chaves et al. [13] is an example of the modulation of a porphyrin activity by the microenvironment. The different activities of TMPyP and other porphyrins are described herein.

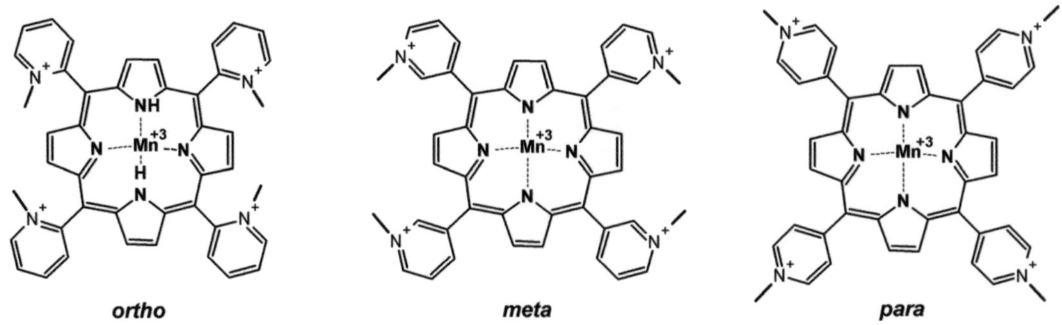

Figure 2. *Ortho, meta* and *para* isomers of MnTMPyP.

2. Biological applications of porphyrins

2.1. Porphyrins in photodynamic therapy (PDT)

2.1.1. A brief historical of PDT

The term PDT—photodynamic therapy—is recent. However, the heliotherapy—the therapeutic exposure to sunlight—was already used more than 4000 years ago by Egyptians, Greeks, and Indians as a treatment for several skin disorders, like psoriasis, vitiligo, cancer and even psychosis [21–29]. Heliotherapy, recently known as phototherapy, employs either UV and visible light with/without an exogenous photosensitizer. The photosensitizer is a molecule which when exposed to light absorbs determined wavelength becomes electronically excited and starts photochemical reactions that can produce a desirable beneficial effect, as in the case of vitamin D synthesis or damage and death, as in the case of tumor and infections treatment [2, 30]. Phototherapy without an exogenous photosensitizer is used in dermatology to treat vitiligo, eczema, neonatal jaundice and vitamin D deficiency, and even some cancer types [30–33]. During 18th and 19th centuries, phototherapy without exogenous photosensitizer was used in France in the treatment of many diseases, including tuberculosis, rheumatism, edema, rickets and paralysis [28, 34]. When an exogenous photosensitizer is used in tandem with the sunlight, this therapy is called photochemotherapy. An example of the exogenous photosensitizer is the psoralen series (**Figure 3**). These molecules are used as active treatments of HIV-associated dermatoses, seborrheic dermatitis, mycosis fungoids, prurigo, palmar and plantar pustulosis, among other diseases [30, 35]. The use of psoralens and ultraviolet light—UV (300–400 nm) was used by ancient Egyptians to treat vitiligo in the past and has been accepted for the treatment of psoriasis (PUVA) and in immunotherapy throughout the world [22, 27, 30, 35, 36].

Photodynamic therapy (PDT) is a non-invasive treatment method that uses light, photosensitizer and molecular oxygen for the treatments of cancer, inflammation, immunological

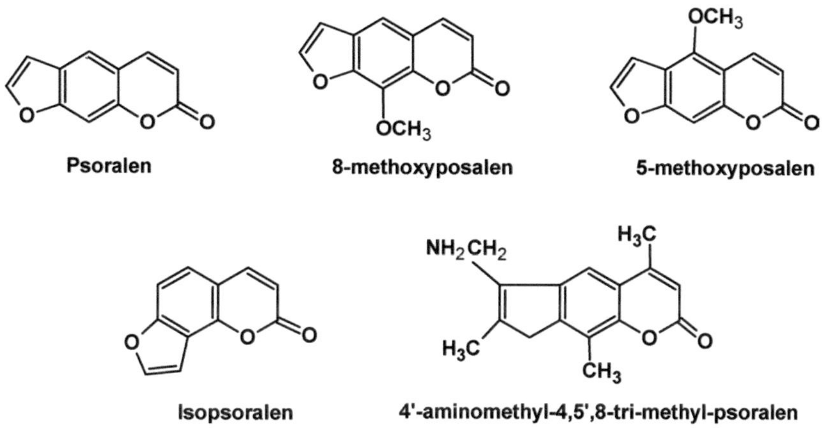

Figure 3. Psoralen series.

diseases and bacterial infections [8, 37–41]. In ancient times, phototherapy was used based on the observation of positive results without a mechanistic knowledge. People using and advocating phototherapy did know the key role of the photosensitizer in this type of treatment. In that times, the photosensitizer role was played by an endogenous biomolecule absorbing sunlight. The domain of the PDT mechanism initiated with the isolation of hematoporphyrin (Hp) (**Figure 4**) [28, 42]. From dried blood cells by Scherer in 1841 followed by the discovery of its fluorescence properties in 1871 [43]. In 1911 and 1913, the side effects of sun exposure after the administration of hematoporphyrin were described by Hausmann and Friedrich Meyer-Bertz. The latter scientist tested on himself the effect of Hp and sun and provided the first scientific communication of human photosensitization [44]. Besides, the powerful cytotoxic effect of phototherapy, another significant finding favoring the consolidation of this type of treatment, was the report of Auler and Banzer showing the affinity of Hp for cancer cells in 1942 [45]. In the following, several other studies led to the development of new range of porphyrinic photosensitizers [28, 43, 46–51].

2.1.2. The PDT mechanism

The Jablonski diagram [52], first proposed by Professor Alexander Jablonski in 1935, has been used to describe the photodynamic processes of photosensitizer molecules used in PDT. The PDT principles are based on the presence of an endogenous or exogenous photosensitizer in the target tissue that can absorb red light to be promoted to a long-lived electronic excited state. In the electronic excited state, the photosensitizer triggers photooxidative events directly or more commonly via energy transfer to molecular oxygen. The quantum yield triplet state generation depends on the molecular structure, and the energy transfer to molecular oxygen competes with other deactivating routes for the excited state [25].

According to **Figure 5**, Jablonski diagram shows that the ground state photosensitizer (S_0) can absorb a photon and be converted to the short-lived excited singlet state (S_n) at different

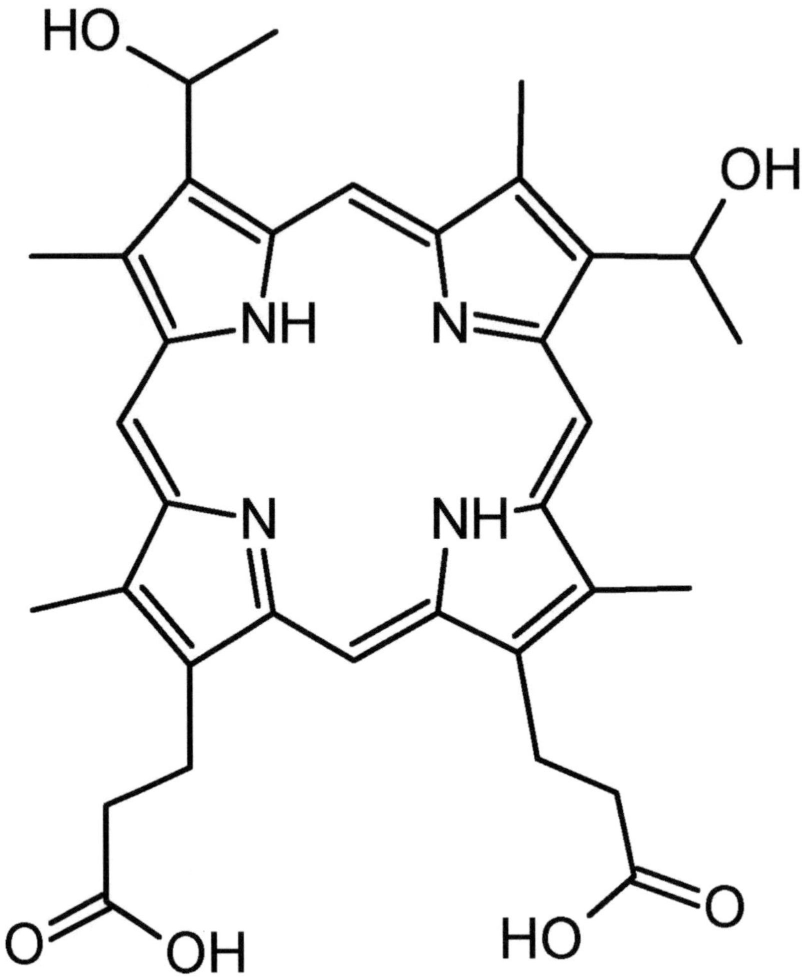

Figure 4. Hematoporphyrin.

vibrational sublevels (S_n'). The S_n state, if n > 1 can lose energy *via* internal conversion (IC) to populate the first excited single state (S_1). In the first singlet excited state, the photosensitizer can return to the ground state via fluorescence and thermal irradiation. Also, the S_1 state of the photosensitizer can undergo intersystem crossing by spin inversion and populate the lower-energy first excited triplet state (T_1), a long-lived state [2, 30, 37, 49]. At this point, two different reaction processes involving molecular oxygen can occur Type I or Type II processes. In the first process, Type I, the photosensitizer in a triplet excited state is reduced with organic substrates by electron exchange. The reduced photosensitizer can react with molecular oxygen ($^3O_2$) to produce reactive oxygen species (ROS) such superoxide anion ($O_2^{-}\cdot$), hydroxyl radical (OH·) and hydrogen peroxide (H_2O_2) [30, 37, 53]. In the second process, Type II, the triplet excited state photosensitizer transfers energy to molecular oxygen, resulting in a long-lived and highly reactive species, the singlet oxygen ($^1O_2$) [37, 49, 54]. Types I and II mechanisms occur concomitantly. However, Type II is the dominant process during PDT [30, 37].

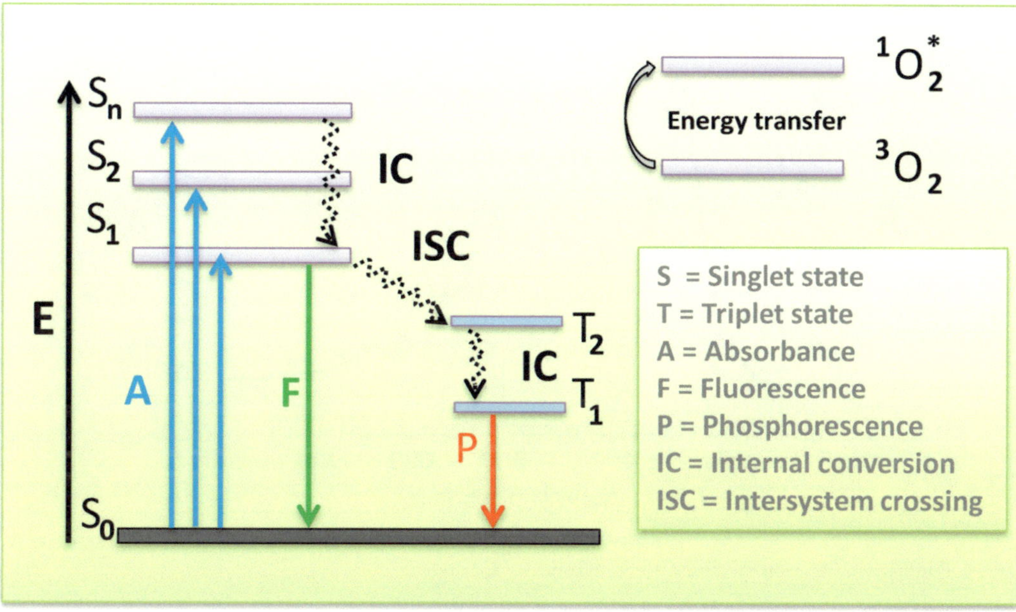

Figure 5. Energy levels of Jablonski diagram for a typical type II photosensitizer and oxygen.

In PDT, singlet oxygen is the principal reactive species. However, as well as others ROS, singlet oxygen has the capacity of damage limited due to its short lifetime (~100 ns in lipid regions of membranes and 250 ns in the cytoplasm) [30, 49, 55], and a diffusion range of approximately 45 nm in the cellular medium [28, 56–58]. The PDT has amino acid residues in proteins, unsaturated lipids, and DNA as the targets for oxidation leading to cell damage [59–61].

2.1.3. Porphyrin as photosensitizers

An ideal photosensitizer needs to have the following characteristics: (1) chemical purity; (2) high yield of singlet oxygen production; (3) high absorption coefficient in the red region of the visible spectrum (680–800 nm). Wavelengths longer than 900 nm should be avoided due to their insufficient energy to excite a dye photosensitizer to the triplet state; (4) efficient accumulation in tumor tissue associated with a rapid clearance in healthy organs; (5) low toxicity in the dark extensive to their metabolites; and (6) small aggregation [8, 30, 49, 62–64].

Porphyrins satisfy most of the desirable properties of photosensitizers, such as high efficiency of singlet oxygen generation, absorption of the higher wavelengths of the electromagnetic spectrum and a relatively higher affinity for malignant cells. Porphyrins have 18π electrons on the aromatic macrocycle that responds for the "*Soret*" band, with a strong absorption band around 400 nm, and Q bands in the 500–700 nm range that constitute the therapeutic window for this photosensitizer (**Figure 6**) [10, 65]. The absorption spectrum of the porphyrins is influenced by ligands and the central metal [66–68].

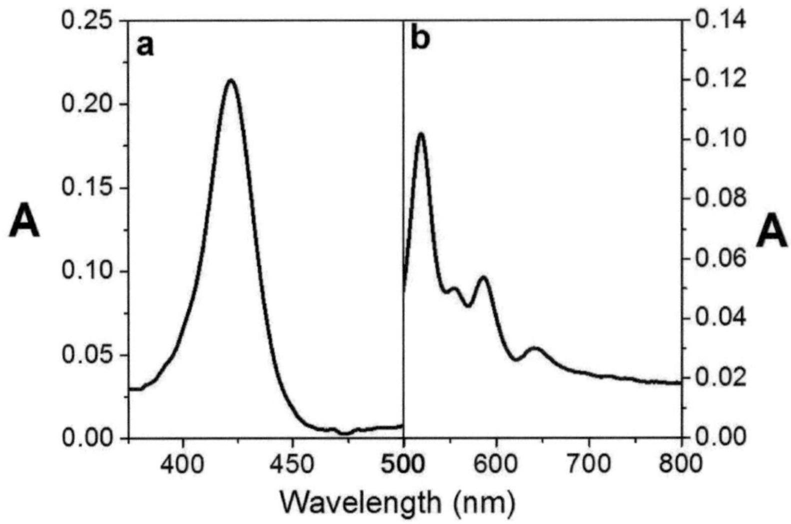

Figure 6. Porphyrin absorption spectrum. a = Soret band; b = Q band.

In the early twentieth century, data of literature described experiments that demonstrated the potential role of Hp in the detection and treatment of cancers; however, one of the major drawbacks was the large doses required to achieve consistent photosensitizer uptake in tumors, which led to inappropriate phototoxicity [45, 69–71]. In 1955, Schwartz et al. [72] demonstrated Hp to be impure and attributed selective fluorescence of malignant tissue after in vivo administration of Hp to a mixture of porphyrins with different properties. Subsequent studies led to the development of a derivative of hematoporphyrin (HpD) by the treatment of crude Hp with acetic and sulfuric acids, which enhanced tumor accumulation. The ability to accumulate selectively in neoplastic tissue using lower doses of HpD than Hp was reported by Lipson and coworkers [73–77]. In 1972, Diamond et al. demonstrated the destructive potential of HpD irradiated with white light on glioma in rats [78]. Six years later, Dougherty et al. reported the partial and complete response of many tumors, including malignant melanomas and carcinomas of the colon, breast, and prostate, treated by photodynamic therapy using HpD as a photosensitizer [79]. In the following, HpD compounds were purified, many of the less active monomers were removed, and the most efficient HpD derivatives were used to produce Photofrin (**Figure 7**).

For a complete study of different porphyrinic photosensitizers [80–109], it is recommended the reviews Josefsen et al. [2], Connor et al. [25], Pushpan et al. [28], and Ethirajan et al. [49]

Among a diversity of porphyrinic photosensitizers, *meso*-tetraphenylporphyrin (TPP) and TMPyP are readily synthesized and metallized, and several derivatives have been studied as a photosensitizer for PDT. The photochemical efficiency of anionic 5,10,15,20-*meso*-tetra(4-sulfonatophenyl)porphyrin (H_2TPPS_4) (**Figure 8A**) was compared with *meso*-tetraphenyl porphyrins with a lower number of sulfonate groups [99, 100] and

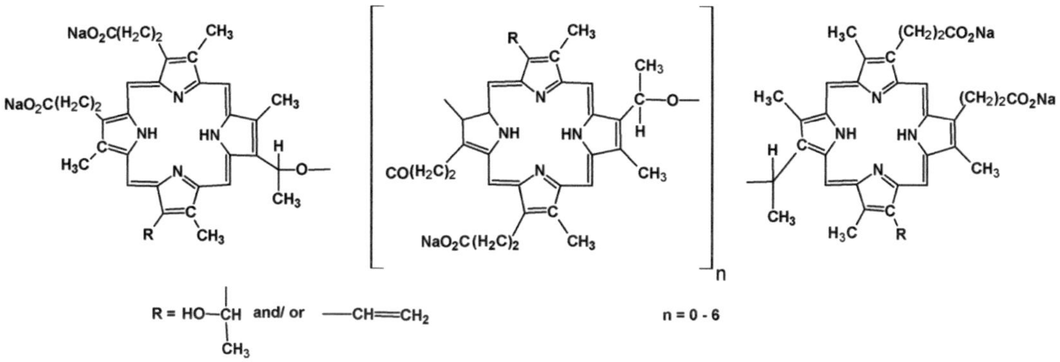

Figure 7. Photofrin.

with 5,10,15,20-tetrakis(4-sulfonatophenyl-21,23-dichalcogenaporphyrin [110] (**Figure 8B**). These studies showed that H_2TPPS_4 is less efficient in PDT than *meso*-tetraphenyl porphyrins with a lower number of sulfonate groups. Also, the replacement of nitrogen atoms of the macrocycle by chalcogens S and Se increased the photodynamic efficiency of the porphyrin in vitro and in vivo studies. Particularly in vivo, these chalcogen derivatives exhibited lower toxicity, morbidity and side effects post administration in animal models.

Regarding TMPyP, the focus of the present study, its efficiency as a photosensitizer is related to its topology. A study comparing photodamage in a mitochondrial membrane model modulated by the topology of TPPS4 and 5,10,15,20-tetrakis(N-methyl pyridinium L)porphyrin (TMPyP) [8] shows that in L-α-phosphatidylcholine/cardiolipin (PC/CL)liposomes (mitochondrial membrane model) both porphyrin can damage the membrane *via* the Type II mechanism. However, the injuries on the lipid membranes promoted by TMPyP were greater than the damages promoted by TPPS4 due to the affinity between TMPyP and this biological

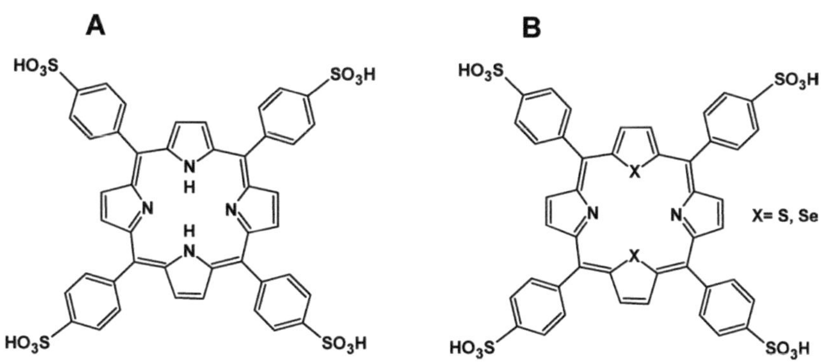

Figure 8. TPP-based photosensitizers. (A) Tetrasulfonated *meso*-tetraphenyl porphyrin; (B) *meso*-tetrakis(4-sulfona tophenyl)-21,23-dichalcogenaporphyrin.

structure [111, 112] that in turn influences the photosensitizer and the generation of long-lived singlet oxygen. In cells, the positively charged TMPyP accumulates in the nucleus and mitochondria and could attack DNA, mitochondrial DNA and cardiolipin. The association of TMPyP with the inner mitochondrial membranes due to the affinity to cardiolipin favors the generation of singlet oxygen in situ with a high efficiency since its concentration is higher in the hydrophobic core of the lipid bilayers. Metalloporphyrins have also been studied as potential sensitizers for PDT. However, the results were less promising than those obtained with the free-base species [113, 114].

2.2. Porphyrins in chemical therapy

The synthetic analogs of porphyrins are widely used in therapy of diseases connected to oxidative stress processes. A quantitative structure-activity relationship (QSAR) studies have been performed to identify the optimal active molecule within a series of analog structure characteristics to diversify the biological action of the compound. The QSAR studies can correlate the physicochemical characteristics that affect the compound's activity in biological systems. These studies assumed that the binding affinity of the compound to the target receptor could determinate the biological activity [115]. The biological effects of two meso-tetrakis porphyrins, TPPS4 (anionic) and TMPyP (cationic) demonstrated that the cationic porphyrin has affinity to the inner mitochondrial membrane [99]. Therefore, in mitochondria, Mn^{3+}TMPyP has been used as an antioxidant against superoxide ions. The replacement of manganese by an iron ion in TMPyP makes this porphyrin a prooxidant agent [116]. Au-porphyrins have been reported as excellent antiproliferative agents, showing cytotoxic effects on cancer cells. Regarding to the mimetic SOD activity of porphyrins, the correlation between the metal-centered reduction potential and the catalytic rate constant for the $O_2^{\bullet-}$ dismutation was found for Fe and Mn porphyrins. The structure-activity relationships have been established over the years by the rate-limiting step of metal reduction of this class of compounds [117]. Modulation of SOD activity has been achieved by decreasing the electron density of the groups at the meso and β-pyrrile positions, thus increasing the Mn^{3+}/Mn^{2+} potential and facilitating its reduction [118–120]. Either the mimetic SOD activity can occur when the $O_2^{\cdot-}$ is directed to the catalytic site by the metal-centered positive charges via electrostatic facilitation [118, 119]. The manganese (III) 5,10,15,20-tetrakis(N-ethylpyridinium-2-yl) porphyrin ($Mn^{3+}TE\text{-}2\text{-}PyP^{5+}$, $E_{½}$ = +228 mV *vs* NHE, log $_{kcat}$ = 7.76) and manganese (III) 5,10,15,20-tetrakis(N-n-hexylpyridinium-2-yl) porphyrin ($MnTnHex\text{-}2\text{-}PyP^{5+}$, $E_{½}$ = +314 mV *vs* NHE, log $_{kcat}$ = 7.48), alkylated manganese (III) 5,10,15,20-tetrakis(2-pyridyl)porphyrin ($MnT\text{-}2\text{-}Pyp^{+}$), combined the thermodynamic and electrostatic optimizations and yielded compounds because they exhibit the $E_{½}$ close to the reduction potential of the SOD enzyme and are excellent mimetics of the SOD activity ($E_{½} \cong$ +300 mV *vs* NHE, kcat $\cong 2 \times 10^9$ M^{-1} s^{-1}) [19, 117–122]. Recently, it has been reported that the para isomer ($E_{½}$ = +60 mV *vs* NHE) of Mn^{3+}TMPyP is less efficient as a SOD mimic relative than the *ortho* isomer ($E_{½}$ = +260 mV *vs* NHE) [12, 19, 123, 124].

In a cell redox balance, the association of Mn^{3+}TMPyP to membrane lipid bilayers can be intrinsically related to the redox potential of the Mn^{2+}/Mn^{3+} couple. In homogeneous systems, Batinić-Haberle et al. [19] had reported the effect of Mn^{3+}TMPyP in a CL-containing inner mitochondrial membrane under pH 11 to 7.8 conditions. The potential values of Mn^{2+}/Mn^{3+} redox process were found to be $E_{1/2}$ =94 mV for *ortho* Mn^{3+}TMPyP and $E_{1/2}$ = 42 and 50 mV, respectively, for *meta* and *para* isomers. However, in a heterogeneous system, Araujo-Chaves et al. [20] have reported that the *para* isomer has the redox potential increased by the association with the negatively charged interface of lipid bilayers. Interestingly, the association of *para* Mn^{3+}TMPyP to PC/PS liposomes at physiological pH exhibited a redox potential of +110 mV *vs* NHE. The shift of the Mn^{2+}/Mn^{3+} $E_{1/2}$ value to a more positive value favors the SOD and peroxidase activities. Theoretical calculations corroborated with these results.

3. Technological applications

Porphyrins free base are extensively applied in solar cells and sensor due to their photophysical characteristics. The intense absorption bands covering a significant range of the visible region of the electromagnetic spectrum and due to the relatively low cost of these compounds as compared with inorganic semiconductors make these molecules appropriate for application in solar cells. These characteristics experimentally observed are consistent with the results obtained by density functional theory (DFT). Therefore, DFT/time-dependent (TD)DFT calculation is a useful strategy for the molecular design of porphyrins with the more appropriate characteristics for application is dye-sensitizer solar cells (DSSCs) [125–129]. As an example, Santhanamoorthi et al. [129] have presented the theoretical study of newly designed porphyrin dyes (1–5) for DSSC applications. In this study, the authors calculated seven different structures of porphyrins and found the best characteristics for use in solar cells for two calculated molecules that were named Dyes 2 and 4. Dyes 2 and 4 presented smaller HOMO-LUMO energy gaps and absorption in Q band significantly stronger. Equally, DFT/TDDFT can be used for conceiving porphyrin derivatives for a diversity of technological applications. Theoretical calculations allow the prediction of the best characteristics for porphyrins to be used in technological applications and optimize the subsequent efforts for the synthesis.

3.1. Porphyrins in solar cells

Solar energy is an important source of energy (~3×10^{24} J $year^{-1}$) that sustains the life on the Earth [130–132], and it can be an alternative to using fossil fuels due to be a clean, inexhaustible and sustainable source of energy [133–139]. The utilization of solar energy as solar fuel or electricity is fundamental for the maintenance of development and live on Earth and has attracted the attention of various members of the scientific community.

O'Regan and Grätzel [140] have discussed dye-sensitized solar cells (DSSC), a viable and promising technology which have low-cost production and high power conversion efficiency [141–148]. To build an efficient system of the solar cell is necessary [149–152] three components: (1) dye (light-absorber); (2) a hole transport agent; and (3) an electron-transport agent. **Figure 9** shows the schematic representation of components and representative operational principles of DSSC.

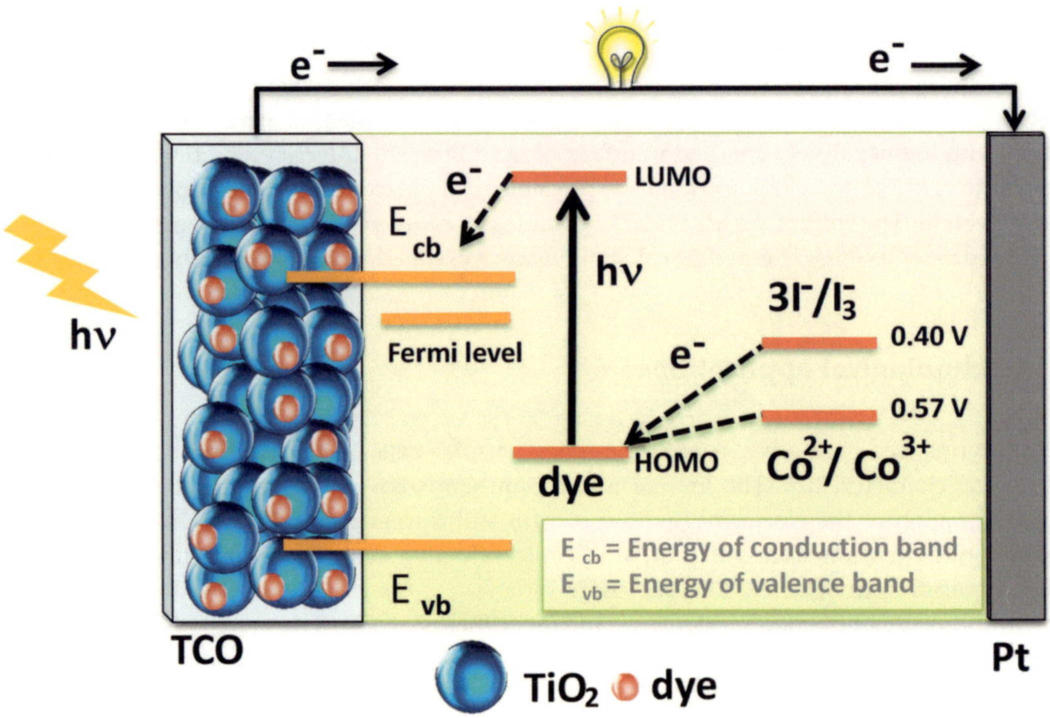

Figure 9. Schematic representation of components and representative operational principles of DSSC.

A typical DSSC device consists of a dye-sensitizer photoanode (TiO_2, anode) and a platinum counter electrode (Pt-coated, cathode) sandwiching an electrolyte that contains a redox mediator (iodine-based or cobalt complexes, redox mediator). Upon light illumination, the photoexcited dye in the LUMO level of sensitizer injects an electron into a conduction band (CB) of TiO_2, and then, the resultant oxidized dye is reduced by I^- species (or Co^{2+} complex). The injected electrons move through an external circuit to the platinized counter electrode. Finally, the I^- species (or Co^{2+} complex) is regenerated to produce the I_3^- species (or Co^{3+} complex) at the surface of the platinized counter electrode, and the circuit is completed [133]. The efficiency of conversion of light to electric power (η) increases when a light-absorbing the dye, and therefore, the choice of a suitable dye is essential to a high η [127, 144, 153–156].

Despite to the intense absorption band, typical porphyrins have poor light-harvesting ability in the Q bands, being necessary the introduction of a push-pull structure [157–160] and the elongation of porphyrin π-conjugated system into *meso* or β-positions to improve the light-harvesting property of porphyrins [158].

Porphyrin also can be used as a dye in thin layers on the porous TiO_2 film. However, this system results in weak absorption of irradiated light, being essential the development of a way to strongly absorb the light in the dye layer. Gold layer can have been used in these systems due to its surface plasmon resonance (SPR) that offers an enhanced optical field with increased short-circuit current, which can be corroborated by theoretical calculations [161].

3.2. Porphyrins in catalysis and sensing

The application of metalloporphyrins in bioinorganic chemistry has attracted interest in catalytic reactions. Synthetic metalloporphyrins are mimetic models inspired two heme proteins: cytochrome P450 (biosynthesis and degradation of biomolecules) and peroxidases as lignin peroxidases (degrades the lignin-cell wall). In 1970, Groves et al. [162] designed the first-generation of metalloporphyrin chlorine (5,10,15,10-tetraphenyl-porphyrinato)iron(III), or [Fe^{3+}TPPCl], activated by iodosylbenzene (PhIO) revealed a catalytic activity in the epoxidation of alkenes and the hydroxylation of alkanes. About 30 years ago, Traylor and Tsuchyia [163] presented the first synthesis of porphyrins with more stability and more efficient catalytic activity due to the introduction of electronegativity and/or bulky auxiliaries groups such as halogen, nitro or sulfonate at the *meso* and/or β-pyrrolic positions, to obtain the second and third generation of porphyrin catalysts. Lately, the metal complexes like *meso*-tetrakis(penta fluorophenyl)porphyrin H_2(TPFPP) represent alternative possibilities to structural modification of porphyrins by nucleophilic substitution of its fluorine atoms [164, 165]. The second generation of porphyrins, especially the manganese (II) and iron(III) porphyrins is the most important representatives as catalysts in the epoxidation of alkenes (cyclohexane, adamantane, or n-hexane). In this case, during the epoxidation reactions, Mn and Fe ions can accept active species from different substrates and oxygen atom donors that result in metal-oxo species formation. In some conditions, the catalytic efficiency of iron(III) porphyrins can be limited due to the presence of some by-products resulted from the epoxidation of alkenes, for example to the allylic oxidation reactions. In anadamant oxidation reaction, the catalytic reaction of manganese porphyrins (MnPor) derived from 5,10,15,20-tetrakis(4-methoxyphenyl)porphyrin had an increased product yield of 1-adamantanol than those obtained with [Fe^{3+}TPPCl] catalyst [166]. The MnPors can exhibit different behaviors regarding the electron-withdrawing substituents in the macrocycle structure. Doro et al. [167] revealed that MnPors had lower catalysis efficient than the second generation of catalyst [Mn^{3+}PFTDCPP]Cl due the high-valence active species caused by the electronegativity of the substituents (fluoro and chlorine) at the *meso*-aryl positions of the macrocycle in [Mn^{3+}PFTDCPP]Cl. Consistently with this observation, Rayati et al. [168] made a comparative catalytic study of two partially brominated MnPs, namely [$Mn^{3+}Br_4$TPP]Cl and [$Mn^{3+}Br_4T_4$(-OME)PP]Cl revealing that the electron-deficient Mnps were a better catalyst than electron-rich MnPs. Lately, new materials of metalloporphyrin catalysts supported on mesoporous silica have shown a high efficiency of stability and reaction conditions. Poltowicz et al. [169] have studied the supported MnTMPyP catalysts on aluminated MCM-41 and SBA-15 mesoporous to investigate the oxidation of cyclooctane with molecular oxygen (as air) without the use of sacrificial co-reductant. Due to the existence diffusion limitations within the pore inner space, the supported MnTMPyP had increased the catalysis activity in the SBA-15 mesoporous because it exhibits increased-size pore. The catalytic activity of porphyrins, including TMPyP, allows the use of these compounds in sensing. Porphyrins can form complexes with almost all metals, and consequently, a broad diversity of catalytic properties can be achieved. The central metal in porphyrins determines the affinity for additional ligands. In general, the complex of Cu^{2+} and Ni^{2+} has low affinity for additional ligands. The Mg^{3+}, Cd^{2+} and Zn^{2+} porphyrins form pentacoordinate complexes with square-pyramidal structure. The metalloporphyrins with (Fe^{2+}, Co^{2+}, Mn^{2+}) in the central position produce distorted octahedral structure with two axial ligands. Metallo *meso* tetrakis porphyrins have been extensively used in the voltammetric determination of oxygen, NO, sugars, organohalides, DNA, alcohols, dopamine and others. Therefore, due to their switchable structures and a diversity of catalytic properties, porphyrins are widely used in analytical chemistry. A diversity of porphyrins can be applied biosensors and as stationary phases in HPLC.

Acknowledgements

We are grateful to FAPESP (2015/17688-0; 2016/15465-7) for the financial support.

Author details

Juliana Casares Araujo Chaves[1], Carolina Gregorutti dos Santos[1], Érica Gislaine Aparecida de Miranda[1], Jeverson Teodoro Arantes Junior[2] and Iseli Lourenço Nantes[1]*

*Address all correspondence to: ilnantes@gmail.com

1 Centro de Ciências Naturais e Humanas (CCNH), Universidade Federal do ABC (UFABC), Brazil

2 Centro de Engenharia, Modelagem e Ciências Sociais Aplicadas (CECS), Universidade Federal do ABC (UFABC), Brazil

References

[1] Battesby AR, Fookes CJR, Matcham GW, McDonald E. Biosynthesis of the pigments of life: formation of the macrocycle. Nature. 1980;**285**:4011–4022. DOI: 10.1038/285017a0

[2] Josefsen LB, Boyle RW. Unique diagnostic and therapeutic roles of porphyrins and phthalocyanines in photodynamic therapy, imaging and theranostics. Theranostics. 2012;**2**:916–966. DOI: 10.7150/thno.4571

[3] Mauzerall D. Porphyrins, Chlorophyll, and Photosynthesis. Vol. 5. Berlin, Heidelberg: Springer; 1977. pp. 117–124. DOI: 10.1007/978-3-642–66505-9_5

[4] Armstrong D. and Stratton R. D. Oxidative Stress and Antioxidant Protection: The Science of Free Radical Biology and Disease. Hoboken, New Jersey: John Wiley & Sons. 2016;**27**:415–470. DOI: 10.1002/9781118832431.

[5] Nantes IL, Crespilho FN, Mugnol KCU, Araujo-Chaves JC, Luz RAS, Nascimento OR, Pinto SMS. Magnetic Circular Dichroism Applied in the Study of Symmetry and Functional Properties of Porphyrinoids. David S. Rodgers. (Org.). Circular Dichroism: Theory and Spectroscopy. 1st ed. New York; Nova Science Publishers. 2010. pp. 321–344

[6] Van Santen R. Catalysis in perspective: Historic review. Beller M, Renken A, Van Santen R, editors. In Catalysis: From Principles to Applications. 1st ed. Weinheim, Germany; Wiley-VCH: 2012. pp. 3–19

[7] Latter MJ, Langford SJ. Porphyrinic molecular devices: Towards nanoscaled processes. International Journal of Molecular Sciences. 2010;**11**:1878–1887. DOI: 10.3390/ijms11041878.

[8] Kawai C, Araújo-Chaves JC, Magrini T, Sanches COCC, Pinto SMS, Martinho H, Daghastanli N, Nantes IL. Photodamage in a mitochondrial membrane model modulated by the topology of cationic and anionic meso-tetrakis porphyrin free bases. Photochemistry and Photobiology. 2014;**90**:596–608. DOI: 10.1111/php.12228

[9] Carvalho CMB, Brocksom TJ, Oliveira KT. Tetrabenzoporphyrins: Synthetic developments and applications. Chemical Society Reviews. 2013;**42**:3302–3317. DOI: 10.1039/C3CS35500D

[10] Robertson CA, Evans DH, Abrahamse H. Photodynamic therapy (PDT): A short review on cellular mechanisms and cancer research applications for PDT. Journal of Photochemistry and Photobiology B. 2009;**96**:1–8. DOI: 10.1016/j.jphotobiol.2009.04.001

[11] Bonnett R. Photosensitizers of the porphyrin and phthalocyanine series for photodynamic theory. Chemical Society Reviews. 1995;**24**:19–33. DOI: 10.1039/CS9952400019

[12] Ferrer-Sueta G, Vitturi D, Batinić -Harbele I, Fridovish I, Goldstein S, Czapski G, Radi R. Reactions of manganese porphyrins with peroxynitrite and carbonate radical anion. The Journal of Biological Chemistry. 2003;**278**:27432–27438. DOI: 10.1074/jbc.M213302200

[13] Araujo-Chaves JC, Yokomizo CH, Kawai C, Mugnol KCU, Prieto T, Nascimento OR, Nantes IL. Towards the mechanisms involved in the antioxidant action of MnIII [meso-tetrakis(4-N-methyl pyridinium) porphyrin] in mitochondria. Journal of Bioenergetics and Biomembranes. 2011;**43**:663–671. DOI: 10.1007/s10863-011-9382-3

[14] Pessoto FS, Inada NM, Nepomuceno MF, Ruggiero AC, Nascimento OR, Vercesi AE, Nantes IL. Biological effects of anionic meso-tetrakis (para-sulfonatophenyl) porphyrins modulated by the metal center. Studies in rat liver mitochondria. Chemico-biological Interactions. 2009;**181**:400–408. DOI: 10.1016/j.cbi.2009.07.012

[15] Richards RA, Hammos K, Joe M, Miskelly GM. Observation of a stable water-soluble lithium porphyrin. Inorganic Chemistry. 1996;**35**:1940–1944. DOI: 10.1021/ic941434w

[16] Tabata T, Andreo P, Shinoda K. An analytic formula for the extrapolated range of electrons in condensed materials. Nuclear Instruments and Methods in Physics Research Section B. 1996;**119**:461–470. DOI: 10.1016/S0168-583X(96)00687-8

[17] Tabata M, Nishimoto J, Ogata A, Kusano T, Nahar N. Metalation of water-soluble octabromoporphyrin with lithium(i), cadmium(II), and mercury(II). Bulletin of the Chemical Society of Japan. 1996;**96**(3):673–677. DOI: 10.1246/bcsj.69.673

[18] Batinić -Haberle I, Liochev SI, Spasojevic I, Fridovich I. A Potent superoxide dismutase mimis: manganese β-octabromo-meso-tetrakis-(N-methylpyridinium-4-yl)-porphyrin. Archives of Biochemistry and Biophysics. 1997;**343**:225–233. DOI: 10.1006/abbi.1997.0157

[19] Batinić -Haberle I, Benov L, Spasojevic I, Fridovich I. The ortho effect makes manganese (III) meso-tetrakis (N-methylpyridinium-2-yl) porphyrin a powerful and potentially useful superoxide dismutase mimic. The Journal of Biological Chemistry. 1998; **273**:24521–24528. DOI: 10.1074/jbc.273.38.24521

[20] Araujo-Chaves J, Kawai C, Melo AFAA, Mugnol KCU, Nascimento OR, Arantes JT, Crespilho FN, Nantes IL. Interaction and reaction of the antioxidant MnIII [meso-tetrakis(4-Nmethyi pyridinium) porphyrin] with the apoptosis reporter lipid phosphatidylserine. Current Physical Chemistry. 2013;**3**(2):187–198. DOI: 10.2174/1877946811303020009

[21] Edelson MF. Light-activated drugs. Scientific American. 1988;**259**:68–75

[22] Sternberg ED, Dolphin D. Porphyrin-based photosensitizers for use in photodynamic therapy. Tetrahedron. 1998;**54**:4151–4202. DOI: 10.1016/S0040–4020(98)00015-5

[23] Nayak CS. Photodynamic therapy in dermatology. Indian Journal of Dermatology, Venereology and Leprology. 2005;**71**:155–160. DOI: 10.4103/0378-6323.16228

[24] Pervaiz S, Olivo M. Art and science of photodynamic therapy. Clinical and Experimental Pharmacology and Physiology. 2006;**33**:551–556. DOI: 10.1111/j.1440-1681.2006.04406.x

[25] O'Connor AE, Gallagher WM, Byrne AT. Porphyrin and Nonporphyrin photosensitizers in oncology: preclinical and clinical advances in photodynamic therapy. Photochemistry and Photobiology. 2009;**85**:1053–1074. DOI: 10.1111/j.1751-1097.2009.00585.x

[26] Spikes JD. The historical development of ideas on applications of photosensitized reactions in health sciences. In: Bergasson RV, JoriG, Land EJ, Truscott TG, editors. In Primary Photo Processes in Biology and Medicine. New York; Plenum Press: 1985. pp. 209–227

[27] Epstein JM. Phototherapy and photochemotherapy. The New England Journal of Medicine. 1990;**32**:1149–1151. DOI: 10.1056/NEJM199004193221609

[28] Pushpan SK, Venkatraman S, Anand VG, Sankar J, Parmeswaran D, Ganesan S, Chandrashekar TK. Porphyrins in photodynamic therapy—a search for ideal photosensitizers. Current Medicinal Chemistry—Anti-Cancer Agents. 2002;**2**:187–207. DOI: 10.2174/1568011023354137

[29] Roelandts R. The history of phototherapy: Something new under the sun?. Journal of the American Academy of Dermatology. 2002;**46**:926–930. DOI: 10.1067/mjd.2002.121354

[30] Ormond AB, Freeman HS. Dye sensitizers for photodynamic therapy. Material. 2013;**6**:817–840. DOI: 10.3390/ma6030817.817-840

[31] Bonnett R. Photosensitizers of the porphyrin and phthalocyanine series for photodynamic therapy. Chemical Society Reviews. 1995;**24**:19–33. DOI: 10.1039/CS9952400019

[32] Gambichler T, Breuckmann FBS, Altmeyer P, Kreuter A. Narrowband UVB phototherapy in skin conditions beyond psoriasis. Journal of the American Academy of Dermatology. 2005;**52**:660–670. DOI: 10.1016/j.jaad.2004.08.047

[33] Paus S, Schmidt-Hubsch T, Wullner U, Vogel A, Klockgether T, Abele M. Bright light therapy in Parkinson's disease: A pilot study. Movement Disorders. 2007;**22**:1495–1498. DOI: 10.1002/mds.21542

[34] Cauvin JF. Des bienfaits de I'insolation. (FThesis) Paris. 1815

[35] Ledo E, Ledo A. Phototherapy, photochemotherapy, and photodynamic therapy: Unapproved uses or indications. Clinics in Dermatology. 2000;**18**:77–86

[36] Schmitt IM, Chimenti S, Gasparro FPJ. Psoralen-protein photochemistry—a forgotten field. Photochemistry and Photobiology B. 1995;**27**:101–107. DOI: 10.1016/1011-1 344(94)07101-S

[37] Swavey S. and Tran M. Porphyrin and Phthalocyanine Photosensitizers as PDT Agents: A New Modality for the Treatment of Melanoma. Recent Advances in the Biology, Therapy and Management of Melanoma. Croatia InTech (Dr. Lester Davids (Ed.)). 2013:Cap.11:253–283. DOI: 10.5772/54940.

[38] Vzorov AN, Dixon DW, Trommel JS, Marzilli LG, Compans RW. Inactivation of human immunodefiency virus type i by porphyrins. Antimicrobial Agents and Chemotherapy. 2002;**46**:3917–3925. DOI: 10.1128/AAC.46.12.3917-3925.2002

[39] Mcmahon KS, Wieman TJ, Moore PH, Fingar VH. Effects of photodynamic therapy using mono-L-aspartyl chlorin e6 on vessel constriction, vessel leakage, and tumor response. Cancer Research. 1994;**15**:5374–5379

[40] Neurath RA, Strick N, Debinath AK. Structural requirements for and consequences of an antiviral porphyrin binding to the V3 loop of the human immunodeficiency virus (HIV- 1) envelope glycoprotein gp120. Journal of Molecular Recognition. 1995;**8**:345–357. DOI: 10.1002/jmr.300080604

[41] Bamfield P. Chromic Phenomena: Technological Applications of Colour Chemistry. Cambridge, UK: The Royal Society of Chemistry; 2001. DOI: 10.1039/9781849731034

[42] Scherer H. Chemical-physiological investigations. Annalen der Chemie und Pharmacie 1841;**40**:1–64

[43] Hoppe- Seyler F. Ueber die Chemische Zusammensetzung des Eiters. Medicinisch-Chemische Untersuchungen. 1871;**4**:486–501

[44] Meyer- Bertz F. Untersuchungen uber due bioloische (photodynamische) wirkung des hamatoporphyrins und anderer derivative des blut-und gallenfarbstoffs. Deutsches Archiv Für Klinische Medizi. 1913;**112**:476–503

[45] Auler H, Banzer G. Untersucheungen uber die rolle der porphyrine bei geschwulstkranken menschen und tieren. Zeitschrift für Krebsforschung und klinische Onkologie. 1942;**53**:65–68

[46] Maganiello LOJ, Figge FH. Cancer detection and therapy II. Journal of Bulletin of the School of Medicine, University of Maryland. 1951;**36**:3–7

[47] Thudichum JL. Tenth Report of the Medical Officer of the Privy Council. London: H. M. Stationary Office; 1867

[48] Rabb O. Ueber die Wirkung fluorizierender Stoffe auf Infusorien. Zeitschrift Fur Biologie. 1900;**36**:524–546

[49] Ethirajan M, Chen y, Joshi P, Pandey RK. The role of porphyrin chemistry in tumor imaging and photodynamic therapy. The role of porphyrin chemistry in tumor imaging and photodynamic therapy. Chemical Society Reviews. 2011;**40**:340–362. DOI: 10.1039/B915149B

[50] von Tappeiner H, Jesionek H. Therapeutische versuche mit fluoreszierenden stoffen. Munch Med. Wochenschr. 1093;**47**:2042–2044

[51] von Tappeiner HA, Jodlbauer A. Die sensibilisierende Wirkung fluorescierender Substanzen. Gesammelte Untersuchungen über die photodynamische Erscheinung., Leipzig, Germany: F. C. W. Vogel; 1907

[52] Jabłoński A. Über den Mechanismus der Photolumineszenz von Farbstoffphosphoren. Zeitschrift für Physik A Hadrons and Nuclei. 1935;**94**:38–46. DOI: 10.1007/BF01330795

[53] Zimcik P, Miletin M. Photodynamic Therapy. In: Lang AR, editor. In Dyes and Pigments: New Research. New York, NY, USA: Nova Science Publishers; 2008. pp. 1–62

[54] Turro NJ. Modern Molecular Photochemistry. California: University Science Books; 1991. pp. 583–593

[55] Figge FHJ. The relationship of pyrrol compounds to carcinogenesis. In: Moulton FR, editor. In AAAS Research Conference on Cancer. 1945; pp. 117–128. DOI: 10.7326%2F0003-4819-27-1-143

[56] Ochsner M. Photophysical and photobiological processes in the photodynamic therapy of tumours. Journal of Photochemistry and Photobiology B: Biology. 1997;**39**:1–18. DOI: 10.1016/S1011-1344(96)07428-3

[57] Rosenthal L, Ben-Hur E. Role of oxygen in the phototoxicity of phthalocyanines. International Journal of Radiation Biology. 1995;**67**(1):85–91. DOI: 10.1080/09553009514550111

[58] Moan JJ. On the diffusion length of singlet oxygen in cells and tissues. Photochemistry and Photobiology B. 1990;**6**:343–347. DOI: 10.1016/1011-1344(90)85104-5

[59] Zamzami N, Susin SA, Marchetti P, Hirsch T, Gomez-Monterrey I, Castedo M, Kroemer G. Mitochondrial control of nuclear apoptosis. The Journal of Experimental Medicine. 1996;**183**(4):1533–1544. DOI: 10.1084/jem.183.4.1533

[60] Gomer CJ, Luna M, Ferraio A, Wong S, Fischer AMR. Photodynamic therapy-mediated oxidative stress can induce expression of heat shock proteins. Cancer Research. 1996;**56**:2355–2360

[61] Hamblin M, Newman ELJ. Photosensitizer targeting in photodynamic therapy II. Conjugates of haematoporphyrin with serum lipoproteins. Journal of Photochemistry and Photobiology B. 1994;**26**:147–157. DOI: 10.1016/1011-1344(94)07036-9

[62] Ravanat JL, Di Mascio P, Martinez GR, Medeiros MHG, Cadet J. Singlet oxygen induces oxidation of cellular DNA. The Journal of Biological Chemistry. 2000;**275**:40601–40604. DOI: 10.1074/jbc.M006681200

[63] Castano AP, Demidova TN, Hamblin MR. Mechanisms in photodynamic therapy: part one-photosensitizers, photochemistry and cellular localization. Photodiagnosis and Photodynamic Therapy. 2004;**1**:279–293. DOI: 10.1016/S1572-1000(05)00007-4

[64] Dolmans DE, Fukumura D, Jain RK. Photodynamic therapy for cancer. Nature Reviews Cancer. 2003;**3**:380–387. DOI: 10.1038/nrc1071

[65] Carvalho CBM, Brocksom TJ, Oliveira KT. Tetrabenzoporphyrins: Synthetic developments and applications. Chemical Society Reviews. 2013;**42**:3302–3317. DOI: 10.1039/c3cs35500d

[66] Oleinick NL, Evans HH. The photobiology of photodynamic therapy: Cellular targets and mechanisms. Radiation Research. 1998;**150**:S146–S156. DOI: 10.2307/3579816

[67] Ben-Dror S, Bronshtein I, Wiehe A, Röder B, Senge MO, Erenberg B. On the correlation between hydrophobicity, liposome binding and cellular uptake of porphyrin sensitizers. Photochemistry and Photobiology. 2006;**82**:695–701. DOI: 10.1562/2005-09-01-RA-669

[68] Berg K, Bommer JC, Moan J. Evaluation of sulfonated aluminum phthalocyanines for use in photochemotherapy—cellular uptake studies. Cancer Letters. 1989;**44**:7–15. DOI: 10.1016/0304-835(89)90101-8

[69] Figge FH, Weiland GS, Manganiello LO. Cancer detection and therapy; affinity of neoplastic, embryonic, and traumatized tissues for porphyrins and metalloporphyrins. Proceedings of the Society for Experimental Biology and Medicine. 1948;**68**(3):640–641. DOI: 10.3181/00379727-68-16580

[70] Peck GC, Mack HP, Holbrook WA. Use of hematoporphyrin fluorescence in biliary and cancer surgery. Annals of Surgery. 1955;**21**:181–188

[71] Rassmussan-Taxdal DS, Ward GE, Figge FH. Fluorescence of human lymphatic and cancer tissues following high doses of intravenous hematoporphyrin. Fluorescence of human lymphatic and cancer tissues following high doses of intravenous hematoporphyrin. Journal of Cancer. 1955;**5**:619–624. DOI: 10.1002/1097-0142(1955)8:1<78::AID-CNCR2820080109>3.0.CO;2-L

[72] Schwartz SK, Absolon K, Vermund H. Some relationships of porphyrins, X-rays and tumours. University of Minnesota Medical Bulletin. 1955;**27**:7–8

[73] Lipson RL, Baldes EJ. The photodynamic properties of a particular hematoporphyrin derivative. Archives of Dermatology. 1960;**82**:208–516. DOI: 10.1001/archderm.1960.01580040026005

[74] Lipson RL, Baldes EJ, Gray MJ. Hematoporphyrin derivative for detection and management of cancer. Cancer. 1967;**20**:2255–2257. DOI: 10.1002/1097-0142(196712)20:12<2255::AID-CNCR2820201229>3.0.CO;2-U

[75] Lipson RL, Baldes EJ, Olsen AM. The use of a derivative of hematoporhyrin in tumor detection. Journal of the National Cancer Institute. 1961;**26**:1–11

[76] Lipson RL, Gray MJ, Blades E. Haematoporphyrin derivative for detection and management of cancer. Cancer. 1966;**20**:2255–2257. DOI: 10.1002/1097-0142(196712)20:12<2255::AID-CNCR2820201229>3.0.CO;2-U

[77] Gray MJ, Lipson RL, Maeck JVS, Parker L, Romeyn D. Use of hematoporphyrin derivative in detection and management of cervical cancer: A preliminary report. American Journal of Obstetrics & Gynecology. 1967;**99**:766–771. DOI: 10.1016/0002-9378(67)90392-4

[78] Diamond I, Granelli SG, McDonagh AF, Nielsen S, Wilson CB, Jaenicke R. Photodynamic therapy of malignant tumours. Lancet. 1972;**2**:1175–1177. DOI: 10.1016/S0140-6736(72)92596-2

[79] Dougherty TJ, Kaufman JE, Goldfarb A, WeishauptKR, Boyle D, Mittleman A. Photoradiation therapy for the treatment of malignant tumors. Cancer Research. 1978;**38**:2628–2635

[80] Usuda J, Kato H, Okunaka T, Furukawa K, Tsutsi H, Yamada K, Suga Y, Honda H, Nagatsuka Y, Ohira T, Tsuboi M, Hirano T. Photodynamic therapy (PDT) for lung cancers. Journal of Thoracic Oncology. 2006;**1**:489–493. DOI: 10.1016/S1556-0864(15)31616-6

[81] Sharman WM, Allen CM, van Lier JE. Photodynamic therapeutics: Basic principles and clinical applications. Drug Discovery Today. 1999;**4**:507–517. DOI: 10.1016/S1359-6446(99) 01412-9

[82] Dougherty TJ, Gomer CJ, Henderson BW, Jori G, Kessel D, Korbelik M, Moan J, Peng Q. Photodynamic therapy. Journal of the National Cancer Institute. 1998;**60**:889–905. DOI: 10.1093/jnci/90.12.889

[83] Silva JN, Filipe P, Morlière P, Mazière J-C, Freitas JP, Cirne de Castro JL, Santus R. Photodynamic therapies: Principles and present medical applications. Bio-Medical Materials and Engineering. 2006;**16**:S147–S154

[84] Allison RR, Mota HC, Sibata CH. Clinical PD/PDT in North America: An historical review. Photodiagnosis and Photodynamic Therapy. 2004;**1**:263–277. DOI: 10.1016/S1572-1000(04)00084-5

[85] Allison RR, Bagnato VS, Cuenca R, Downie GH, Sibata CH. The future of photodynamic therapy in oncology. Future Oncology. 2006;**2**:53–71. DOI: 10.2217/14796694.2.1.53

[86] Phillips D. Chemical mechanisms in photodynamic therapy with phthalocyanines. Progress in Reaction Kinetics and Mechanism. 1997;**22**:175–300

[87] Wohrle D, Hirth A, Bogdahn-Rai T, Schnurpfeil G, Shopova M. Photo-dynamic therapy of cancer: second and third generations of photosensitizers. Russian Chemical Bulletin. 1998;**47**:807–816. DOI: 10.1007/BF02498146

[88] Levy JG, Jones CA, Pilson LA. The preclinical and clinical development and potential application of benzoporphyrin derivative. International Photodynamic Therapy. 1994;**1**:3–5

[89] Aveline BM, Hasan T, Redmond RW. The effects of aggregation, protein binding and cellular incorporation on the photophysical properties of benzoporphyrin derivative monoacid ring A (BPDMA). Journal of Photochemistry and Photobiology B: Biology. 1995;**30**:161–169

[90] Stables GI, Ash DV. Photodynamic therapy. Cancer Treatment Reviews. 1995;**21**:311–323. DOI: 10.1016/0305-7372(95)90035-7

[91] Berdugo M, Bejjani RA, Valamanesh F, Savoldelli M, Jeanny JC, Blanc D, Ficheux H, Scherz A, Salomon Y, BenEzra D, Behar-Cohen F. Evaluation of the new photosensitizer stakel (WST-11) for photodynamic choroidal vessel occlusion in rabbit and rat eyes. Investigative Ophthalmology & Visual Science. 2008;**49**:1633–1644. DOI: 10.1167/iovs.07–0767

[92] Koudinova NV, Pinthus JH, Brandis A, Brenner O, Bendel P, Ramon J, Eshhar Z, ScherzA, Salomon Y. Photodynamic therapy with Pd-Bacteriopheophorbide (TOOKAD): Successful in vivo treatment of human prostatic small cell carcinoma xenografts. International Journal of Cancer. 2003;**104**:782–789. DOI: 10.1002/ijc.11002

[93] Trachtenberg J, Bogaards A, WeersinkRA, Haider MA, Evans A, McCluskey SA, Scherz A, Gertner MR, Yue C, Appu S, Aprikian A, Savard J, Wilson BC, Elhilali M. Vascular targeted photodynamic therapy with palladium-bacteriopheophorbide photosensitizer for recurrent prostate cancer following definitive radiation therapy: Assessment of safety and treatment response. Journal of Urology. 2007;**178**:1974–1979. DOI: 10.1016/j.juro.2007.07.036

[94] Trachtenberg J, Weersink RA, Davidson SR, Haider M. A, Bogaards A, Gertner MR, Evans A, Scherz A, Savard J, Chin JL, Wilson BC, Elhilali M. Vascular-targeted photodynamic therapy (padoporfin, WST09) for recurrent prostate cancer after failure of external beam radiotherapy: A study of escalating light doses. BJU International. 2008;**102**:556–562. DOI: 10.1111/j.1464-410×.2008.07753.x

[95] Razum NJ, Snyder AB, Doiron DR. SnET2: Clinical update. Proceedings of SPIE. 1996;**2675**:43–46. DOI: 10.1117/12.237549

[96] Selman SH, Keck RW. The effect of transurethral light on the canine prostate after sensitization with the photosensitizer tin (II) etiopurpurin dichloride: a pilot study. Journal of Urology. 1994;**152**:2129–2132

[97] MacDonald IJ, Dougherty TJ. Basic principles of photodynamic therapy. Journal of Porphyrins and Phthalocyanines. 2001;**5**:105–129. DOI: 10.1002/jpp.328

[98] Bonnett R, Martinez G. Photobleaching of photosensitisers used in photodynamic therapy. Tetrahedron. 2001;**57**:9513–9547. DOI: 10.1016/S0040-4020(01)00952-8

[99] Detty MR, Gibson SL, Wagner SJ. Current clinical and preclinical pho-tosensitizers for use in photodynamic therapy. Journal of Medicinal Chemistry. 2004;**47**:3897–3915. DOI: 10.1021/jm040074b

[100] Morgan AR, Garbo GM, Keck RW, Selman SH. New photosensitizers for photodynamic therapy: combined effect of metallopurpurin derivatives and light on transplantable bladder tumors. Cancer Research. 1998;**48**:194–198

[101] Ali H, van Lier JE. Metal complexes as photo- and radio-sensitizers. Chemical Reviews. 1999;**99**:2379–2450. DOI: 10.1021/cr980439y

[102] Mang TS, Allison R, Hewson G, Snider W, Moskowitz R. A phase II/III clinical study of tin ethyl etiopurpurin (Purlytin)-induced photodynamic therapy for the treatment of recurrent cutaneous metastatic breast cancer. The Cancer Journal From Scientific American. 1998;**4**:378–384

[103] Primbs GB, Casey R, Wamser K, Snyder WJ, Crean DH. Photodynamic therapy for corneal neovascularization. Ophthalmic Surgery, Lasers. 1998;**29**:832–838

[104] Kaplan MJ, Somers RG, Greenberg RH, Ackler J. Photodynamic therapy in the management of metastatic cutaneous adenocarcinomas: Case reports from phase 1/2 studies using tin ethyl etiopurpurin (SnET2). Journal of Surgical Oncology. 1998;**67**:121–125. DOI: 10.1002/(SICI)1096-9098(199802)67:2<121::AID-JSO9>3.0.CO;2-C

[105] Selman SH, Albrecht D, Keck RW, Brennan P, Kondo S. Studies of tin ethyl etiopurpurin photodynamic therapy of the canine prostate. Journal of Urology. 2001;**165**:1795–1801. DOI: 10.1016/S0022-5347(05)66416-6

[106] Rifkin R, Reed B, Hetzel F, Chen K. Photodynamic therapy using SnET2 for basal cell nevus syndrome: A case report. Clinical Therapeutics. 1997;**19**:639–641. DOI: 10.1016/S0149-2918(97)80089-6

[107] Hsi RA, Kapatkin A, Strandberg J, Zhu T, Vulcan T, Solonenko M, Rodriguez C, Chang J, Saunders M, Mason N, Hahn S. Photodynamic therapy in the canine prostate using motexafin lutetium. Clinical Cancer Research. 2001;**7**:651–660

[108] Kessel D, Thompson P, Saatio K, Nantwi KD. Tumor localization and photosensitization by sulfonated derivatives of tetraphenylporphine. Photochemistry and Photobiology. 1987;**45**:787–790. DOI: 10.1111/j.1751-1097.1987.tb07883.x

[109] Berg K, Western A, Bommer J, Moan J. Intracellular localization of sulfonated meso-tetraphenylporphines in a human carcinoma cell line. Photochemistry and Photobiology. 1990;**52**:481–487. DOI: 10.1111/j.1751-1097.1990.tb01789.x

[110] Stilts CE, Nelen MI, Hilmey DG, Davies SR, Gollnick SO, Oseroff AR, Gibson SL, Hilf R, Detty MR. Water-soluble, core-modified porphyrins as novel, longer-wavelength-absorbing sensitizers for photodynamic therapy. Journal of Medicinal Chemistry. 2000;**43**:2403–2410. DOI: 10.1021/jm000044i

[111] Nepomuceno MF, Tabak M, Vercesi AE. Opposite effects of Mn(III) and Fe(III) forms of meso-tetrakis(4-N-methyl pyridiniumyl) porphyrins on isolated rat liver mitochondria. Journal of Bioenergetics and Biomembranes. 2002;**34**:41–47. DOI: 10.1023/A:1013818719932

[112] Inada NM, da Silva AR, Jorge RA, BoreckyJ, Vercesi AE. Irradiated cationic meso-porphyrin rat liver mitochondria induces larger damage to isolated than the anionic form. Archives of Biochemistry and Biophysics. 2007;**457**:217–224. DOI: 10.1016/j.abb.2006.10.025

[113] Sommer S, Rimington C, Moan J. Formation of metal complexes of tumor-localizing porphyrins. FEBS Letters. 1984;**17**:267–271. DOI: 10.1016/0014–5793(84)81138-2

[114] Kennedy JC, Pottier RH. New trends in photobiology: Endogenous protoporphyrin IX, a clinically useful photosensitizer for photodynamic therapy. Journal of Photochemistry and Photobiology B. 1992;**6**:275–292. DOI: 10.1016/1011-1344(92)85108-7

[115] Henderson BW, Bellmier DA, Greco WR, Sharm A, Pandey RY, Vaughan LA, Weushauot KR, Dougherty TJ. An in vivo quantitative structure-activity relationships for a congeneric series of pyropheophorbide derivatives as photosensitizers for photsynamic therapy. Cancer Research. 1997;**57**:4000–4007

[116] Salvemini D, Riley DP, Cuzzocrea S. SOD mimetics are coming of age. Nature Reviews Drug Discovery. 2002;**1**:367–374. DOI: 10.1038/nrd796

[117] Rebouças JS, Spasojevic I, Batinić-Harbele I. Pure manganese (III) 5,10,15,20-tetrakis(4.bonzoic acid)porphyrin (MnTBAP) is not a superoxide dismutase mimic in aqueous systems: a case of structure-activity relationship as a watchdog mechanism in experimental therapeutics and biology. Journal of Biological Inorganic Chemistry. 2008;**13**(2):289–302. DOI: 10.1007/s00775-007-0324-9

[118] Vance CK, Miller AF. A simple proposal that can explain the inactivity of metal-substituted superoxide dismutases. Journal of the American Chemical Society. 1998;**120**(3):461–467. DOI: 10.1021/ja972060j

[119] Ellerby RM, Cabelli DE, Graden JA, Valentine JS. Copper-zinc superoxide dismutase: why not pH dependent. Journal of the American Chemical Society. 1996;**118**(28):6556–6561. DOI: 10.1021/ja953845x

[120] Goldstein S, Fridovich I, Czapski G. Kinetic properties of Cu, Zn-superoxide dismutase as a function of metal content—order restored. Free Radical Biology & Medicine. 2006;**41**(6):937–941. DOI: 10.1016/j.freeradbiomed/2006/05/026

[121] Saba H, Batinić -Harbele I, Munusamy S, Mitchell T, Lichti C, Megyesi J, MacMillan-Crow LA. Manganese porphyrin reduces renal injury and mitochondrial damage during ischemia/reperfusion. Free Radical Biology & Medicine. 2007;**42**(10):1571–1578. DOI: 10.1016/j.freeradbiomed.2007.02.016

[122] Zhao Y, Chaiswing L, Oberley TD, Batinić -Haberle I, St. Clair W, Epstein CJ, St. Clair D. A mechanism-based antioxidant approach for the reduction of skin carcinogenesis. Cancer Research. 2005;**65**(4):1401–1405. DOI: 10.1158/0008-5472.CAN-04-3334

[123] Haberle I, Rebouças JS, Spasojević I. Superoxide dismutase mimics: Chemistry, pharmacology, and therapeutic potential. Antioxidants & Redox Signaling. 2010;**13**(6):877–918. DOI: 10.1089/ars.2009.2876

[124] Nagami H, Umakoshi H, Shimanouchi T, Kuboi R. Variable SOD-like activity of liposome modified with Mn(II)–porphyrin derivative complex. Biochemical Engineering Journal. 2004;**21**(3):221–227. DOI: 10.1016/j.bej.2004.05.006

[125] Dong H, Zhou X, Jiang C. Molecular design and theoretical investigation on novel porphyrin derivatives for dye-sensitized solar cells. Theoretical Chemistry Accounts. 2012;**131**(2):1–11. DOI: 1007/s00214-012-1102-5

[126] Mugnol KCU, Martins MVA, Nascimento EC, Nascimento OR, Crespilho FN, Arantes JT, Nantes IL. Interaction of Fe^{3+}meso-tetrakis (2,6-dichloro-3-sulfonatophenyl)porphyrin with cationic bilayers: magnetic switching of the porphyrin and magnetic induction at the interface. Theorical Chemistry Accounts. 2011;**130**:829–837. DOI 10.1007/s00214-011-1055-0

[127] Nazeeruddin MK, De Angelis F, Fantacci S, Selloni A, Viscardi G, Liska P, Ito S, Bessho T, Gratzel M. Combined experimental and DFT-TDDFT computational study of photoelectrochemical cell ruthenium sensitizers. Journal of the American Chemical Society.2005;**127**(48):16835–16847. DOI: 10.1021/ja052467l

[128] Balanay MP, Kim DH. DFT/TD-DFT molecular design of porphyrin analogues for use in dye-sensitizer solar cells. Physical Chemistry Chemical Physics. 2008;**10**(33):5121–5127. DOI: 10.1039/b806097e

[129] Santhanamoorthi N, Lo C-M, Jiang J-C. Molecular design of porphyrins for dye-sensitized solar cells: a DFT/TDDFT study. The Journal of Physical Chemistry. 2013;**4**(3):524–530. DOI: 10.1021/jz302101j

[130] Grätzel M. Solar energy conversion by dye-sensitized photovoltaic cells. Inorganic Chemistry. 2005;**44**(20):6841–6851. DOI: 10.1021/ic0508371

[131] Grätzel M. Photoelectrochemical cells. Nature. 2001;**414**:338–344. DOI: 10.1038/35104607

[132] Araújo-Chaves JC, Tofanello A, Yokomizo CH, Carvalho-Jr WM, Souza FL, Nantes IL. Cytochrome c as an electron acceptor of nanostructured titania and hematite semiconductors. Journal of Energy Challenge and Mechanics. 2014;**1**:86–94

[133] Higashino T, Imahori H. Porphyrins as excellent dyes for dye-sensitized solar cells: recent developments and insights. Dalton Transactions. 2015;**44**(2):448–463. DOI: 10.1039/C4DT02756F

[134] Panda MK, Ladomenou K, Coutsolelos AA. Porphyrins in bio-inspired transformations: Light-harvesting to solar cell. Coordination Chemistry Reviews. 2012;**256**:2601–2627. DOI: 10.1016/j.ccr.2012.04.041

[135] Hoffert MI, Caldeira K, Benford G, Criswell DR, Green C, Herzog H, Jain AK, Kheshgi HS, Lackner KS, Lewis JS, Lightfoot HD, Manheimer W, Mankins JC, Mauel ME, Perkins LJ, Schlesinger ME, Volk TT, Wigley ML. Advanced technology paths to global climate stability: energy for a greenhouse planet. Science. 2002;**298**:981–987. DOI: 10.1126/science.1072357

[136] Blankenship RE, Tiede DM, Barber J, Brudvig GW, Fleming G, Ghirardi M, Gunner MR, Junge WD, Kramer M, Melis A, Moore TA, Moser CC, Nocera DG, Nozik AJ, Ort DR, Parson WW, Prince RC, Sayre RT. Comparing photosynthetic and photovoltaic efficiencies and recognizing the potential for improvementol. Science. 2011;**332**:805–809. DOI: 10.1126/science.1200165

[137] Grätzel M, Janssen RA, Mitzi JDB, Sargent EH. Materials interface engineering for solution-processed photovoltaics. Nature. 2012;**488**:304–312. DOI: 10.1038/nature11476

[138] Shih PM, Zarzycki J, Niyogi KK. Kerfeld CA. Introduction of a synthetic CO_2-fixing photorespiratory bypass into a cyanobacterium. The Journal of Biological Chemistry. 2014;**289**:9493–9500. DOI: 10.1074/jbc.C113.543132

[139] Falkowski PG, Raven JA. Aquatic Photosynthesis. Princeton: University Press; 2007

[140] O'Regan B, Grätzel M. A low-cost, high-efficiency solar cell based on dye-sensitized colloidal TiO_2 films. Nature. 1991;**353**:737–740. DOI: 10.1038/353737a0

[141] Grätzel M. Recent advances in sensitized mesoscopic solar cells. Accounts of Chemical Research. 2009;**42**(11):1788–1798. DOI: 10.1021/ar900141y

[142] Knodler R, Sopka J, Harbach F, Grunling HW. Photoelectrochemical cells based on dye sensitized colloidal TiO_2 layers. Solar Energy Materials and Solar Cells. 1993;**30**(3):277–281. DOI: 10.1016/0927-0248(93)90147-U

[143] Hagfeldt A, Boschloo G, Sun L, Kloo L, Pettersson H. Dye-sensitized solar cells. Chemical Reviews. 2010;**110**(11):6595–6663. DOI: 10.1021/cr900356p

[144] Nazeeruddin MK, Kay A, Rodicio I, Humphry-Baker R, Mueller E, Liska P, Vlachopoulos N, Grätzel M. Conversion of light to electricity by cis-X2bis(2,2′-bipyridyl-4,4′-dicarboxylate)ruthenium(II) charge-transfer sensitizers (X = Cl-, Br-, I-, CN-, and SCN-) on nanocrystalline titanium dioxide electrodes. Journal of the American Chemical Society. 1993;**115**(14):6382–6390. DOI: 10.1021/ja00067a063

[145] Hagfeldt A, Grätzel M. Molecular photovoltaics. Journal of the American Chemical Society. 2000;**33**(5):269–277. DOI: 10.1021/ar980112j

[146] Grätzel M. Perspectives for dye-sensitized nanocrystalline solar cells. Progress in Photovoltaics. 2000;**8**(1):171–185. DOI: 10.1002/(SICI)1099-159X(200001/02)8:1<171::AID-PIP300>3.0.CO;2-U

[147] Gao F, Wang Y, Shi D, Zhang J, Wang M, Jing X, Humphry-Baker R, Wang P, Zakeeruddin SM, Grätzel M. Enhance the optical absorptivity of nanocrystalline TiO_2 film with high molar extinction coefficient ruthenium sensitizers for high performance dye-sensitized solar cells. Journal of the American Chemical Society. 2008;**130**(32):10720–10728. DOI: 10.1021/ja801942j

[148] Tan LL, Huang JF, Shen YL, Xiao M, Liu JM, Kuang DB, Su CY. Highly efficient and stable organic sensitizers with duplex starburst triphenylamine and carbazole donors for liquid and quasi-solid-state dye-sensitized solar cells. Journal of Materials Chemistry A. 2014;**2**:8988–8994

[149] Zhou H, Fan T, Zhang D. An insight into artificial leaves for sustainable energy inspired by natural photosynthesis. ChemCatChem. 2011;**3**(3):513–528. DOI: 10.1002/cctc.201000266

[150] Liu L-N, Duquesne K, Oesterhelt F, Sturgis JN, Scheuring S. Forces guiding assembly of light-harvesting complex 2 in native membranes. Proceedings of the National Academy of Sciences of the United States of America. 2011;**108**(23):9455–9459. DOI: 10.1073/pnas.1004205108

[151] Lott GA, Perdomo-Ortiz A, Utterback JK, Widom JR, Aspuru-Guzik A, Marcus AH. Conformation of self-assembled porphyrin dimers in liposome vesicles by phase-modulation 2D fluorescence spectroscopy. Proceedings of the National Academy of Sciences of the United States of America. 2011;**108**(40):16521–16526. DOI: 10.1073/pnas.1017308108

[152] Gust D, Moore TA, Moore AL. Realizing artificial photosynthesis. Faraday Discuss. 2012;**155**:9–26. DOI: 10.1039/C1FD00110H

[153] Péchy P, Renouard T, Zakeeruddin S, Humphry-Baker MR, Comte P, Liska P, Cevey L, Costa E, Shklover V, Spiccia L, Deacon GB, Bignozzi CA, Grätzel M. Engineering of efficient panchromatic sensitizers for nanocrystalline TiO_2-based solar cells. Journal of the American Chemical Society. 2001;**123**(8):1613–1624. DOI: 10.1021/ja003299u

[154] Zeng W, Cao Y, Bai Y, Wang Y, Shi Y, Zhang M, Wang F, Pan C, Wang P. Efficient dye-sensitized solar cells with an organic photosensitizer featuring orderly conjugated ethylenedioxythiophene and dithienosilole blocks. Chemistry of Materials. 2010;**22**(5):1915–1925. DOI: 10.1021/cm9036988

[155] Tsao HN, Burschka J, Yi C, Kessler F, Nazeeruddin MK, Grätzel M. Influence of the interfacial charge-transfer resistance at the counter electrode in dye-sensitized solar cells employing cobalt redox shuttles. Energy & Environmental Science. 2011;**4**(12):4921–4924. DOI: 10.1039/C1EE02389F

[156] Xu M, Zhang M, Pastore M, Li R, De Angelis F, Wang P. Joint electrical, photophysical and computational studies on D-π-A dye sensitized solar cells: The impacts of dithiophene rigidification. Chemical Science. 2012;**3**(4):976–983. DOI: 10.1039/C2SC00973K

[157] Yella A, Lee H-W, Tsao HN, Yi C, Chandiran AK, Nazeeruddin MK, Diau EWG, Yeh C-Y, Zakeeruddin SM, Grätzel M. Porphyrin-sensitized solar cells with cobalt (II/III)–based redox electrolyte exceed 12 percent efficiency. Science. 2011;**334**(6056):629–634. DOI: 10.1126/science.1209688

[158] Lu HP, Tsai CY, Yen WN, Hsieh CP, Lee CW, Yeh CY, Diau EWG. Control of dye aggregation and electron injection for highly efficient porphyrin sensitizers adsorbed on semiconductor films with varying ratios of coadsorbate. The Journal of Physical Chemistry C. 2009;**113**(49):20990–20997. DOI: 10.1021/jp908100v

[159] Lee CW, Lu HP, Lan CM, Huang YL, Liang YR, Yen WN, Liu YC, Lin YS, Diau EWG, Yeh CY. Novel zinc porphyrin sensitizers for dye-sensitized solar cells: synthesis and spectral, electrochemical, and photovoltaic properties. Chemistry - A European Journal. 2009;**15**(6):1403–1412. DOI: 10.1002/chem.200801572

[160] Hsieh C-P, Lu H-P, Chiu C-L, Lee C-W, Chuang S-H, Mai C-L, Yen W-N, Hsu S-J, Diau EW-G, Yeh C-Y. Synthesis and characterization of porphyrin sensitizers with various electron-donating substituents for highly efficient dye-sensitized solar cells. Journal of Materials Chemistry. 2010;**20**(6):1127–1134. DOI: 10.1039/B919645E

[161] Baba A, Wakatsuki K, Shinbo K, Kato K, Kaneko F. Increased short-circuit current in grating-coupled surface plasmon resonance field-enhanced dye-sensitized solar cells. Journal of Materials Chemistry. 2011;**21**(41):16436–16441. DOI: 10.1039/C1JM12935J

[162] Groves JT, Nemo TE, Myers RS. Hydroxylation and epoxidation catalyzed by iron-porphine complexes. Journal of the American Chemical Society. 1979;**101**(4):1032–1033. DOI: 10.1021/ja00498a040

[163] Traylor TG, Tsychyia S. Perhalogenated tetraphenylhemins: Stable catalysts of high turnover catalytic hydroxylations. Inorganic Chemistry. 1987;**26**:1338–1339. DOI: 10.1021/ic00255a028

[164] Silva S, Pereira PMR, Silva P, Paz FAA, Faustino MAF, Cavaleiro JAS, Tomé J. Porphyrin and phthalocyanine glycodendritic conjugates: synthesis, photophysical and photochemical properties. Chemical Communications. 2012;**48**(30):3608–3610. DOI: 10.1039/C2CC17561D

[165] Costa JIT, Tomé J, Cavaleiro J. 5,10,15,20 -Tetrakis(pentafluorophenyl)porphyrin: A versatile platform to novel porphyrinic materials. Porphyrins Phtalocyanines. 2011;**15**(11): 1116. DOI: 10.1142/S1088424611004294

[166] Nakagaki S, Ferreira GKB, Ucoski GM, Castro KADF. Chemical reactions catalyzed by metalloporphyrin-based metal-organic frameworks. Molecules. 2013;**18**(6):7279–7308. DOI: 10.3390/molecules18067279

[167] Doro FG, Smith AG, Assis MD. Oxidation of alkanes and alkenes by iodosylbenzene and hydrogen peroxide catalysed by halogenated manganese porphyrins in homogeneous solution and covalently bound to silica. Journal of Molecular Catalysis A: Chemical. 2000;**164**:97–108. DOI: 10.1016/S1381–1169(00)00352-6

[168] Rayati S, Zavaki S, Valinejad H. Oxidation of hydrocarbons with tetra-n-butylammonium peroxy monosulfate catalyzed by β-tetrabromo-meso-tetrakis(4-methoxyphenyl)- and β -tetrabromo-meso-tetraphenylporphyrinatomanganese(III). Turkish Journal of Chemistry. 2014;**38**(4):611–616. DOI: 10.3906/kim-1310-35

[169] Poltowicz J, Pamin K, Matachowski L, Serwicka EM, Mokaya R, Xia Y, Olejniczak Z. Oxidation of cyclooctane over Mn(TMPyP) porphyrin-exchanged Al,Si-mesoporous molecular sieves of MCM-41 and SBA-15 type. Catalysis Today. 2006;**14**:287–292. DOI: 10.1016/j.cattod.2006.02.015

Chapter 2

Preparation and Structural Characterization of Metallophthalocyanine Particles Embedded in a Polymer Matrix

María Elena Sánchez-Vergara,
Angelina Romo Ubeda, Enrique Garibay Ochoa and
José Ramón Alvarez-Bada

Additional information is available at the end of the chapter

http://dx.doi.org/10.5772/67576

Abstract

In this work, thin-film deposition of FePc particles nucleated and grown in gels was carried out in air by spin coating. The surface morphology and structure of these films were analysed by scanning electron microscopy (SEM) and Fourier transform infrared (FTIR) spectroscopy. The optical parameters have been investigated using spectrophotometric measurements of transmittance in the wavelength range of 200–1100 nm. The absorption spectra recorded in the UV-Vis region for the deposited samples showed a single band, namely the B or Soret band in the region between 285 and 305 nm. The dependence of the Tauc and Cody optical gaps associated with the thickness of the film was determined and found to be around 4.2 eV from direct transitions and 3.8 eV from non-direct transitions. The films' electric properties and their dependence in the presence of radiation of several wavelengths were evaluated. At lower voltages, ohmic conduction is evident, while space-charge limited conductivity (SCLC) governed by an exponential trap distribution is to be found at higher voltages.

Keywords: thin films, spin coating, metallophthalocyanines, optical properties, electrical properties

1. Introduction

Photoconducting agents and other photoelectronic compounds embedded in polymer films as nanocomposite films have attracted considerable attention, as they exhibit many useful optical and electrical properties. Because of their large chemical and structural stability, as well as their optical and electrical properties, metallic phthalocyanines (MPcs) have been introduced into polymeric matrices as nanoparticles. A polymeric matrix composite (PMC) is a compound material consisting of a polymeric primary phase, or matrix, which is embedded in a secondary phase based mainly on matrix-reinforcing fibres and particles. The polymeric matrix enhances material stability, as it limits the introduction of environmental oxygen or water, which could reduce the potential usefulness of the MPcs. Nanostructuring also permits two other goals: to achieve optical homogeneity of the polymeric composite medium and to take advantage of specific properties of MPcs in their crystalline form. MPcs are usually ordered in crystalline arrangements, as their aromatic rings stack neatly. Due to the strength of π bonds, MPcs can be accommodated in a large number of different structures, which depend on the substituents they have. The type of structure determines the physical properties of a specific MPc, as well as its applications. The main modes of MPc molecular organization that may be observed are: (i) crystals, which can be in the alpha or beta allotropic forms (the beta polymorph being thermodynamically more stable). The two types are distinguished by the angle formed between the symmetry axis and the stacking direction. Alpha and beta crystals form angles of 26.5 and 45.8°, respectively. (ii) Liquid crystals, where Pcs are substituted by flexible lipophilic chains, which allow the formation by substituents of a quasi-liquid medium surrounding the in-plane aromatic nuclei, which overlap in columns distributed over two-dimensional positions with hexagonal or tetragonal symmetries. (iii) Thin films are solid structures whose thicknesses can be neglected for many physical purposes. In applications involving interaction with electromagnetic waves, thin-film thickness must be of the same order as the wavelength of the interacting disturbance. Thin films represent the Pcs arrangement most commonly considered for electronic applications. (iv) Skewer-structured polymers are obtained by polymerizing MPcs through bridge ligands; due to the variety of ligands that may be used and their properties, the distance between molecules can be controlled rather well and, thanks to the rigidity of the unidirectional connection in this type of structures, very good electronic and optical properties can be obtained from the material.

The purpose of this work is to report the generation of MPc crystals, their dispersion into a polymeric matrix and the evaluation of their optical and electrical properties in thin-film form. In this study, a polystyrene polymeric matrix was used. The materials thus obtained were characterized by different methods, including infrared (IR) and ultraviolet-visible (UV-Vis) spectroscopy, as well as scanning electron microscopy (SEM). First nanoparticles were synthesized in a molecular solution obtained from a supersaturated MPc solution. Second a solid composite was prepared by introducing pre-grown colloidal MPc particles into a polymeric matrix in a spin coating process. Spin coating leads to the production of uniform, flat, high-quality films or coatings. This process involves the application of a certain amount of nanoparticles suspended in a polymer and previously solved in an organic solvent. A small amount of the fluid is put on a substrate attached to a plate that is made to rotate at high speed, so that the resulting centripetal force spreads the suspension until the desired film thickness is achieved for the composite material. This process has four stages: deposition, centrifugation, de-centrifugation and evaporation. The evaporation of the fourth stage represents the main thinning mechanism for the film. After the film is deposited, it is annealed for 10 min at 90°C to accelerate matrix polymerization.

As some polymeric materials have conductivities similar to those of metals, they represent an important research area for the next generation of organic electronic devices. Conductivities in some polymers, such as poly(3,4-ethylenedioxythiophene) (PEDOT), are comparable to those of indium oxide or tin, while showing significant optical transmission. In this work, the electrical conductivity of the thin films was evaluated by means of a four-point technique. The films' electric properties and their dependence in the presence of radiation of several wavelengths were evaluated in order to determine whether this type of PMC films may have applications in the construction of electronic and optoelectronic flexible devices, such as OLEDs, photovoltaic devices and visual information devices. Additionally, the optical activation energies were evaluated by the Cody and Tauc methods from the transmittance values of the films at different thicknesses [1, 2].

2. Research method

2.1. Crystallization process

To carry out the crystallization of MPcs embedded in a polymeric matrix, the gel crystallization method was used, where a very viscous medium that favours slow crystallization is used to mix the constituent phases, mainly by diffusion. In this method, crystal growth in the gel takes place by diffusion-controlled mass transport. This procedure minimizes the sedimentation and convection effects of traditional crystallization by evaporation methods. One must take into account that the crystallization mechanism consists of three steps, i.e. solution supersaturation, formation of crystalline nuclei and crystal growth. The gel is a means to transport molecules or ions (precipitant agents, shock absorbers), with no or almost no chemical reactivity to molecules and ions that diffuse through their three-dimensional polymeric network. Gels can be classified, according to their preparation method, as chemical or physical. Chemical gels are those obtained by poly-addition processes, like those achieved from neutralization of sodium metasilicate, or by poly-condensation processes, such as those obtained from the hydrolysis reaction of tetramethoxysilane. The physical gels, including agar and agarose, are defined as those where the gelation process is carried out by the variation of some physical parameter, like temperature.

For the current study, tetramethoxysilane gel at 10% volume, with 50% of ethanol for crystallization in FePc capillary tubes, was used. Before introducing the solved gel into the capillary, this tube must be carefully washed with detergent, followed by double-distilled water and then acetone, and finally dried with warm air. The introduction of gel into the capillary is carried out by the application of air pressure with a syringe, taking care to avoid the formation of bubbles in the gel. The gel must occupy the central 4-cm section of the capillary. After the dispersion has gelled (a process which takes about 4 weeks), MPc is added through the ends of the capillary, travelling a distance of 3 cm of length. These MPcs, previously dissolved, must be added in the same way as the gel, by means of air pressure with the help of a syringe, while taking care not to form bubbles. The capillary is then sealed at the two ends and kept at a constant temperature of 22°C, until the product is formed. The conformation of the system used for gel crystallization can be shown in **Figure 1**, where the diagram of the tube used

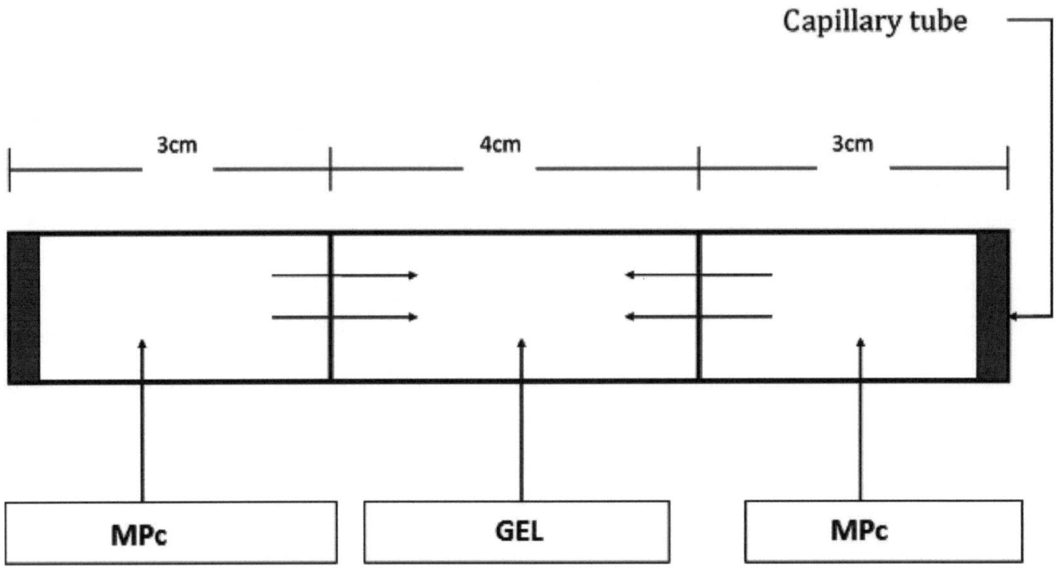

Figure 1. Capillary system used for crystallization.

for the crystallization is divided into three parts, as shown in the figure: one in the middle, where the gel was initially placed and the two ends where the dissolved MPc was placed before the MPc molecules migrated to the gel zone, where they nucleated and grew. This gel-based technique provides continuous control over the crystal or particle growth process, since it becomes possible to increase the growth rate by adding a larger amount of reagents through the ends of the capillary. Moreover, it also reduces the risk of damage to the crystal or the particle that could occur because of physical instabilities in the experimental arrangement, as it avoids the direct manipulation of the grown crystals.

2.2. Thin-film deposition and characterization

Most of the advanced devices manufactured today depend, at some point of their fabrication, on the synthesis and growth of films or thin layers. For this work, thin-film deposition of FePc particles nucleated and grown in gels was carried out in air by spin coating. The material was deposited onto a Corning 7059 glass, quartz, (100) single-crystalline silicon (c-Si) 200 Ω-cm wafers and ITO-coated glass slides. The quartz and Corning glass substrates were ultrasonically degreased in warm methanol and dried under a nitrogen atmosphere. The silicon substrates were chemically etched with a *p-etch* solution and dried under a nitrogen atmosphere. The composition of the solution was selected to have an FePc: polystyrene ratio of 1:3 in chloroform. The solution was spin coated on the substrates in a two-step process: 2500 rpm for 30 s, followed by annealing at 393 K for 10 min. These processes, spin coating and annealing, were repeated to obtain a suitable thickness. The thicknesses of the films obtained in the present study are shown in **Table 1**. We also report the determination of optical parameters related to the main transitions in the UV-Vis region, as well as the fundamental energy gap calculations for these films. Devices consisting of polystyrene matrix film were placed onto Corning

Sample	Film thickness (nm)	Direct Cody optical gap (eV)	Indirect Cody optical gap (eV)	Direct Tauc optical gap (eV)	Indirect Tauc optical gap (eV)
Thin Film 1	29	5.4	5.4	5.4	5.1
Thin Film 2	35	5.3	5.3	5.3	4.7
Thin Film 3	52	5.3	5.2	5.3	4.7
Thin Film 4	75	5.3	5.2	5.3	4.3
Thin Film 5	99	5.3	5.1	5.3	4.3
Thin Film 6	122	5.3	4.7	5.3	4.2
Thin Film 7	348	4.3	4.3	4.2	3.8

Table 1. Characteristic parameters of the FePc/polystyrene films.

glass substrates with a contact conductor of indium tin oxide (ITO) by spin coating. After the deposition, in order to diffuse MPc particles into the polystyrene matrix, the films were heat treated at 393 K for 10 min. The electric conductivity at 298 K of the device was evaluated with a four-point probe; for these measurements, the substrates were ITO-coated glasses with silver strips acting as electrodes. The strips were deposited by the painting process, the current due to hole-injection from positively-biased ITO was measured and the current due to hole-injection from silver was measured by reversing the polarity of the bias voltage [3].

2.3. Instruments

For the preparation of the thin films, a *Best Tools Smart Coater* 200, operating at 400 W, 110 V and 50/60 Hz, was used. FT-IR measurements were obtained with a Nicolet iS5-FT spectrophotometer using KBr pellets for the powders and silicon wafers as substrates for the thin films. Film thickness values were determined by profilometry in a quartz substrate with a *Bruker* profilometer, model DEKTAK XT, with STYLUS, LIS 3, 2 μm RADIUS-Type B. For SEM, a ZEISS EVO LS 10 scanning electron microscope was coupled to a microanalysis system and operated at a voltage of 20 kV and a focal distance of 25 mm, using thin films on a glass substrate. The size and distribution of dispersed particles were observed using a *JEOL* JEM2010 transmission electron microscope (TEM), LaB_6 cathode at 200 kV, 105 μA. UV-Vis spectroscopy was carried out in a *Unicam* spectrophotometer, model *UV300*, with a quartz substrate. Electric characterization was performed with a programmable voltage source, an auto-ranging pico-ammeter *Keithley* 4200-SCS-PK1 and a sensing station with a *Next Robotix* lighting controller circuit.

3. Results and discussion

The capillaries with FePc at the ends and tetramethoxysilane in the centre were allowed to stand at 22°C for 2 weeks. Subsequently, the generated particles were extracted from the capillary within the gel and were observed by SEM. **Figure 2a** and **b** show, at different magnifications, the FePc particles embedded in tetramethoxysilane. Despite being very small, they showed several structures-amorphous particles, regular particles and needles. In all cases, there was a heterogeneous distribution of particles inside the gel. The particles were removed from the tetramethoxysilane, washed and dried in a vacuum. The use of this technique demonstrated its applicability to the *in situ* formation of nanometric-size particles inside the gel. A preliminary TEM study of the nanometric FePc particles was also performed. **Figure 2c** shows a high-resolution bright field image of the FePc sample, where particles ranging in size between 2.8 and 20 nm can be seen. The shape of the particles is irregular, although some quasi-spherical forms can be discerned. A heterogeneous dispersion of the nanoparticles can also be seen. Among the advantages of using this technique for reinforcing particles in the manufacture of composite materials are that a very small sample can be used and the continuous manipulation of particles can be avoided; furthermore, it permits a continuous control of the growth process. It is difficult to determine the crystalline arrangement from TEM imaging in real space, so a wider characterization by IR spectroscopy was required. IR spectroscopy was specifically used to identify the structural nature of FePc, given that the IR spectrum depends on the crystal structure [4]. MPcs are known to have different polymorphs which are strongly identified by the IR absorption technique [4, 5]. It has been reported that the α-form of MPc can be characterized by a band around 720 cm^{-1}, while the β-form can be characterized by a band at a greater wave number at approximately 778 cm^{-1} [4–7]. In **Table 2**, it can be observed that FePc particles were present in the α and β crystalline structures.

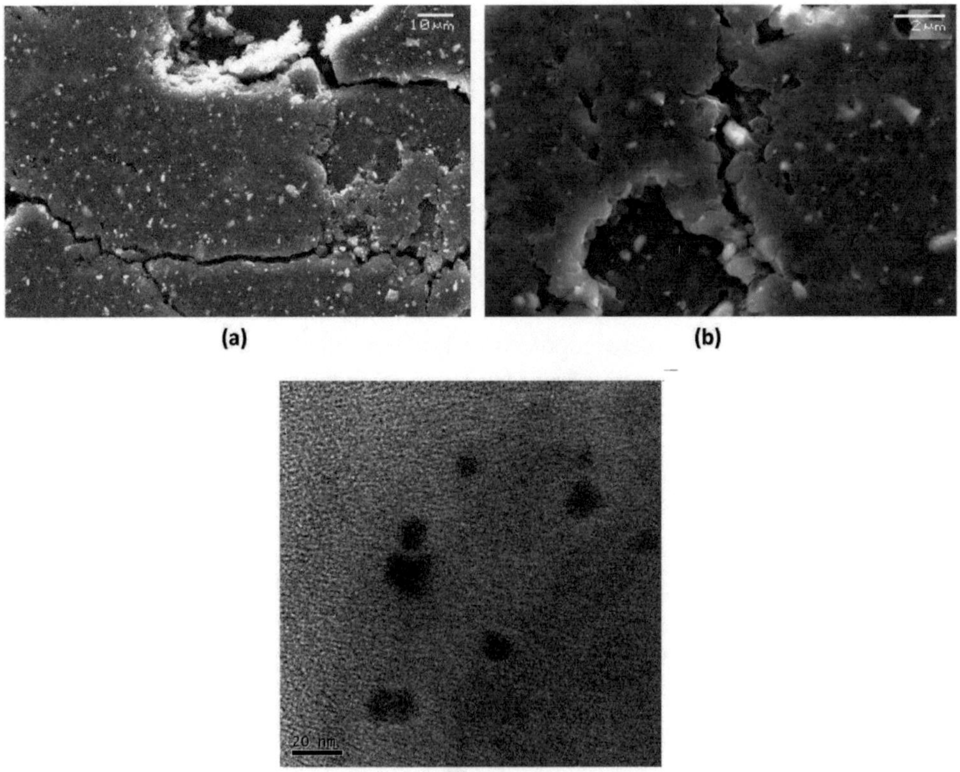

Figure 2. Gel with FePc particles (a) 1000×, (b) 7000× and (c) HRTEM micrographs.

Sample	ν (C—C) cm^{-1}	ν (C=N) cm^{-1}	ν (C—H) cm^{-1}	α-form cm^{-1}	β-form cm^{-1}
FePc (particle)	1609	1336	1164, 1119, 750	724	771
Thin Film 1	1607	1331	1163, 1119, 754	-	769
Thin Film 2	1609	1336	1164, 1119, 750	724	771
Thin Film 3	1603	1330	1164, 1117, 754	720	770
Thin Film 4	1604	1330	1163, 1119, 754	720	771
Thin Film 5	1604	1331	1166, 1117, 755	721	771
Thin Film 6	1603	1331	1165, 1119, 755	719	769
Thin Film 7	1603	1331	1165, 1116, 754	719	769

Table 2. Characteristic FT-IR bands for particles and thin films (cm^{-1}).

IR spectroscopy was also used in this study to ascertain the presence of the more representative bonds in the FePc compound and to determine whether significant chemical changes took place in this compound during gel nucleation and growth. **Table 2** shows the characteristic bands of the FePc particles deposited in the gel. The band appearing at 1605 ± 4 cm^{-1} was assigned to the C=C stretching vibration for pyrrole. The peak responsible for carbon-nitrogen stretching and bending occurs at 1332 ± 4 cm^{-1}. The peaks located at 1164 ± 2, 1117 ± 2 and 753 ± 2 cm^{-1} are due to the interaction of carbon atoms with the peripheral-ring hydrogen atoms [8–10]. As mentioned above, spin coating and annealing were carried out to produce the thin films. IR spectroscopy was performed in these films in order to verify that no chemical changes occurred in the FePc when interacting with the polymeric matrix. The results reported in **Table 2** indicate that the MPc did not experience any chemical changes during the deposition; on the other hand, in the thinnest film, the crystalline phase α is not observed. It is worth mentioning that the signals in the MPc film show slight changes in location. This occurs because, in thin films deposited by any method, internal stress affects intramolecular angles and bonding energies. Nevertheless, no significant changes occurred in these films, so we may conclude that the production of thin films from the FePc-polystyrene composite by the spin coating and annealing method is appropriate.

The films obtained by spin-coating were analysed by SEM. **Figure 3** shows the presence of the two phases-polymeric matrix and reinforcement. During the annealing, polymerization of polystyrene generated the needles shown in the images, while the FePc appears as irregular conglomerates. It is possible to observe that the MPc particles have been embedded in the matrix homogeneously, i.e. the particles are not agglomerated or separated, which in turn indicates that polystyrene is an appropriate matrix for this kind of films.

Optical absorption measurements are widely used to characterize the electronic properties of the thin films through the determination of parameters describing the electronic transitions, such as the band gap [11]. Additionally, the absorption spectra of different polymorphs of some Pc compounds show significant differences among each other [7, 12]. MPcs have two typical absorption bands, namely the *Q*-band in the visible region and the *B* or Soret-band in the near-ultraviolet

Figure 3. SEM images for spin-coated films (a) 83×, (b) 500× y (c) 1000×.

region [13–17]. The Q-band absorption is responsible for the characteristically intense blue/green colour of the FePc and this band has been interpreted in terms of π-π^* excitation between bonding and antibonding molecular orbitals [7, 18]. The electronic spectrum of the FePc particles obtained in tetramethoxysilane (**Figure 4a**) shows the characteristic Q-band absorption in the 578–750 nm region. The Soret-band of FePc arising from the deeper π levels → LUMO transitions is observed in the UV region at about 400–463 nm. On the other hand, the optical transmittance spectra of the thin-films deposited on quartz were recorded from 200 to 1100 nm and are shown in **Figure 4b**. Differences in the transmittance of the films under examination can be attributed to differences in thickness (see **Table 1**) according to Beer's law [19]. When the thickness of the film increases, its transmittance diminishes. The UV-Vis spectra of FePc-polystyrene thin films exhibited a characteristic B-band in the region between 285 and 305 nm. The observation of a single peak in the Soret band resembles that observed for CoPc, NiPc and other Pc thin films [20, 21]. This may imply that the splitting structure of this peak could be affected by the orbital overlap of the Pc ring with the central metal [21], although this effect could also be attributed to the presence of the polymeric matrix which, while protecting the FePc from oxygen and environmental humidity, also alters its optical properties in the visible region of the spectrum.

Considering the above results, we further apply the Cody and the Tauc models for the determination of the band gaps of the thin films [7, 22, 23]. The Cody model provides an effective option for the determination of the optical band of thin films in terms of its thickness. It uses the dependence between the photon energy (hv) and the absorption coefficient (α). The optical gap associated with the thin films is determined by extrapolating the linear trend observed in the spectral dependence of $(\alpha/hv)^n$ on hv. Here, n is a number characterizing the transition process, depending upon the nature of the electronic transitions responsible for the absorption; for direct transitions, $n = ½$, and, for indirect transitions, $n = 2$. The intersection with the x-axis of this linear extrapolation corresponds to the Cody optical gap for a given thickness of the film [22, 23]. The Cody optical gaps Eg_i and Eg_d for both transitions were obtained from the curves corresponding to those shown in **Figure 5** for the film with the largest thickness (*Thin Film 7*), which was of 348 nm.

For this film, the optical gap value is similar for both transitions, direct and indirect (see **Table 1**); apparently, the high concentration of FePc related to the highest thickness could be the cause of the similar values, but this could also be related to the fact that 4.3 eV is the lower (indirect) gap of the films under examination and may be quantitatively close to the direct gap for that particular film. On the other hand, the Tauc model argues that the optical gap associated with the thin film is determined through an extrapolation of the linear trend observed in the spectral

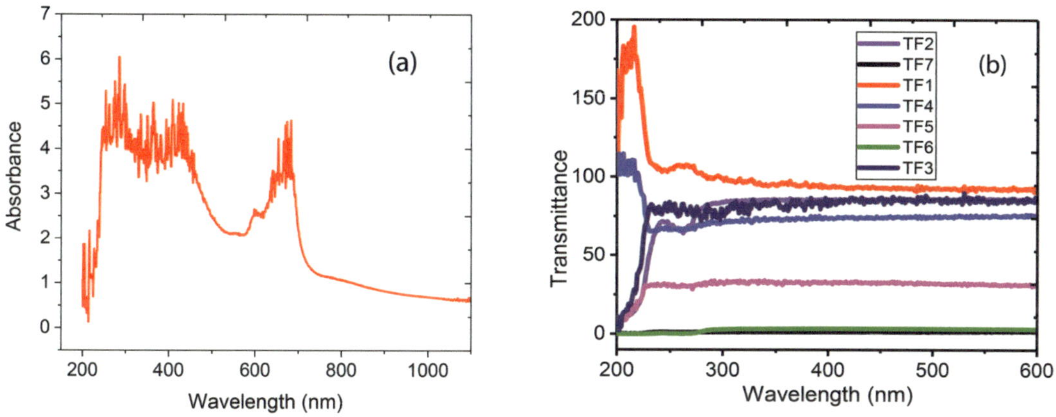

Figure 4. UV-Vis spectroscopy for: (a) FePc and (b) thin films.

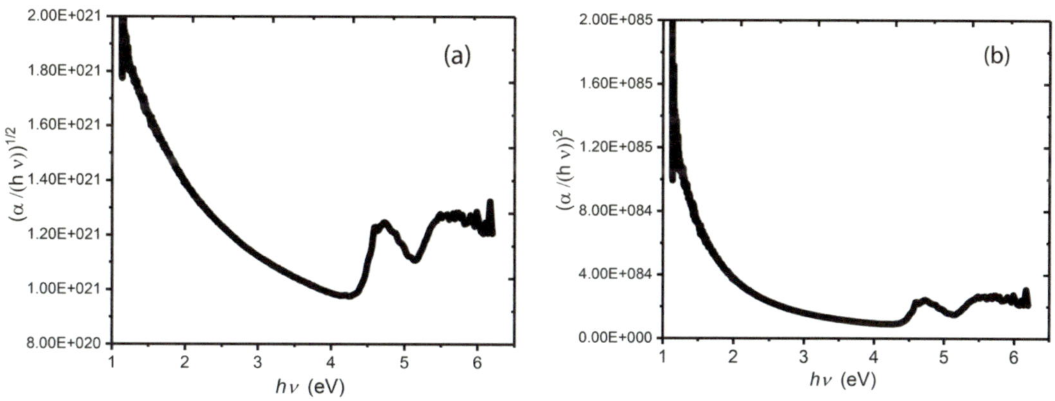

Figure 5. Plot of (a) $(\alpha/h\nu)^{1/2}$ and (b) $(\alpha/h\nu)^2$ versus photon energy $h\nu$ of *Thin Film 7*.

dependence of $(\alpha h\nu)^n$ over a limited range of $h\nu$ [1, 2]. The Tauc optical gaps for Eg_i and Eg_d were obtained from the curves corresponding (see **Table 1**) and they are shown in **Figure 6** for the film with the largest thickness (*Thin Film 7*). According to **Table 1** for the thicker film the smaller gap is obtained. At this thickness, the concentration of FePc is sufficient to decrease the gap and increase the overlap between Pc molecules. As the stacking between molecules increases, the electron flux increases significantly with respect to films with small thickness. On the other hand, for each of the remaining films, the indirect transition is the predominant one, with significantly lower values than the direct transition; this may be expected because of the mainly amorphous characteristics of the films and their effect on orbital overlap, despite FePc showing some α or β crystalline forms. It is important to mention that the variations in optical gaps obtained for the different films are of low significance. This may be attributed to the similar morphology of these systems, which differ only in the quantity and size of the FePc particles and the arrangement of their molecules in the polymeric matrix. Additionally, the gap depends on the number of electrons of the metal in the Pc ring [7, 19], which is the same for all these films.

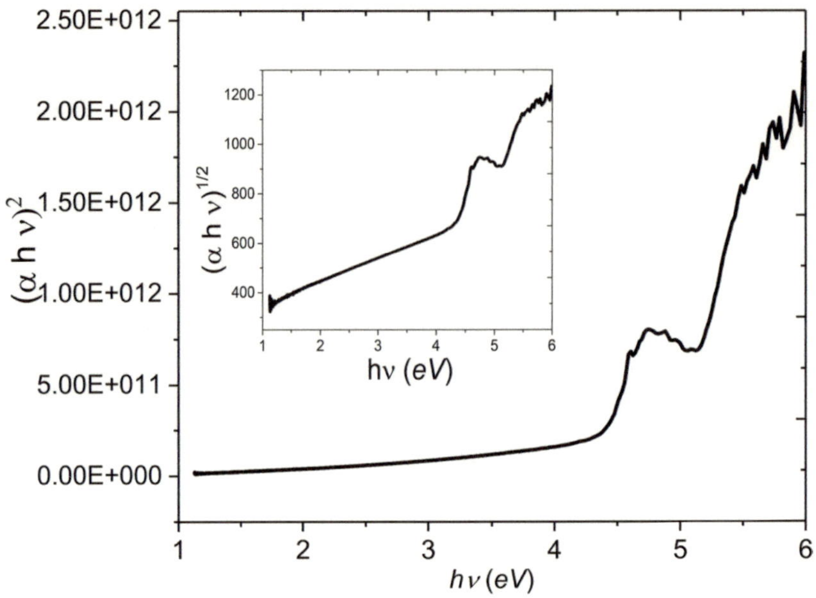

Figure 6. Plot of **($\alpha h\nu$)$^{1/2}$** and **($\alpha h\nu$)^2** versus photon energy *hν* of *Thin Film 7*.

Finally, in order to evaluate the electrical properties of the thin films, the four-point technique was employed, using the glass substrate with an ITO conducting contact. This study was performed on the sample labelled *Thin Film 7*, which was the one having the lowest optical gap. The film had a surface area of 2.16 cm^2. **Figure 7** shows the *I-V* characteristics of *Thin Film 7* under different illumination types (yellow light, white, blue, orange, green, infrared, UV and dark [no light]). Regardless of the wavelength of the incident radiation, the thin film follows the same behaviour. At lower voltages (around 10 V), ohmic conduction is evident, while space-charge limited conductivity (SCLC) governed by an exponential trap distribution is found at higher voltages. On the other hand, the *I-V* characteristics display symmetric

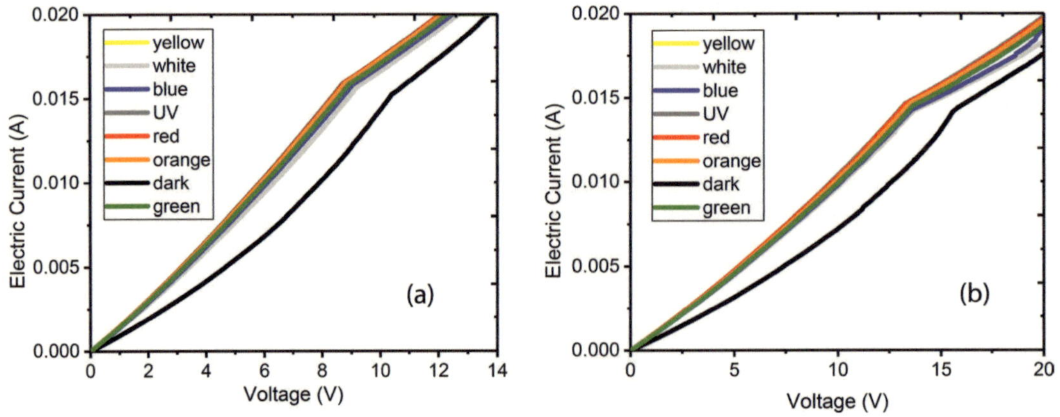

Figure 7. *I-V* characteristics of *Thin Film 7:* (a) ITO is positively biased and (b) ITO is negatively biased.

http://dx.doi.org/10.5772/67576

behaviour, both when (a) the current due to hole injection from positively biased ITO was measured and also when (b) the current due to hole injection from silver was measured by reversing the polarity of the bias voltage. This can be explained by a negligible energy barrier at the *ITO/FePc-polystyrene* and *FePc-polystyrene/Ag* interfaces leading to a SCL bulk current when either the ITO or silver electrode is positively biased [24–26].

4. Conclusions

Different types of particles and crystalline polymorphs of FePc can be obtained with tetramethoxysilane. This blend of structures can be used to produce thin films of a polystyrene matrix in a FePc matrix-reinforcing base by spin coating. Upon examination of the resulting films by SEM, a homogeneous particle distribution is found within the polystyrene matrix. IR spectral analysis confirms that FePc is rich in α and β polymorphs. None of the MPc samples suffers chemical degradation during the thin-film deposition and annealing processes. The UV-Vis spectra of the particles in tetramethoxysilane show two well-defined absorption bands, namely, the Soret and the Q-bands. The exact position of these bands depends on their particular structure, metal complexation, and peripheral substituents. However, only the Soret band appears in the UV-Vis spectra of the thin films, which can be attributed to the presence of the polymeric matrix. The optical gap was calculated from the Cody and Tauc models and the information obtained from the absorption spectra indicates that these films absorb light on either side of the blue-green region. Since these FePc compounds absorb light on either side of the blue-green spectrum, they could be used as photosensitive materials in practical applications. The electrical conductivity of the films was evaluated and ohmic characteristics were found at low voltages, while an SCLC-type behaviour can be observed at higher voltages. Bias inversion in the *I-V* measurements does not have a significant effect on the thin-film electric transport characteristics.

Acknowledgements

The authors wish to thank the technical support of M.I. Mariel Leyva-Esqueda (Universidad Anáhuac) for technical help. María Elena Sánchez-Vergara gratefully acknowledges the financial support of Universidad Anáhuac México under grant INNADBSEVM140129141.

Author details

María Elena Sánchez-Vergara[1]*, Angelina Romo Ubeda[1], Enrique Garibay Ochoa[2] and José Ramón Alvarez-Bada[1]

*Address all correspondence to: elena.sanchez@anahuac.mx

1 School of Engineering, The Anahuac University Mexico, State of Mexico, Mexico

2 Technological Institute of Morelia, José María Morelos y Pavón, Morelia, Michoacán, Mexico

References

[1] Mok T.M., O'Leary S.K. The dependence of the Tauc and Cody optical gaps associated with hydrogenated amorphous silicon on the film thickness: αl experimental limitations and the impact of curvature in the Tauc and Cody plots. Journal of Applied Physics. 2007;102:113525. DOI: 10.1063/1.2817822.

[2] O'Leary S. K., Lim P. K. On determining the optical gap associated with an amorphous semiconductor: a generalization of the Tauc model. Solid State Communications. 1997;104(1):17–21. DOI: 10.1016/s0038-1098(97)00268-8.

[3] Mahapatro A.K., Ghosh S. Charge carrier transport in metal phthalocyanine based disordered thin films. Journal of Applied Physics. 2007;101:034318-1–034318-5. DOI: 10.1063/1.2434946.

[4] Karan S., Basak D., Mallik B. Persistence in photoconductivity and optical property of nanostructured copper (II) phthalocyanine thin films. Current Applied Physics. 2010;10(4):1117–1122. DOI: 10.1016/j.cap.2010.01.011.

[5] Wang J.B., Li W.L., Chu B., Lee C.S., Su Z.S., Zhang G., Wu S.H., Yan F. High speed responsive near infrared photodetector focusing on 808 nm radiation using hexadecafluoro–copper–phthalocyanine as the acceptor. Organic Electronics. 2011;12(1):34–38. DOI: 10.1016/j.orgel.2010.09.015.

[6] Neghabi M., Zadsar M., Ghorashi S.M.B. Investigation of structural and optoelectronic properties of annealed nickel phthalocyanine thin films. Materials Science in Semiconductor Processing. 2014;17:13–20. DOI: 10.1016/j.mssp.2013.08.002.

[7] El-Nahass M.M., Farag, A.M., Abd El-Rahman K.F., Darwish A.A.A. Dispersion studies and electronic transitions in nickel phthalocyanine thin films. Optics & Laser Technology 2005;37(7):513–523. DOI: 10.1016/j.optlastec.2004.08.016.

[8] Berhanu S., Tariq F., Jones T., McComb D.W. Three-dimensionally interconnected organic nanocomposite thin films: implications for donor–acceptor photovoltaic applications. Journal of Materials Chemistry. 2010;20(37):8005–8009. DOI: 10.1039/c0jm01030h.

[9] Farag A.A.M., Yahia I.S., Alfaify S., Bilgiçli A., Kandaz M., Yakuphanoğlu F. Optical dispersion parameters based on single-oscillator model and optical absorption of nanocrystalline metal phthalocyanine films: a comparison study. Superlattices and Microstructures. 2013;60:83–100. DOI: 10.1016/j.spmi.2013.04.018.

[10] Liu L.Y., Wan L., Cao L., Han Y.Y., Zhang W.H., Chen T.X., Guo P.P., Wang K., Xu F.Q. Assistance of partially reduced MoO_3 interlayer to hole-injection at iron phthalocyanine/ITO interface evidenced by photoemission study. Applied Surface Science. 2013;271:352–356. DOI: 10.1016/j.apsusc.2013.01.200.

[11] Yang Y., Samas B., Kennedy V.O., Macikenas D., Chaloux B.L., Miller J.A., Speer R.L., Protasiewicz J., Pinkerton A.A., Kenney M.E. Long, directional interactions in cofacial

silicon phthalocyanine oligomers. The Journal of Physical Chemistry A. 2011;115(45):12474–12485. DOI: 10.1021/jp2019445.

[12] Lbova A.K., Vasiliev M.P., Gutmann E.S. Phthalocyanine and polystyrene film nanocomposites. Russian Journal of Physical Chemistry A. 2011;85(3):457–561. DOI: 10.1134/S0036024411030216.

[13] Liu Z.T., Kwok H.S., Djurišić A.B. The optical functions of metal phthalocyanines. Journal of Physics D: Applied Physics. 2004;37(5):678–688. DOI: 10.1088/0022-3727/37/5/006.

[14] Ottaviano L., Di Nardo S., Lozzi L., Passacantando M., Picozzi P., Santucci S. Thin and ultra-thin films of nickel phthalocyanine grown on highly oriented pyrolitic graphite: an XPS, UHV-AFM and air tapping-mode AFM study. Surface Science. 1997;373(2–3):318–332. DOI: 10.1016/s0039-6028(96)01179.

[15] Guo L., Ma G., Liu Y., Mi J., Qian S., Qiu L. Optical and non-linear optical properties of vanadium oxide phthalocyanine films. Applied Physics B. 2014;74(3):253–257. DOI: 10.1007/s003400200801.

[16] Djurišić A.B., Kwong C.Y., Lau T.W., Guo W.L., Li E.H., Liu Z.T., Kwok H.S., Lam L.S.M., Chan W.K. Optical properties of copper phthalocyanine. Optics Communications. 2002;205(1–3):155–162. DOI: 10.1016/s0030-4018(02)01311-1.

[17] Andzelm J., Rawlett A.M., Orlicki J.A., Snyder J.F., Baldridge K.K. Optical properties of phthalocyanine and naphthalocyanine compounds. Journal of Chemical Theory and Computation. 2007;3(3):870–877. DOI: 10.1021/ct700017b.

[18] Nitschke C., O'Flaherty S.M., Kroell M., Strevens A., Maier S., Rüther M.G., Blau W.J. Preparation and nonlinear optical properties of phthalocyanine nanocrystals. In: Organic Photonic Materials and Devices; 25 January 2003; San Jose, CA. James V., Grote G., Kaino T., Editors, Proceedings of SPIE. 2003, Vol. 4991, pp. 124–131.

[19] Seoudi R., El-Bahy G.S., El Sayed Z.A. Ultraviolet and visible spectroscopic studies of phthalocyanine and its complexes thin films. Optical Materials. 2006;29(2–3):304–312. DOI: 10.1016/j.optmat.2005.10.002.

[20] El-Nahass M.M., Sallam M.M.A. Optical properties of thermally evaporated metal-free phthalocyanine (h 2 pc) thin films. International Journal of Modern Physics B. 2005;19(27):4057–4071. DOI: 10.1142/s0217979205032632.

[21] El-Nahass M.M., Abd-El-Rahman K.F., Darwish A.A.A. Fourier-transform infrared and UV-Vis spectroscopes of nickel phthalocyanine thin films. Materials Chemistry and Physics. 2005;92(1):185–189. DOI: 10.1016/j.matchemphys.2005.01.008.

[22] Cody G., Tiedje T., Abeles B., Brooks B., Goldstein Y. Disorder and the optical-absorption edge of hydrogenated amorphous silicon. Physical Review Letters. 1981;47(20):1480–1483. DOI: 10.1103/physrevlett.47.1480.

[23] Cody G.D. Chapter 2 The Optical Absorption Edge of a-Si:H. Hydrogenated Amorphous Silicon-Optical Properties. Semiconductors and Semimetals. 1984;21(Part B):11–82. DOI: 10.1016/s0080-8784(08)62910-5.

[24] Anthopoulos T.D., Shafai T.S. SCLC measurements in nickel phthalocyanine thin films. Physica Status Solidi (A). 2000;181(2):569–574. DOI: 10.1002/1521-396X (200010).

[25] Gravano S., Hassan A.K., Gould R.D. Effects of annealing on the trap distribution of cobalt phthalocyanine thin films. International Journal of Electronics. 1991;70(3):477–484. DOI:10.1080/00207219108921297.

[26] Hassan A.K., Gould R.D. The interpretation of current density-voltage and activation energy measurements on freshly prepared and heat treated nickel phthalocyanine thin films. International Journal of Electronics. 1993;74(1):59–65. DOI:10.1080/00207219308925813.

Chapter 3

Electrochemical Investigation of Porphyrin and Its Derivatives at Various Interfaces

Xiaoquan Lu and Samrat Devaramani

Additional information is available at the end of the chapter

http://dx.doi.org/10.5772/67637

Abstract

This chapter describes the electrochemistry of the porphyrins at solid-liquid and liquid-liquid interfaces. The fundamental electrochemical approach toward the porphyrin molecules in estimating their HOMO and LUMO energy levels is given. Various factors such as the effect of central metal ion, the periphery of the aromatic ring and axial ligands on the redox potentials of porphyrins have been discussed. Electrochemical sensing application of porphyrin molecules is described with few examples in brief. Much focus has been given on the electrochemistry of the self-assembled monolayer (SAM) of thiol-porphyrins on the gold electrode. Structural characterization and charge transfer across the SAM using cyclic voltammetry and electrochemical impedance spectroscopy are discussed. Theory and methodologies developed to study photoinduced charge transfer kinetics of porphyrin molecules using scanning electrochemical microscope at the solid-liquid and liquid-liquid interface have been described. Use of porphyrin molecules as luminophores in electrochemiluminescence sensing applications and the mechanisms involved are described through representative examples.

Keywords: porphyrin, electrochemistry, interface, SECM, ECL

1. Introduction

In the natural photosynthesis process, chlorophyll converts incident light into chemical energy with nearly 100% quantum yield through many complex steps. This excellent phenomenon inspired many scientists to study porphyrin derivatives and their metallated forms extensively for many decades and continue to be so. Substantial information has been gathered on the synthesis, structural characterization, and dependence of their property on the structure and applications of porphyrins [1]. Porphyrins can be tailored by modifying the aromatic ring at the β and *meso* positions of the pyrrole and by metallating the tetradentate core of the porphyrin ring with almost all the transition metal ions. Thereby electronic properties of the porphyrins such as redox process, light absorption property, energy, and electron transfer capabilities can be amended [2]. Hence, porphyrins have witnessed their participation in the wide range of applications in various fields such as photovoltaics, artificial photosynthesis, photodynamic therapy, catalysis, and enzymatic systems.

As mentioned above, crafting the redox potentials of the porphyrins by modifying the periphery or the core of the aromatic ring remains the key strategy behind its multifunctional behavior. Most of such compounds are electroactive, exhibit multiple redox couples, and have been investigated for their electrochemical properties, generally, in nonaqueous solvents. Various factors such as a type of metal ion and its oxidation state present at the core, nature of the macrocyclic aromatic ring, and an axial ligand attached to the metal ion will affect the electrochemistry of the molecule.

Porphyrins exhibit outstanding absorption of electromagnetic radiation in the visible region. Upon light illumination, electrons present in the HOMO will get excited to LUMO of the porphyrin. Photoexcitation followed by various relaxation processes and charge separation is shown in **Figure 1**. Long-lived radical ion pairs of porphyrins can be observed by stabilizing the charge separated states. Generally, the basic electrochemistry of the porphyrins is related to its electron donating or accepting behavior in the ground state. Electrochemistry of porphyrins under conditions similar to that of photovoltaic devices, artificial photosynthetic systems involves the other states depicted in **Figure 1**. In the following sections, we discuss the fundamental and applied electrochemistry of the porphyrins and its derivatives. Without going for the exhaustive citation of all the reported literature, representative examples have been chosen to support our discussion.

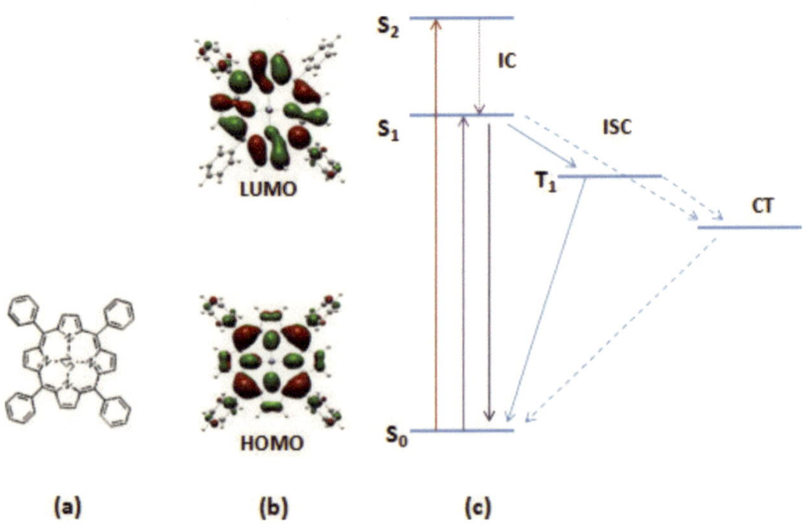

Figure 1. Representation of the molecular structure (a), HOMO and LUMO (b) of zinc tetraphenylporphyrin (ZnTPP) and the photoexcitation process followed by various relaxation events (c). Reprinted with permission from Ref. [65]. Copyright 2015 Elsevier Ltd.

2. HOMO and LUMO energy levels of porphyrins

Electrochemical techniques such as cyclic voltammetry and differential pulse voltammetry are generally used to estimate the HOMO and LUMO energy levels of the organic compounds. Oxidation onset potential, that is, the energy required to take out the first electron from the HOMO of the molecule will give the HOMO energy level of the molecule under study in eV versus the reference electrode used. In a similar way, reduction onset potential, that is, the energy required to add the first electron to LUMO of the molecule will give information about the energy level of the LUMO of the molecule. Ferrocene (Fc) or other common references used as an internal standard to complete the calculation by using following equations. Such electrochemical studies generally carried out in organic solvents with a suitable electrolyte.

$$E_{HOMO} = -(E^{ox}{}_{onset} + E_{1/2} \text{ of reference}) \text{ (eV)} \tag{1}$$

$$E_{LUMO} = -(E^{red}{}_{onset} + E_{1/2} \text{ of reference}) \text{ (eV)} \tag{2}$$

By knowing HUMO and LUMO energy levels, one can calculate the energy gap (E_g) between them.

2.1. Effect of metal ion

Cheng et al. calculated the HOMO and LUMO energy levels of the tetraphenylporphyrin (TPP) and Cu, Zn, Ni, Pd, and Pt metallated porphyrins (MTPP) [3]. Cyclic voltammograms (CVs) were recorded for TPP and MTPP in acetonitrile using tetra-n-butylammonium hexafluorophosphate as an electrolyte. Quinoxalinoporphyrin and its zincated form were studied in chlorobenzene with tetrabutylammonium tetrafluoroborate as an electrolyte versus Fc/Fc^+ by recording the CVs. E_g values calculated from the electrochemical method were comparable to those obtained from the electronic spectra [4]. In both the examples mentioned above, expected change in the energy levels of HOMO and LUMO of the porphyrin molecules after metallation was obtained in the electrochemical results. A linear relationship between the electronegativity of the divalent central metal ion and the first ring-centered oxidation and reduction potential was observed [5].

2.2. Effect of modifying the periphery of aromatic ring

Factors such as electron donating or withdrawing nature of the substituent, where it has been located on the ring and its number will affect the oxidation and reduction potentials. Influence of π-extension of the aromatic ring on the electrochemistry has been reported for platinum (II) porphyrin derivatives [6]. Dependence of the oxidation and reduction half-wave potentials of the porphyrins on the planarity of the molecule has been discussed by Shelnutt et al. in their review with various examples [7].

2.3. Effect of axial ligand

Coordination of nitrogenous bases is generally used in the axial ligation of metalloporphyrins. Type of ligand participated in the axial ligation can manipulate the oxidation and reduction half-wave potentials of the metalloporphyrins. Kadish et al. have discussed this in detail with exemplifying a large number of ligands possessing nitrogen as a donor atom with iron and cobalt porphyrins [8]. Basic electrochemistry of metalloporphyrins has been discussed in detail with numerous examples by Kadish et al. in a series of book volumes and also in reviews [9].

3. Porphyrins at solid-liquid interface

3.1. Sensors

Because of the multifunctional property and their interaction with the various analyte molecules, porphyrins deliver different signal outputs. Porphyrin molecules have been used for sensing applications through optical, electrochemical, different spectral modes. Electrochemical sensing methods developed using porphyrin molecules by our group is discussed here briefly. Porphyrin monolayer was used to electrochemically sense the phosphate anion based on the hydrogen bonding interaction. Upon hydrogen bonding of PO_4^{2-} with ^{-}NH, ease of charge transfer between the redox mediator and monolayer on the electrode was increased. Taking the advantage of this, decrease in the charge transfer resistance, R_{ct} and increase in the magnitude of the normalized current of the approach curves recorded by SECM were measured to sense the phosphate anion [10]. The same strategy has been used to quantify the porphyrin molecules in the pheophytin samples obtained from the spinach leaves. In this method, a gold electrode was modified with phosphate monolayer and used for the electrochemical sensing of porphyrin. R_{ct} value with increased concentration of porphyrin was found to be linear in the 1.0×10^{-7} M to 5.0×10^{-5} M concentration range. The detection limit, 3.0×10^{-8} M was superior to that of an optical method which was parallelly done [11]. The electrochemical sensing of m-dinitrobenzene (m-DNB) was demonstrated based on the same concept. Hydroxyl group(s) present at the periphery of the porphyrin ring form hydrogen bond with the nitro group of the m-DNB. Further, the benzene ring of the analyte will orient parallel to the macrocyclic π-ring of the porphyrin to result in the charge transfer interactions. As expected, increase in the number of hydroxyl groups on the porphyrin ring lead to the improved differential pulse stripping voltammetric analytical signal [12]. Taking the advantage of hydrogen bonding and π-π interaction between the analyte molecules and porphyrin macrocyclic ring, simultaneous determination of hydroquinone (HQ), catechol (CA), and resorcinol (RC) was proposed using 5,10,15,20-tetrakis(4-hydroxyphenyl)porphyrin (THPP)-CNT composite. HOMO and LUMO levels of the three analytes calculated from the density functional theory. HOMO energy level of the hydroquinone found to be highest and that of resorcinol is least. It is understood that higher the energy level of HOMO, it is easier to oxidize the molecule. The strength of the hydrogen bonding, an extent of π-π interaction between the THPP and the three analytes is different due to a difference in the charge distribution. Hence the affinity and oxidation potentials of the HQ, CA, and RC on the THPP-CNT-modified electrode resulted in the well-separated and sensitive peaks (**Figure 2**) which are not possible in case of bare and CNT-modified electrode [13]. Composites of porphyrin with carbon substrates such as graphene and fullerene have been reported for the electro and photoelectrochemical sensing of hydroquinone and m-DNB [14, 15].

http://dx.doi.org/10.5772/67637

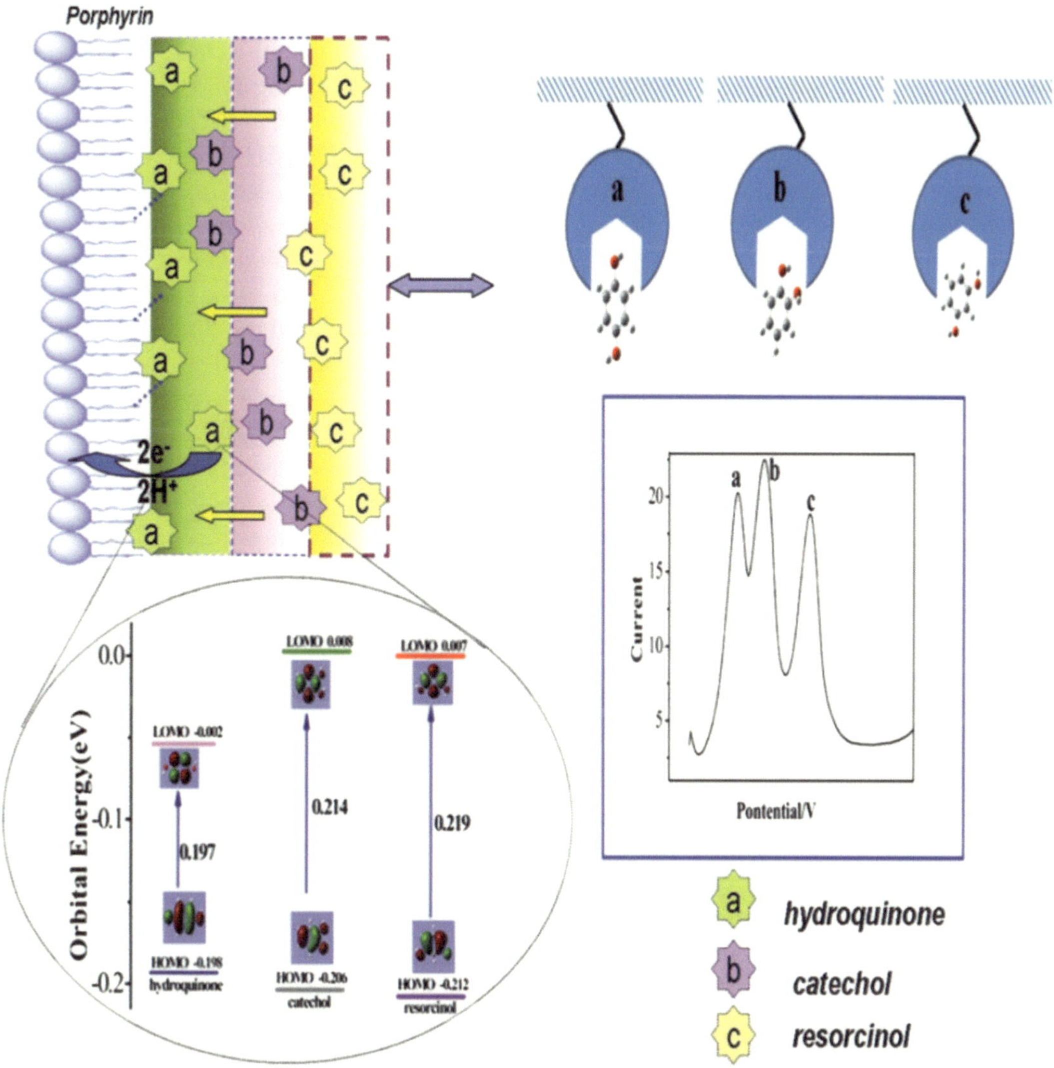

Figure 2. Representation of the density of the electron atmosphere of the dihydroxybenzene isomers and the interaction between porphyrin and dihydroxybenzene isomers. Adapted from Ref. [13] with permission from The Royal Society of Chemistry.

3.2. Monolayers of porphyrin derivatives

Organosulfur compounds are well studied for the formation of self-assembled monolayers (SAMs) on the gold substrate. SAMs have been studied for their effect on the interfacial properties. A molecule which involved in the formation of SAM can be divided into three parts. Head group of the molecule will interact with the gold substrate, a free end of the molecule can be considered as a tail group and the thickness and structure of the SAM will be decided by the spacer or linker moiety present between the head and tail. SAMs provide an ideal system for the electrochemical study of heterogeneous charge transfer. Effect of the length of linker molecule on the adsorption kinetics of 5-[p-(mercaptoalkoxy)-phenyl]-10,15,20-triphenylporphyrin molecules denoted as $H_2TPPO(CH_2)nSH$ was studied by varying the n from 3 to 12 [16]. Cyclic voltammetry and electrochemical impedance spectroscope were used to observe the time dependence of the surface coverage and orientation of $H_2TPPO(CH_2)nSH$ on the gold electrode. Adsorption rate constant was found to decrease with the increase in the length of the linker molecule. Though the bulky porphyrin molecules are present at the terminal, adsorption steps were similar to that of bare alkanethiols. The monolayers formed as a result of interaction between thiol and gold substrate are compact. Still, there will be imperfections in the form of pinholes. Hence, molecules or ions will reach electrode surface through them to result in a charge transfer. Hence, it is important to have information about such imperfections. The same set of different alkyl length thiol-porphyrin molecules was used to study the surface coverage, size, and distribution of pinholes present in the monolayer. The size of the pinholes estimated using EIS was ranged between 4 and 6 μM with 40–70 μM separation between them. Randles equivalent circuit and pore size distribution model were used in the analysis of monolayer structure [17]. Electrocatalytic activity of the metalloporphyrins depends on their orientation in the film produced on the electrode [18, 19]. Cobalt tetraphenylporphyrin (CoTPP) monolayers were prepared on the gold electrode using two different linker molecules such as a 3-mercaptopropionic acid (MPA), 4-mercaptopyridine (MPY). Free base porphyrin, tetra-[*p*-(3-mercaptopropyloxy)-phenyl]-porphyrin (TMPP) was first assembled to result in monolayer and then metallated with cobalt. The orientation of the porphyrins with respect to a gold electrode in all the three cases was different. Further, the second layer of CoTPP was prepared using imidazole as an axial ligand. Effect of orientation of porphyrin molecules in SAMs on the dioxygen reduction was studied in perchloric acid. Mono- and multi-layers of CoTPP prepared using MPY exhibited the highest catalytic activity [20]. The cofacial arrangement of porphyrins on the electrode found to be more effective for oxygen reduction [21, 22]. Electron transfer across the thiol-TPP and thiol-CoTPP monolayers was examined on a gold electrode in the aqueous solution. Direct electron transfer was blocked when the compact monolayer of the thiol-porphyrin was present. With the decrease in the density of the thiol-porphyrin in the monolayer, electron transfer was observed. Different potentials were applied to promote the charge transfer across the monolayer and the electron transfer rate constants were calculated using the cole-cole plot [23].

Generally, porphyrin monolayer formed by assembling each porphyrin molecule on the gold electrode through one thio- or thioacetate-group, that is, through one clip. Studies on the formation of SAMs using multi clips are seldom [24, 25]. Our group investigated the formation of SAM of tetra[*p*-(3-mercaptopropyloxy)phenyl]porphyrin (PPS_4) with four clips. Dense SAM was formed in the case of porphyrins with four clips compared to that with one clip. Charge transfer through the SAM by tunneling mechanism and the thickness of the films were taken into consideration to propose the arrangement of porphyrin molecules in the SAM. Considering the arrangement of porphyrins as shown in **Figure 3**, the thickness of the porphyrin film in the case of four clips should be less compared to that of one clip. Therefore greater tunneling current can be expected in the case of four clipped porphyrins. But the contradictory electrochemical results were observed, hence the arrangement of porphyrins with four clips was proposed as shown in **Figure 3d**. Gold electrode modified with the PPS_4 monolayer can behave as nanometer scaled photoswitches [26].

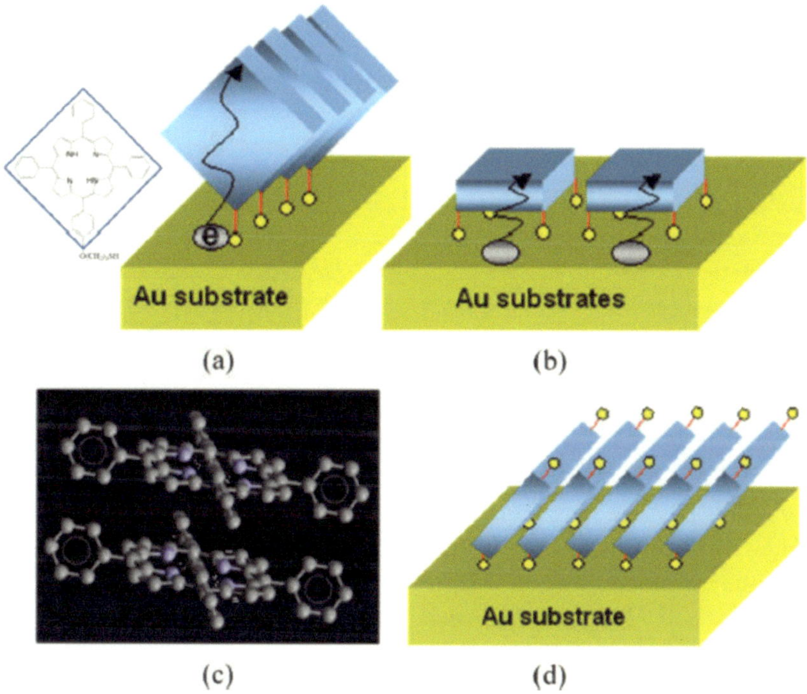

Figure 3. Representation of the possible arrangement of porphyrin molecules bearing one clip (a) and four clips (b and d) to form SAM on Au Surface. Skeletal structures of the porphyrin molecules (c). Adapted with permission from Ref. [26]. Copyright 2010 American Chemical Society.

A monolayer of base porphyrin 5-[p-(mercaptopropyloxy)-phenyl]-10,15,20-triphenylporphyrin, H_2MPTPP and its Co and Ni metallated forms were produced on the gold electrode to study their interaction with the DNA at the electrode/electrolyte interface. The magnitude of interaction was understood by calculating the heterogeneous rate constant values from SECM and EIS. Electrostatic attraction between DNA and Ni-MPTPP found highest and it was least in the case of H_2MPTPP [27]. Based on the strong interaction between iron porphyrin and DNA, a composite film prepared out of this combination was used for electrochemical sensing of p-nitrophenol [28].

Moving on from electrochemical properties of a monolayer of porphyrins, multilayered films were constructed. Gold nanoparticles (AuNPs) and 5,15-di-[p-(6-mercaptohexyl)-phenyl]-10,20-diphenylporphyrin (trans-PPS_2) were used as inorganic and organic materials, respectively, to form hybrid multilayer film on the gold electrode. Electrochemical property particularly heterogeneous charge transfer constant, k_{eff} was deduced at different stages of multilayer formation using EIS and SECM. Irrespective of the number of layers on the gold electrode, the material present at the film/electrolyte interface influence the charge transfer resistance of the multilayer. Low charge transfer resistance was observed when the AuNPs are present as the final layer, whereas the trans-PPS_2 at the interface resulted in the high charge transfer resistance. SECM images were recorded by exposing the AuNPs or trans-PPS_2 to the interface. In the case of AuNPs' layer at the interface greater extent of normalized current was seen compared to that of trans-PPS_2 confirmed the charge transfer blocking behavior of porphyrin layers. As a conclusion, charge transfer between the gold electrode and the AuNPs decrease with the increase in a number of layers of trans-PPS_2 between them [29].

3.3. SECM

So far, electrochemistry of porphyrin molecules at the solid-liquid interface studied using basic techniques such as CV and EIS was discussed. Deducing charge transfer constants using CV is simple and straightforward. But, the factors such as resistive potential drop and double-layer charging current pose an ambiguity on the reliability of the results obtained. SECM measures the steady state current using micro- or submicro-size of the tip. Hence, the measured current will also be very small that in turn minimizes the influence of resistive potential drop and double-layer charging current. Hence, SECM emerged as a versatile experimental technique to study the adsorption kinetics of films on various substrates, charge transfer kinetics across the thin films [30, 31]. Very few studies have also been reported on the investigation of porphyrin films using SECM [32, 33]. Understanding and optimizing the long-range charge transfer across the nanometer thickness films is of prime technological importance in various research fields. Theoretical and experimental approaches have been developed to study such cases using SECM [34].

SAMs of thiol-porphyrins, $H_2TPPO(CH_2)nSH$ with varied alkyl chain length were formed on the gold electrode to investigate the electron transfer between the electrode and the redox mediator, $[Fe(CN)_6]^{3-}$ present in the electrolyte. Three pathways for the electron transfer were proposed. (I) Mediated electron transfer, in this case, product formed at the tip will be regenerated by the bimolecular reaction. That is the film is also redox active. (II) Tunneling through the film, in this case, film is electro-inactive. Hence, the product formed at the tip of the SECM is regenerated at the electrode by tunneling through the film. (III) Imperfections of the SAM such as pinholes and defects give a way for charge transfer. Theoretical approximations were deduced for all the three situations to calculate the SECM tip current. Experimental approach curves were recorded for the SAMs of different thickness and fitted with the theoretically simulated curves to extract the heterogeneous charge transfer constant, k_{eff}, values for all the three cases. SECM investigation of the electron transfer in porphyrin systems through the bimolecular process closely resembles the charge transfer in photosystem II [35]. Highly ordered monolayers can be obtained by first assembling the alkanethiols on the gold electrode as a template then introducing the porphyrin molecules on to it. When the alkanethiols were used as templates, the surface coverage of the electrode and electron transfer significantly altered. Also, porphyrin molecules stood perpendicular to the electrode. k_{eff} value was significantly improved after introducing the cobalt ions into free porphyrin bases of such monolayer [36]. Electroactive zinc porphyrin films were produced on the transparent electrode by electropolymerization using the bipyridinium as a pendant molecule between the two molecules. The permeability of the film and the charge transport within the film was studied under conditions similar to photovoltaic devices. Four different organic redox mediators were used to record the approach curves. Lateral charge transportation between the adjacent redox active sites was also probed [37]. Micropatterning of transparent electrodes (substrate) was done using zinc porphyrin molecular squares to produce a film composed of meso- and micro-porous material. SECM was used to study the molecular sieving and permeability of these films. Cavity size of the sieves was controlled by modifying the porphyrinic squares and the permeability was examined by using the redox mediators with smaller in size compared to that of the cavity. It was observed from the results that the steric property of the redox ion needed for charge compensation also played a significant role in deciding the permeability of the redox species in addition to its size. SECM tip current is observed only if the redox mediators generated at tip successfully accesses the underlying substrate through the cavity of the film and diffuse back to the tip. Based on this substrate generation/tip collection mode of the SECM, imaging of the film was done to understand its sieving ability [38]. Complete removal of thiol-porphyrin SAMs from the electrode is also as important as producing the perfect SAMS to get the clean electrode surface. It is difficult to mechanically wash off the SAMs from the electrode surface. But it can be done electrochemically by applying the sufficient negative potential to the modified electrode [39].

In all the above examples, the charge transfer across the porphyrin film/electrolyte interface was originated as a result of the applied potential. But, the photo-excitation of the porphyrins followed by charge transfer in the dyads and triads systems is well known and widely adopted for the construction of photovoltaic devices and artificial photosynthetic systems. Hence, a model system to study the photoelectrochemical properties of the porphyrin molecules is of prime importance. Our group proposed a novel experimental methodology to study the photoinduced charge transfer kinetics of the porphyrin films using SECM [40]. Porphyrin coated on the transparent electrode will be excited by illuminating light from the bottom to result in the porphyrin cation. The reduced form of the redox mediator at the SECM tip will diffuse toward the film and reduce the porphyrin cation back to its original form by undergoing oxidation. Redox mediator oxidized at the film will travel back to the tip, thereby diffusion cone of redox mediator was generated between the SECM tip and porphyrin-coated electrode (**Figure 4**). Hence, the tip current was increased as it moves close to substrate, that is, positive feedback. Through this bimolecular reaction, SECM tip was used to capture the photoinduced charge by performing the probe approach curve experiment. Such diffusion cone does not form when the light was not illuminated according to the above mechanism. Therefore, negative feedback was obtained when the probe approach curve was recorded (**Figure 5**).

Theoretical equations were proposed for tip current (I_T) to fit the experimentally obtained approach curves to directly extract the k_{eff} values [41].

$$I_T^k = I_S^k \left(1 - \frac{I_T^{ins}}{I_T^{cond}}\right) + I_T^{ins} \quad (3)$$

$$I_S^k = \frac{0.78377}{L\left(1 + \frac{1}{\Lambda}\right)} + \frac{0.68 + 0.3315 exp^{\left(\frac{-1.0672}{L}\right)}}{1 + \left(\frac{11}{\Lambda + 7.3}\right)/(110 - 40L)} \quad (4)$$

Where, I_T^{ins} and I_T^{cond} denote the tip current in case of insulating and conducting substrates, respectively. In this case, the transparent electrode is a conducting substrate, hence

$$I_T^{cond} = 0.68 + \frac{0.78377}{L} + 0.3315\ \exp^{\left(\frac{-1.0672}{L}\right)} \quad (5)$$

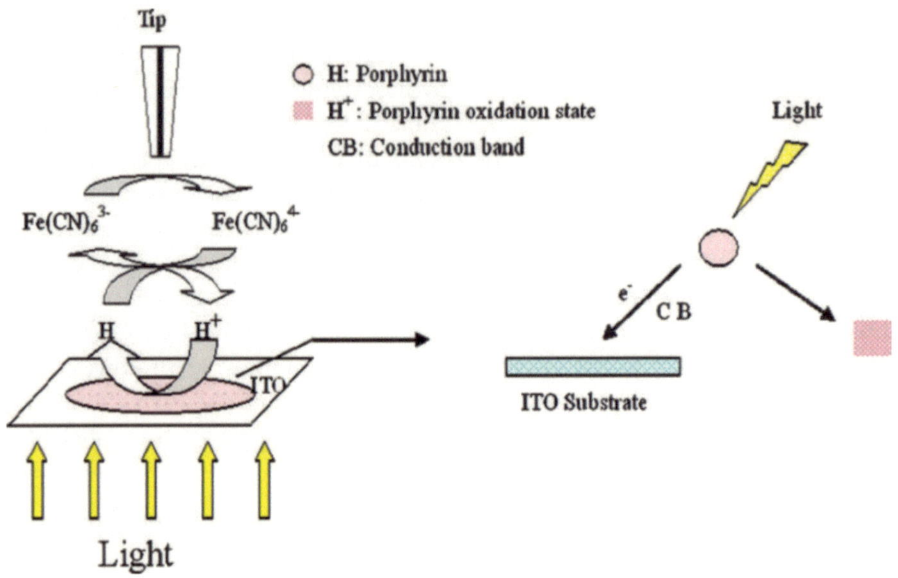

Figure 4. Cartoon represents the photoinduced excitation of porphyrin coated on the transparent electrode followed by the bimolecular reaction due to the presence of tip reduced species. Adapted from Ref. [40] with permission from The Royal Society of Chemistry.

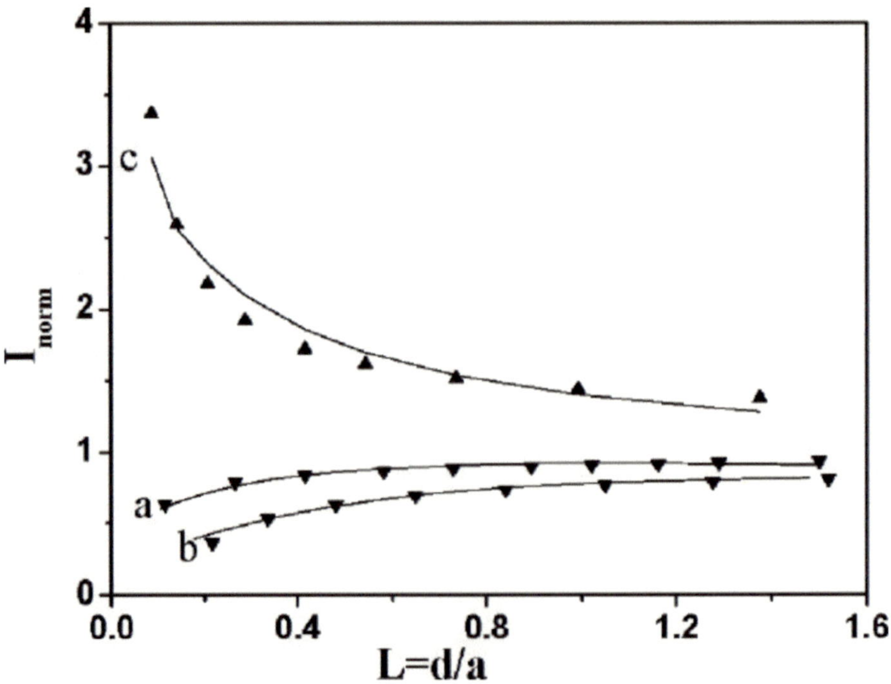

Figure 5. Experimental probe approach curves (dotted lines) fitted with the theoretical ones (solid lines) for bare ITO electrode (a), ITO coated with porphyrin in the absence of light (b) and in the presence of light (c). Adapted from Ref. [40] with permission from The Royal Society of Chemistry.

http://dx.doi.org/10.5772/67637

where $\wedge = \frac{d\, k_{eff}}{D}$, k_{eff} is the heterogeneous charge transfer constant, d is the radius of the electroactive part of the SECM tip and D is the diffusion coefficient of the mediator used. One will arrive with the value of $\wedge$ after fitting experimental approach curves with theoretical ones. Then by substituting the value of d, D and $\wedge$ in the above equation k_{eff}(cm s^{-1}) value can be obtained.

Above methodology was adopted to study the influence of various parameters on the PCT kinetics of zinc porphyrin across the solid/liquid interface using benzoquinone (BQ) as a redox mediator. The family of approach curves was recorded by varying the parameters such as wavelength, the intensity of the light source and for the different concentration of the mediator. The favorable condition for the PCT resulted in the greater k_{eff} value [42]. A simple model was constructed using the combination of AuNPs, porphyrin and CNT to mimic the natural photosynthesis system. Core-shell structured composite of AuNP-porphyrin was adsorbed on the vertically aligned CNT on the ITO electrode. The presence of AuNPs at the center of the vesicle structure diminishes the recombination of photogenerated charges and facilitates interfacial charge transfer. CNTs will successfully transfer the received electrons to ITO electrode. These effects were photoelectrochemically studied by recording the approach curves using benzoquinone as a redox mediator. PCT kinetics of this model was found to be dependent on the concentration of electrochemically active benzoquinone. This behavior resembles the role of plastoquinone in natural photosynthesis [43]. Porphyrin molecules loaded on the TiO_2 nanowire array grown on the ITO electrode by hydrothermal method. Then, change in the PCT kinetics with respect to the length of nanowire array was tested using SECM. With the increase in the length of the nanowire, k_{eff} value also becomes greater, may be due to greater amount porphyrin loading [44].

4. Electrochemiluminescence (ECL) of porphyrin

Electrochemiluminescence (ECL) involves a conversion of electrical energy into radiative energy. Fundamental principles, various luminophore systems, applications, and recent advances of ECL have been discussed in detail elsewhere [45, 46]. Polypyridyl complexes are the excessively studied luminophores so far. Because of the rich photo and electrochemical properties, porphyrins have also been used as luminophores in ECL. There are two well-established mechanisms through which ECL can be produced. First one is the annihilation mechanism: In this, a potential of the electrode alternatively pulsed between the two values to produce the oxidized and reduced species of the luminophore. These electrogenerated species at the vicinity of the electrode will interact with each other to produce the excited species, which will return back to the ground state by emitting the radiation [47].

$$\mathrm{TPP} \rightarrow \mathrm{TPP}^{+\bullet} + e^- \tag{6}$$

$$\mathrm{TPP} + e^- \rightarrow \mathrm{TPP}^{-\bullet} \tag{7}$$

$$\mathrm{TPP}^{+\bullet} + \mathrm{TPP}^{-\bullet} \rightarrow \mathrm{TPP}^* + \mathrm{TPP} \tag{8}$$

$$\mathrm{TPP}^* \rightarrow \mathrm{TPP} + h\nu \tag{9}$$

Another way of generating the ECL is by coreactant mechanism: in this, coreactant species upon oxidation or reduction will generate an intermediate, which will further react with the luminophore to cause the excitation. For example, oxalate ion upon oxidation produces the strong reductant. Hence, it is also called as "oxidative-reductive" coreactant [48].

$$C_2O_4^{2-} - e^- \rightarrow [C_2O_4^{-\bullet}] \rightarrow CO_2^{-\bullet} + CO_2 \quad (10)$$

ECL luminophore present in the system also undergoes oxidation at the same potential. For example, tetrakis(3-sulfonatomesityl)porphyrin (H_2TSMP) [49].

$$H_2TSMP - e^- \rightarrow H_2TSMP^{+\bullet} \quad (11)$$

Then the reaction takes place between the oxidized porphyrin and $CO_2^{-\bullet}$ produced from the coreactant to result in the excited porphyrin, which will emit radiation.

$$H_2TSMP^{+\bullet-} + CO_2^{-\bullet-} \rightarrow H_2TSMP^* \quad (12)$$

$$H_2TSMP^* \rightarrow H_2TSMP + hv \quad (13)$$

There is another category of coreactant referred to as "reductive-oxidative," that is, reduction of the coreactant will produce the strong oxidant species. Peroxydisulfate ($S_2O_8^{2-}$) can be mentioned as an example.

$$S_2O_8^{2-} + e^- \rightarrow SO_4^{-\bullet} + SO_4^{2-} \quad (14)$$

Let us consider luminophore, meso-tetra(4-sulfonatophenyl)porphyrin (TSPP) which is also undergoing reduction to produce radical anion under the same conditions [50].

$$TSPP + e^- \rightarrow TSPP^{-\bullet} \quad (15)$$

$$TSPP^{-\bullet} + SO_4^{-\bullet} \rightarrow +SO_4^{2-} \quad (16)$$

$$TSPP^* \rightarrow TSPP + hv \quad (17)$$

Based on the above mechanisms, ECL sensors have been developed from our group to quantify pheophorbide, Cu^{2+}, meso-tetra(4-carboxyphenyl) porphyrin [50–52]. ECL behavior of ruthenium and zinc porphyrins has been electrochemically investigated [53, 54]. Different porphyrin molecules have been studied in combination with clay, carbon nitride, and graphene to improve the intensity of the ECL signal and to achieve the applicability [55–57].

5. Porphyrins at liquid-liquid interface

Investigation of the charge transfer process at the liquid-liquid interface, that is, interface between the two immiscible electrolytes (ITIES) has got significance because of its mimicking nature of various fields such as phase transfer catalysis, biomembranes, and drug delivery systems. One of the important outcomes from ITIES studies is the dependence of charge transfer on the driving force. Three important charge transfer processes have been studied at the ITIES.

- Ion transfer
- Electron transfer across the ITIES
- Ion transfer across the ITIES facilitated by the complexing agent

More detailed information about theory, techniques used to study, and applications of ITIES have been discussed in detail in the review by Pekka and Hubert [58]. Generally, in ITIES system, one out of the two phases will be rich of redox species and can be considered metal-like. The potential drop across the ITIES is referred to as the difference of the Galvani potentials of the organic and aqueous phases, $\Delta^{\circ}_{\omega}\varnothing$. At low overpotentials, the exponential dependence of electron transfer (ET) rate constant on the $\Delta^{\circ}_{\omega}\varnothing$ can be expected (Butler-Volmer theory). Whereas at very high overpotentials, the ET rate constants will level off to follow Marcus theory [59].

The theory has been developed to split the complex multistep ET reaction into several one-step processes and the ET rate constant of each step can be calculated. Taking ZnTPP/$[Fe(CN)6]^{4-}$ as a model system, effect of concentration of the species in the two phases, a thickness of the thin layer on the multistep ET processes was studied using thin layer cyclic voltammetry (TLCV). Experimentally obtained results were found to be in good correlation with the theoretical simulations [60]. Owing to the close resemblance of iron porphyrin with the heme, ET kinetics of the various substituents bearing iron porphyrin was studied by constructing the artificial membrane in the form of ITIES (**Figure 6**). Effect of substituent with different electron affinity on the ET kinetics was understood by recording the approach curves using SECM. Good agreement between the experimentally obtained rate constant values and the electronic structure and molecular orbital energies calculated by the density functional theory was observed. More the number of electron donating substituent, more easily iron porphyrin will tend to lose the electron(s) at the ITIES [61].

Encouraged by these results, ET kinetics of iron porphyrin substituted with a range of electron accepting and donating groups was investigated by choosing the benzoquinone as a redox mediator. Dependence of the ET kinetics at the nitrobenzene-water ITIES of the porphyrins was envisaged using SECM. Both Butler-Volmer and Marcus inverted region ET kinetics were observed with the increase of low and high overpotentials as driving force, respectively [62]. We extended our study for the complex two-step electron transfer process, that is, zinc porphyrin substituted with the electron withdrawing groups. As expected, the oxidation potentials of the zinc porphyrin were positively shifted in the voltammograms recorded by TLCV. But, the ET kinetic data for the three different zinc porphyrin at the ITIES were not in line with the theoretical approximation [63]. Consecutive ET kinetics of zinc porphyrin was investigated by extending its π-conjugation using phenyl, naphthyl, and pyrenyl groups. ET rate estimated from the TLCV experiments were found to be slow if the substituted molecule is small and that of larger molecules was fast [64]. From the SECM experiments, it was revealed that, increase in the driving force lead to the slowdown of the ET rate. This was explained by taking the stereostructures of the molecules into consideration. The structure of the molecule has a dominating effect on the ET rate over that of Galvani potential of the ITIES [65].

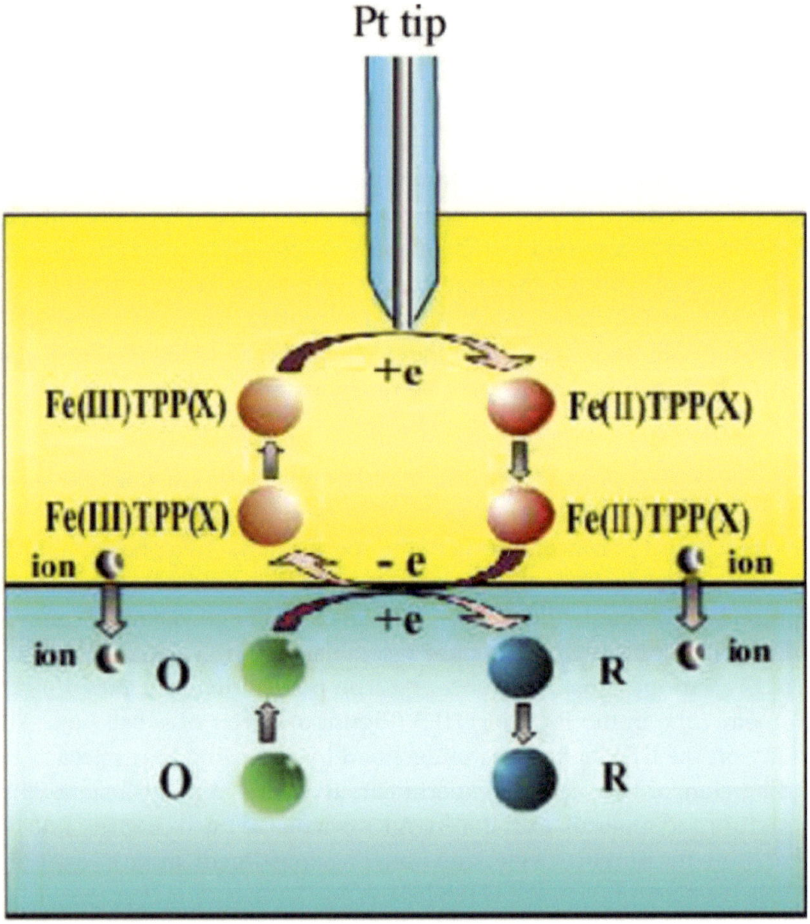

Figure 6. Schematic of the bimolecular redox reaction at the ITIES between the iron porphyrin taken in the nitrobenzene and $[Fe(CN)_6]^{3-}$ present in the aqueous phase. Reprinted from Ref. [61]. Copyright 2010 Elsevier Ltd.

6. Conclusions

Desired changes in the electronic properties of the porphyrin molecules achieved by tailoring the macrocyclic ring are investigated in detail using electrochemical techniques. Thiol-derivatized porphyrins are used to produce the self assembled monolayer (SAM) on the gold electrode to study their behavior at the solid/liquid interface. The SAMs composed of porphyrin molecules have been electrochemically investigated for the arrangement of molecules in it, imperfections and charge transfer across it. Simple and straightforward electrochemical methodologies have been developed to estimate the heterogeneous charge transfer constant at the solid/liquid and liquid/liquid interfaces using scanning electrochemical microscope (SECM). Porphyrin molecules have also been used in the electrochemical and electrochemiluminescence sensing applications.

Author details

Xiaoquan Lu* and Samrat Devaramani

*Address all correspondence to: luxq@nwnu.edu.cn

Key Laboratory of Bioelectrochemistry and Environmental Analysis of Gansu Province, College of Chemistry and Chemical Engineering, Northwest Normal University, Lanzhou, PR China

References

[1] Stéphanie D, Julien T, Valérie H. Multiporphyrinic cages: architectures and functions. Chem. Rev. 2014;**114**:8542–8578. doi:10.1021/cr400673y

[2] Gema de la T, Giovanni B, Michael S, Anita H, Dirk MG, Tomas T. A voyage into the synthesis and photophysics of homo- and heterobinuclear ensembles of phthalocyanines and porphyrins. Chem. Soc. Rev. 2013;**42**:8049–8105. doi:10.1039/C3CS60140D

[3] Chao H, Qingguo H, Changmin D, Liqi S, Defeng Z, Yanyan F, Huimin C, Jiangong C. Turn on fluorescence sensing of vapor phase electron donating amines via tetraphenylporphyrin or metallophenylporphrin doped polyfluorene. Chem. Commun. 2010;**46**:7536–7538. doi:10.1039/C0CC01972K

[4] Wallace WHW, Tony K, Doojin V, Chao Y, David JJ, Maxwell JC, Andrew BH. A porphyrin-hexa-peri-hexabenzocoronene-porphyrin triad: synthesis, photophysical properties and performance in a photovoltaic device. J. Mater. Chem. 2010;**20**:7005–7014. doi:10.1039/C0JM00311E

[5] Karl MK, Eric VC. Electrochemistry of porphyrins and related macrocycles. J. Solid State Electrochem. 2003;**7**:254–258. doi:10.1007/s10008-002-0306-3

[6] Ping C, Olga SF, Zhongping O, Sergei AV, Karl MK. Alectrochemistry of platinum(II) porphyrins: effect of substituents and π-extension on redox potentials and site of electron transfer. Inorg. Chem. 2012;**51**:6200–6210. doi:10.1021/ic3003367

[7] John AS, Xing-Zhi S, Jian-Guo M, Song-Ling J, Walter J, Craig JM, Craig JM. Nonplanar porphyrins and their significance in proteins. Chem. Soc. Rev. 1998;**27**:31–42. doi:10.1039/A827031Z

[8] Karl MK, Erin VC, Guy R. Electrochemistry of metalloporphyrins in non-aqueous media. In: Karl M Kadish, Kevin MS, Roger G, editors. The porphyrin handbook, Vol 8. San Diego, CA: Academic Press; 2000, pp. 1–114

[9] Kadish KM, Guy R, Erin VC, Gueletti L. Metalloporphyrins in nonaqueous media: database of redox potentials. In: Karl MK, Kevin MS, Roger G, editors. The porphyrin handbook, Vol 9. San Diego, CA: Academic Press; 2000, pp. 1–219

[10] Fupeng Z, Xiaoquan L, Jiandong Y, Xiaoyan W, Hui S, Shaohua Z, Zhonghua X. Selective anion sensing through a self-assembled monolayer of thiol-end-functionalized porphyrin. J. Phys. Chem. C. 2009;**113**:13166–13172. doi:10.1021/jp9003278

[11] Xiaoquan L, Dongxia Z, Zhengen S, Bowan W, Bingzhang L, Xibin Z, Zhonghua X. A valuable visual colorimetric and electrochemical biosensor for porphyrin. Biosensors Bioelectron. 2011;**27**:172–177. doi:10.1016/j.bios.2011.06.043

[12] Xiaoquan L, Yanli Q, Zhonghua X, Bowan W, Hetong Q, Dong L. Determination of explosives based on novel type of sensor using porphyrin functionalized carbon nanotubes. Colloids Surf. B Biointerfaces. 2011;**88**:396–401. doi:10.1016/j.colsurfb.2011.07.020

[13] Yanli Q, Zhonghua X, Haicai S, Xibing Z, Jie D, Xiuhui L, Xiaoquan L. A high-performance and simple method for rapid and simultaneous determination of dihydroxybenzene isomers. Analyst. 2012;**137**:944–952. doi:10.1039/C1AN15945C

[14] Yaqi H, Zhonghua X, Hongxia H, Ruixia A, Xiuhui L, Xiaoquan L. Photoelectrochemical sensing for hydroquinone based on porphyrin-functionalized Au nanoparticles on graphene. Biosens. Bioelectron. 2013;**47**:45–49. doi:10.1016/j.bios.2013.02.034

[15] Xiaoquan L, Duoliang S, Jianmin Y, Baomei H, Xibin Z. Determination of m-dinitrobenzene based on novel type of sensor using thiol-porphyrin mixed monolayer-tethered polyaniline with intercalating fullerenols. Talanta. 2013;**115**:457–461. doi:10.1016/j.talanta.2013.06.002

[16] Guofang Z, Xiuhui L, Jiandong Y, Xiujuan L, Xiaoquan L. Study of the adsorption kinetics of thiol-derivatized porphyrin on the surface of gold electrode. J. Electroanal. Chem. 2007;**605**:81–88. doi:10.1016/j.jelechem.2007.03.020

[17] Xiaoquan L, Huiqing Y, Guofang Z, Jiandong Y. Study of the size and separation of pinholes in the self-assembled thiol-porphyrin monolayers on gold electrodes. Thin Solid Films. 2008;**516**:6476–6482. doi:10.1016/j.tsf.2008.02.036

[18] James EH, Timothy AP, Chun-hsien C, Kevin WH, Roychelle SI, Wei O, Richard WL, Royce WM. Electrocatalytic activity of an immobilized cofacial diporphyrin depends on the electrode material. Langmuir. 1997;**13**:2143–2148. doi:10.1021/la960980u

[19] Katsuaki S, Miwa T, Hiroshi F, Minoru S, Hiroshi S, Tetsuhiko Y, Kohei U. Formation and characterization of thiol-derivatized zinc (II) porphyrin monolayers on gold. Thin Solid Films. 1996;**273**:250–253. doi:10.1016/0040-6090(95)06790-6

[20] Guofang Z, Huiqing Y, Jiandong Y, Ruixue Z, Xiaoquan L. Study of orientation mode of cobalt-porphyrin on the surface of gold electrode by electrocatalytic dioxygen reduction. J. Mol. Catal. A Chem. 2007;**269**:46–52. doi:10.1016/j.molcata.2006.11.041

[21] Karl MK, Laurent F, Fabien B, Jean-Michel B, Claude PG, Roger G. Cobalt(IV) corroles as catalysts for the electroreduction of O_2: reactions of heterobimetallic dyads containing a face-to-face linked Fe(III) or Mn(III) porphyrin. J. Inorg. Biochem. 2006;**100**:858–868. doi:10.1016/j.jinorgbio.2006.01.010

[22] Christopher JC, Zhi-Heng L, Chunnian S, Fred CA, Daniel GN. Targeted proton delivery in the catalyzed reduction of oxygen to water by bimetallic pacman porphyrins. J. Am. Chem. Soc. 2004;**126**:10013–10020. doi:10.1021/ja049115j

[23] Qin W, Fupeng Z, Wenting W, Xinghua X, Xiuhui L, Fanfu M, Yanyan S, Chen Y, Xiaoquan L. Direct electron transfer of thiol-derivatized tetraphenylporphyrin assembled on gold electrodes in an aqueous solution. J. Phys. Chem. C. 2009;**113**:9359–9367. doi:10.1021/jp803725x

[24] Kin-ya T, Lianhe Y, Lingyun W, David FB, Jonathan SL. Synthesis of cyclic hexameric porphyrin arrays. Anchors for surface immobilization and columnar self-assembly. J. Org. Chem. 2003;**68**:8199–8207. doi:10.1021/jo034861c

[25] Amir AY, Dennis S, Vladimir LM, Robert SL, Jonathan SL, Francisco Z, David FB. Characterization of self-assembled monolayers of porphyrins bearing multiple thiol-derivatized rigid-rod tethers. J. Am. Chem. Soc. 2004;**126**:11944–11953. doi:10.1021/ja047723t

[26] Jiandong Y, Minrui L, Hongxiang L, Yanlian Y, Yoshiaki K, Chen W, Keiichi T, Xiaoquan L, Wenping H. Characterization and application of self-assembly porphyrin with four "clips" on gold surface. J. Phys. Chem. C. 2010;**114**:12320–12324. doi:10.1021/jp1020643

[27] Yan Z, Xiaoquan L, Tianlu L, Yina C, Xiuhui L, Limin Z. Studies on interaction of porphyrin and its complexes with DNA at interface on gold electrode modified by thiol-porphyrin self-assembled monolayer. J. Solid State Electrochem. 2007;**11**:1303–1312. doi:10.1007/s10008-007-0291-7

[28] Shen-Ming C, Sz-Vin C. The interaction of water-soluble iron porphyrins with DNA films and the electrocatalytic properties for inorganic and organic nitro compounds. Electrochim. Acta. 2003;**48**:4049–4069. doi:10.1016/S0013-4686(03)00562-0

[29] Xiaoquan L, Fupeng Z, Hui S, Xiaoyan W, Zhonghua X. Investigation of the electrochemical behavior of multilayers film assembled porphyrin/gold nanoparticles on gold electrode. Electrochim. Acta. 2010;**55**:3634–3642. doi:10.1016/j.electacta.2009.11.004

[30] Fardad F, Allen JB, Michael VM. Voltammetric and scanning electrochemical microscopic studies of the adsorption kinetics and self-assembly of *n*-alkanethiol monolayers on gold. Isr. J. Chem. 1997;**37**:155–163. doi:10.1002/ijch.199700019

[31] Céline C, Frédéric K, Allen JB. Cyclic voltammetric and scanning electrochemical microscopic study of menadione permeability through a self-assembled monolayer on a gold electrode. Langmuir. 2002;**18**:8134–8141. doi:10.1021/la0258906

[32] Jalal G, Fanny H, Bruno F, Philippe H. Scanning electrochemical microscopy investigations of monolayers bound to p-type silicon substrates. Anal. Chem. 2006;**78**:6019–6025. doi:10.1021/ac060058h

[33] Xiaoquan L, Limin Z, Minrui L, Xiaoqiang W, Yan Z, Xiuhui L, Guofang Z. Electrochemical characterization of self-assembled thiol-porphyrin monolayers on gold electrodes by SECM. Chem. Phys. Chem. 2006;**7**:854–862. doi:10.1002/cphc.200500492

[34] Biao L, Allen JB, Michael VM, Stephen E. Creager. Electron transfer at self-assembled monolayers measured by scanning electrochemical microscopy. J. Am. Chem. Soc. 2004;**126**:1485–1492. doi:10.1021/ja038611p

[35] Wenting W, Xiujuan L, Xiaoyan W, Hui S, Xiuhui L, Xiaoquan L. Comparative electrochemical behaviors of a series of SH-terminated-functionalized porphyrins assembled on a gold electrode by scanning electrochemical microscopy (SECM). J. Phys. Chem. B. 2010;**114**:10436–10441. doi:10.1021/jp1026064

[36] Wenting W, Yaqi H, Chunming W, Xiaoquan L. Comparative electrochemical investigations on series of SH-terminated-functional porphyrins. Electrochim. Acta. 2012;**65**:244–250. doi:10.1016/j.electacta.2012.01.049

[37] Yann L, Delphine S, Laurent R, Philippe H. SECM investigations of immobilized porphyrins films. Langmuir. 2010;**26**:14983–14989. doi:10.1021/la101294s

[38] Mary EW, Joseph TH. Scanning electrochemical microscopy assessment of rates of molecular transport through mesoporous thin-films of porphyrinic "molecular squares". J. Phys. Chem. B. 2001;**105**:8944–8950. doi:10.1021/jp010881b

[39] Zhonghua X, Hetong Q, Jie D, Bowan W, Xiuhui L, Xiaoquan L, Xibin Z. A novel method to remove self-assembled monolayer of porphyrin from the gold surface by cyclic voltammetry. J. Adhesion Sci. Technol. 2012;**26**:1521–1529. doi:10.1163/156856111X618371

[40] Wenting W, Duoliang S, Yong Y, Chunming W, Yaqi H, Xiaoquan L. A novel method for dynamic investigations of photoinduced electron transport using functionalized-porphyrin at ITO/liquid interface. Chem. Commun. 2011;**47**:6975–6977. doi:10.1039/c1cc10842e

[41] Michael VM. Theory. In: Allen JB, Michael VM, editors. Scanning electrochemical microscopy. New York: Marcel Dekker; 2001, pp. 145–199

[42] Xiaoquan L, Yaqi H, Wenting W, Jie D, Hongxia H, Ruixia A, Xiuhui L. A novel platform to study the photoinduced electron transfer at a dye-sensitized solid/liquid interface. Colloids Surf. B Biointerfaces. 2013;**103**:608–614. doi:10.1016/j.colsurfb.2012.11.013

[43] Xingming N, Liang M, Shouting Z, Dongdong Q, Duoliang S, Yaqi H, Xiaoquan L. Construction of a porphyrin-based nanohybrid as an analogue of chlorophyll protein complexes and its light-harvesting behavior research. Phys. Chem. C. 2016;**120**:919–926. doi:10.1021/acs.jpcc.5b11246

[44] Yuan J, Dong-Dong Q, Yan-Ru F, Hui-Xia G, Shi-Xia W, Xingming N, Xiaoquan L. Investigation of photoinduced electron transfer on TiO_2 nanowire arrays/porphyrin composite via scanning electrochemical microscopy. RSC Adv. 2015;**5**:56697–56703. doi:10.1039/c5ra08485g

[45] Mark MR. Electrochemiluminescence (ECL). Chem. Rev. 2004;**104**:3003–3036. doi:10.1021/cr020373d

[46] Lianzhe H, Guobao X. Applications and trends in electrochemiluminescence. Chem. Soc. Rev. 2010;**39**:3275–3304. doi:10.1039/b923679c

[47] Toby RL, Mark MR. Electrogenerated chemiluminescence of the platinum (II) octaethylporphyrin/tri-n-propylamine system. Inorg. Chim. Acta. 2005;**358**:2141–2145. doi:10.1016/j.ica.2004.12.017

[48] Israel R, Allen JB. Electrogenerated chemiluminescence. 37. Aqueous ecl systems based on ru(2,2′-bip~ridine)$_3^{2+}$ and oxalate or organic acids. J. Am. Chem. Soc. 1981;**103**:512–516. doi:10.1021/ja00401a031

[49] Fang-Chung C, Jinn-Hsuan H, Chin-Yu C, Oliver SY, Tong-Ing H. Electrogenerated chemiluminescence of sterically hindered porphyrins in aqueous media. J. Electroanal. Chem. 2001;**499**:17–23. doi:10.1016/S0022-0728(00)00439-3

[50] Jing Z, Samrat D, Duoliang S, Xiaoquan L. Electrochemiluminescence behavior of meso-tetra (4-sulfonatophenyl)porphyrin in aqueous medium: its application for highly selective sensing of nanomolar Cu^{2+}. Anal. Bioanal. Chem. 2016;**408**:7155–7163. doi:10.1007/s00216-016-9655-0

[51] Di L, Baomei H, Li W, Ahmed M, Shixia W, Xiaoquan L. Cathodic electrochemiluminescence of meso-tetra(4-carboxyphenyl) porphyrin/potassium peroxydisulfate system in aqueous media. Electrochim. Acta. 2015;**151**:42–49. doi:10.1007/s00216-016-9655-0

[52] Canty P, Väre L, Håkansson M, Spehar AM, Papkovsky D, Ala-Kleme T, Kankare J, Kulmala S. Time-resolved electrochemiluminescence of platinum(II) coproporphyrin. Anal. Chim. Acta. 2002;**453**:269–279. doi:10.1016/S0003-2670(01)01413-1

[53] Angela B, Mark MR. Coreactant electrogenerated chemiluminescence of ruthenium porphyrins. Inorg. Chim. Acta. 2009;**362**:1974–1976. doi:10.1016/j.ica.2008.09.022

[54] Guang-Yao Z, Sheng-Yuan D, Xue-Ji Z, Dan S. Cathodic electrochemiluminescence of singlet oxygen induced by the electroactive zinc porphyrin in aqueous media. Electrochim. Acta. 2016;**190**:64–68. doi:10.1016/j.electacta.2015.12.228

[55] Shengyuan D, Tingting Z, Xubo J, Ying W, Peng X, Dan S, Xueji Z. Detection of zinc finger protein (EGR1) based on electrogenerated chemiluminescence from singlet oxygen produced in a nanoclay-supported porphyrin environment. Anal. Chem. 2015;**87**:9155–9162. doi:10.1021/acs.analchem.5b01318

[56] Shengyuan D, Peixin Y, Xubo J, Dan S, Xueji Z. Carbon nitride nanosheet-supported porphyrin: a new biomimetic catalyst for highly efficient bioanalysis. ACS Appl. Mater. Interfaces. 2015;**7**:543–552. doi:10.1021/am506645h

[57] Li W, Jiasi W, Lingyan F, Jinsong R, Weili W, Xiaogang Q. Label-free ultrasensitive detection of human telomerase activity using porphyrin-functionalized graphene and electrochemiluminescence technique. Adv. Mater. 2012;**24**:2447–2452. doi:10.1002/adma.201200412

[58] Pekka P, Hubert HG. Liquid/liquid interfaces, electrochemistry at. Encyclopedia of analytical chemistry. New Jersey: John Wiley & Sons, Ltd; 2006–2012.

[59] Biao L, Michael VM. Potential-independent electron transfer rate at the liquid/liquid interface. J. Am. Chem. Soc. 1999;**121**:8352–8355. doi:10.1021/jp013543j

[60] Xiaoquan L, Ping S, Dongna Y, Bowan W, Zhonghua X, Xibing Z, Ruiping S, Li L, Xiuhui L. Heterogeneous consecutive electron transfer at graphite electrodes under steady state. Anal. Chem. 2010;**82**:8598–8603. doi:10.1021/ac1016997

[61] Xiaoquan L, Junying M, Ruiping S, Mina N, Fanfu M, Jie D, Xiaoyan W, Hui S. Substituent effects of iron porphyrins: structural, kinetic, and theoretical studies. Electrochim. Acta. 2010;**56**:251–256. doi:10.1016/j.electacta.2010.08.088

[62] Xiaoquan L, Wenting G, Ruiping S, Xiuhui L. Investigation of electrochemical properties of metalloporphyrin species at the liquid/liquid interface by switching substitutes on the porphyrin ring. Electroanalysis. 2012;**24**:2341–2347. doi:10.1002/elan.201200396

[63] Xiaoquan L, Yao L, Ping S, Bowan W, Zhonghua X, Xiuhui L, Xibing Z. Investigation of the consecutive electron transfer of metalloporphyrin species containing different substituents at the liquid/liquid interface by thin-layer cyclic voltammetry. J. Phys. Chem. C. 2012;**116:**16660–16665. doi:10.1021/jp3061043

[64] Yanru F, Yu H, Shixia W, Yuan J, Duoliang S, Xiaoquan L. Investigation of interfacial consecutive electron transfer and redox behaviors of zinc-tetraarylporphyrins. Electrochim. Acta. 2016;**190**:419–425. doi:10.1016/j.electacta.2015.12.167

[65] Yanru Fan, Yu Huang, Yuan Jiang, Xingming Ning, Xuemei Wang, Duoliang Shan, Xiaoquan Lu. Comparative study on the interfacial electron transfer of zinc porphyrins with meso-p-extension at a 2^n pattern. J. Colloid Interface Sci. 2016;**462**:100–109. doi:10.1016/j.jcis.2015.09.063

Chapter 4

Deposited Transition Metal-Centered Porphyrin and Phthalocyanine Molecules: Influence of the Substrates on the Magnetic Properties

Heike C. Herper, Barbara Brena,
Sumanta Bhandary and Biplab Sanyal

Additional information is available at the end of the chapter

http://dx.doi.org/10.5772/intechopen.68224

Abstract

The field of molecular spintronics has gained much attention since molecules with-magnetic centers form natural magnetic units, which do not suffer from the size limitations of conventional electronics, opening a new path towards miniaturization. To fabricate devices, the molecules have to be deposited on a substrate. The key questions are the interaction of the molecules with the substrate and the control of the magnetic properties. Considering molecule-substrate hybrid interfaces as building blocks for spintronic devices, a deep understanding of the electronic structure and the coupling mechanisms is central to future applications. The orientation and reconstruction of the substrates can strongly affect the electronic and magnetic characteristics of the adsorbed molecule and drastically change the properties of the free molecules. In this chapter, we will discuss the interaction of transition metal-centered porphyrins and phthalocyanines with different types of substrates, for example, ferromagnetic transition metals or graphene sheets, in the framework of state-of-the-art density functional theory methods plus insights gained from X-ray absorption/X-ray magnetic circular dichroism experiments. The goal is to give an insight into the relevant processes on the atomic scale and to present possible routes to tailor magnetic properties in molecule-substrate hybrid structures.

Keywords: magnetic molecule, density functional theory, graphene, XAS, XMCD, spin-dipole moment, spin switching

1. Introduction

Single molecular magnets play an important role in realizing the device concepts of molecular nanospintronics. In this context, spin-valves based on molecular magnets have been proposed where the exchange coupling between the magnetic center in the molecule and the magnetic electrodes dictates the magnetoresistance [1]. Among the class of molecular magnets, organometallic molecules exhibit quite complex properties due to their low dimensionality and inherent confinement effects combined with an interplay between crystal field, Coulomb interaction, spin-orbit coupling, and orbital-dependent hybridization with ligands. They exhibit enormous prospects in the context of molecular electronics/spintronics exploring the subtle balance between different energy scales. Bistability, that is, realizing two different states in the molecule which can be accessed and manipulated with external means, forms the basis of device realization. In molecular magnets, the bistability can be achieved in terms of inherent spin state or through magnetic coupling between molecule and surface while adsorbed. As will be discussed in the following sections, the choice of surface plays a crucial role in exploring both the possibilities. The subtle balance of ligand field, Coulomb energy, and Hund's exchange makes molecules with "metallic core" formed out of transition metals particularly interesting. In the molecules of our interest for this chapter, porphyrins and phthalocyanines, Fe, Co, and Mn as metal core respond to the spin crossover feasibilities. The other exotic features, for example, Kondo effect, tunable magnetic coupling, spin-orbit coupling, and orbital-dependent hybridization with ligands, also appear in this class of molecules. The description with local density approximation (LDA) within density functional theory (DFT) thus becomes inadequate and leads to a large underestimation of the highest occupied molecular orbital (HOMO)-LUMO gap. The treatment of the electron correlation hence is essential and plays an extremely significant role in determining electronic configurations, magnetic anisotropy, etc., along with the spin state. The most popular method, DFT+U, includes electron correlation in Hartree-Fock manner staying within a single particle theory where U defines the Coulomb interaction. This explicit inclusion certainly improves the situation of HOMO-LUMO gap and provides a reasonable account for spin-state, electronic configurations, and the bonding situations in the case of adsorption on different surfaces [2, 3]. The biggest advantage is to be able to simulate large systems and, therefore, is mostly used for the results we have presented here. But the method has its own limitation for finding a single Slater determinant ground state leading to the overestimation of correlation effect. A more sophisticated method has been recently adapted that includes the many-body treatment of electron correlation. This is achieved by the so-called DFT++ method, which treats an interacting Hamiltonian within Anderson's impurity model on top of the DFT Hamiltonian [4].

Although similar in structure, as shown in **Figure 1**, porphyrins and phthalocyanines behave a bit differently when adsorbed on magnetic substrates, resulting in different magnetic properties. In the examples that follow, a model system for the porphyrin molecules, that is, a porphine, is used. The porphine, a model system for the theoretical study of single-molecule magnets (SMM), maintains the same central macrocycle ring but lacks the various types of peripheral ligands that stabilize the porphyrins [5].

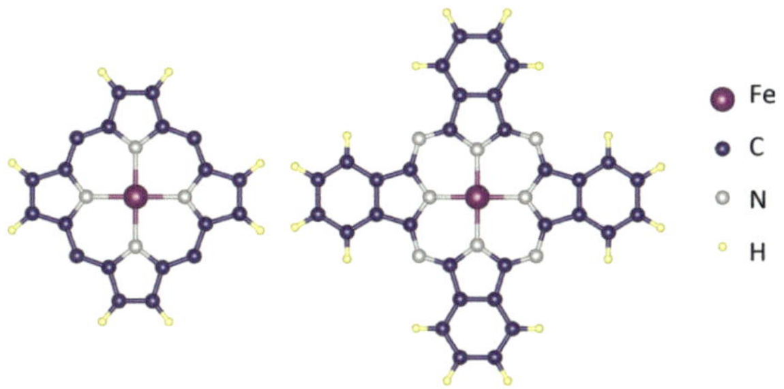

Figure 1. Porphine (left) and phthalocyanine (right) structures.

In iron porphyrin (FeP), Fe^{2+} in the center of the molecule is subjected to a square planar crystal field, along with a strong hybridization predominantly between Fe-$d_{x^2-y^2}$ and N-$p_{x/y}$ orbitals. In a strong crystal field scenario, electrons arrange themselves within the orbitals except for $d_{x^2-y^2}$ leading to an intermediate spin state, S = 1. However, a change in the ligand field lowers the cost of electron occupation in $d_{x^2-y^2}$ while simultaneously gains energy owing to Hund's exchange. As a result, the molecule exhibits a high spin state, S = 2. The energy landscapes corresponding to these two distinct spin states are however separated by an energy barrier. The estimated energy barrier within LDA+U is 0.81 eV [6]. In a square planar crystal field, structural change modifying ligand field boils down to the effective change of Fe-N bond length. In FeP, the Fe-N bond length corresponding to S = 1 spin state is 2.0 Å, while it needs to be stretched to 2.04 Å to achieve S = 2 spin state. The strain required for spin switching is however dependent on the impurity atom as well as the organic ligands. However, it is not only the gross description of spin state which should be the focus of the study. An accurate description of the electronic configuration is also required that in turns dictates crucial properties, such as magnetic anisotropy energy or spin-dipole moment. An explicit treatment of the electron correlation within many-body framework is adapted to describe the electronic structure of the molecules in addition to the DFT calculations, namely, DFT++ method. The method describes noninteracting part of the system within DFT, while an interacting Hamiltonian is treated within many-body framework, acquiring best parts of both the methods.

2. Surface effects

The deposition of molecules on suitable surfaces is an integral part of the device realization. The intrinsic gas-phase properties of molecules, however, are often diminished in that process due to structural deformation or strong chemical bonding. The focus of the chapter is to explore the feasibilities of manipulation of the molecular properties, employing surface molecular interactions yet retaining their key properties intact. Traditional electronic functions, such as memories, modulators, rectifiers, switches, transistors, and wires rely on the bistable nature associated with the charge/spin-state or molecule-surface coupling. Branching out toward spintronics, the intrinsic spin of molecules and/or magnetic coupling with a magnetic surface is exploited. In the following sections, we will explore the effects of different kinds of surfaces in controlling both the spin-state of a molecule and magnetic coupling with it.

2.1. Graphene as a reversible spin manipulator

Since the discovery in 2003, graphene has made an enormous impact on the advancement of "beyond silicon" electronics [7], as well as featuring exotic properties in multidisciplinary fields, such as gas sensing, batteries, drug delivery, understanding of high-energy experiments, and many more. We exploited its new role as an ideally ultrathin, robust 2D surface for the molecular adsorption of flat molecules like metal porphyrin (MP) or phthalocyanine (MPc) [6]. The synthesis of pristine graphene, either by exfoliation or chemical synthesis, often leaves several forms of defects on it which depending on its kind can change local structure or the whole layer structure. A "monovacancy" defect, created with a missing carbon atom, for example, has a local structural modification, while a "Stone-Wales" defect, where a much stronger reconstruction happens forming a pentagon-heptagon pair, leads to a strong ripple in the whole graphene layer. From the adsorption perspective, these distortions play crucial role and will be unveiled during the course of following discussions. However, with the present-day techniques, it is possible to create specific defects [8], which impart an access to the manipulation of molecular properties.

The adsorption of molecules or adatoms is enhanced in the presence of defects owing to the unsaturated bonds compared to that on pristine graphene with perfectly sp^2-bonded carbon network. We have investigated the adsorption scenario of FeP molecule on pristine and defected graphene. The feasibility of spin-state manipulation can be brought in only with a specific defect, "divacancy", where a pair of adjacent C atoms is missing. The binding energy of the FeP molecule on the "divacancy" site is 0.28 eV, which demonstrates its chemical stability. The molecule features almost similar structural properties as of the gas-phase molecule. However, the Fe atom experiences a vertical shift of 0.05 Å compared to that in flat gas-phase molecule. Along with that, the local defect structure modifies the square planar crystal field, experienced by the central Fe atom in the free molecule. In the electronic structure, the d_π-degeneracy which is inherent to the square planar crystal field is broken in the adsorbed molecule although the spin state remains unchanged. The schematic representation of the Fe-d orbital occupations is shown in the left part of **Figure 2**. The six electrons in Fe^{2+} ion are distributed as four and two in two different spin channel, resulting in an intermediate spin state ($S = 1$). The highest occupied molecular orbital (HOMO) is predominantly composed of Fe-d_z orbital, as seen in the isosurface plot in **Figure 2** (left inset). It can also be noted that the spin degeneracy is broken for the C atoms around the divacancy defect, which also contributes to the HOMO.

The surface component of the composite is exerted with a tensile strain. With as much as 1% strain, one brings sufficient change in the defect site that can affect the molecular spin state. The reconstruction in local defect structure is now prohibited with resulting stretching and unsaturated bonds, which in turn brings in sufficient mechanical strain in the molecule. The molecule switches to the high spin ($S = 2$) state. The crystal field, separating the $d_{x^2-y^2}$ and rest of the Fe-d orbitals, is reduced in the stretched molecule. The intra-atomic Hund's exchange overpowers the crystal filed barrier and Coulomb energy cost, and hence the spin crossover arises. The right part of **Figure 2** shows corresponding orbital occupation in the high spin state of the stretched molecule. As expected, the HOMO is now predominantly composed of Fe-$d_{x^2-y^2}$ and in-plane N-p orbitals, shown with an isosurface plot of magnetization density in the right inset of **Figure 2** [6].

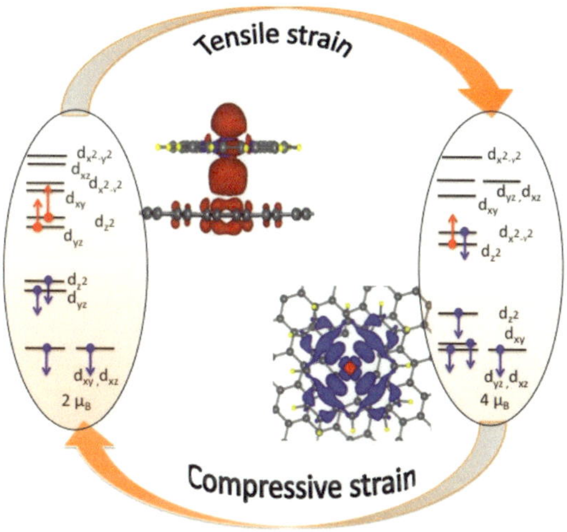

Figure 2. Magnetization density isosurfaces for FeP on (left) 0% and (right) 1% strained graphene. The isosurfaces have been plotted for an energy window of 0.4 eV below the Fermi levels in both cases. The upper (lower) plots are for spin-up (spin-down) densities. The energy levels with the d-orbital character of FeP are shown in the extreme left and right for 0 and 1% strained graphene, respectively. Data from Ref. [6].

The device realization relies on the feasibility of the controlled manipulation. This is the key achievement in this composite system. The strain engineered spin crossover is "reversible." The hysteresis effect can be brought in by applying a compressive strain (2%) to the graphene lattice. The intermediate spin state S = 1 is revived from the high spin state under compressive strain. This allows us to envisage the composite as an ultrathin spintronic device having logic states assigned as "0" and "1" corresponding to S = 1 and S = 2 spin states, respectively, which can be influenced by external strain. The strain engineered spin crossover, however, is very specific to the type of the defect in the graphene lattice and the transition metal atom in the porphyrin molecule. With a reasonable amount of strain in pristine graphene or with "monovacancy," it is not feasible to induce spin crossover in FeP.

2.2. Magnetic substrate manipulation of the spin state by chemisorption

The adsorption of iron porphyrin (FeP) on single crystal magnetic surfaces of Co and Ni shows a possible pathway to tune the molecular spin through deposition on a magnetic surface [2, 3]. The FeP molecule in the gas phase is characterized by an intermediate spin state (S = 1) [9]. In this spin configuration, the six 3d electrons of the Fe ion are distributed, four in the majority spin channel and two in the minority spin channel. A strong interaction between the molecule and the substrate can alter the electronic structure of the molecule and consequently can even affect the magnetic moment. This is even more likely to happen if the molecule adsorbs flatly, maximizing for all its constituent atoms the possibility of hybridization with the substrate. This is precisely the case of FeP on the Co and Ni substrates studied, where partially filled 3d shells can further favor a strong interfacial interaction.

DFT studies of the adsorption of FeP were performed on surfaces of different crystallographic symmetries like Ni(111) with hexagonal unit cell, Ni(110) with a rectangular unit cell, and

Ni(001) and Co(001) with square unit cells [2, 3]. The optimization of the adsorption geometries started from high-symmetry positions, like the top position, where the Fe ion is located on top of a Co or Ni atom; the bridge position, where the Fe sits in-between two metal atoms; and the hollow position where the Fe sits in-between more than two metal atoms. Different orientations of the molecule with respect to the surface major axes were also tested. In all these cases, the geometry relaxation of the hybrid interfaces resulted in short adsorption distances of about 2.0 Å or less, with a robust molecule-metal interaction indicating chemisorption. In all cases, a ferromagnetic coupling was observed. The relaxation of FeP on all these surfaces appears to be driven by the hybridization of the N atoms with the underlying Ni or Co metal atoms [2, 3]. In fact, in the case in which the morphology made it possible, the relaxations resulted in configurations where the N atoms were minimizing their distance with the Ni or Co atoms. Thus, N atoms are in many cases sitting on top positions or in proximity of a top position. To position the N atoms in this way brings about in many cases a geometrical stretching of the molecule and specifically an elongation of the intramolecular bonds between the Fe and the N atoms. The Fe-N bond lengths increased from the calculated bond length of about 1.9 Å in the gas phase [1] to values larger than 2.0 Å in the strained positions. The increase of the Fe-N bond length is directly responsible for a modification in the Fe 3d ligand field.

The different symmetries of these three substrates produce slightly different molecule-surface hybridization. On the squared (001) surfaces of both Ni and Co, FePc can arrange itself in such a way that the distances of Fe from all the four N atoms are basically the same, for example, if the Fe ion sits on the high-symmetry top or hollow sites on the surface, in these cases, the overall symmetry of the molecule is maintained (panel A in **Figure 3**), but a stretching of the bond lengths is generally observed. On the rectangular and on the hexagonal surfaces, the Fe-N bond lengths are increased asymmetrically (panels B and C in **Figure 3**). However, in all cases, a similar mechanism is observed: the elongation of the Fe-N bond lengths in FeP above about 2.04 Å causes analogous alterations of the Fe 3d crystal field, bringing about a high spin state with S = 2 on the Fe. The Fe-N bond length elongations reach up to the value of 2.13 Å

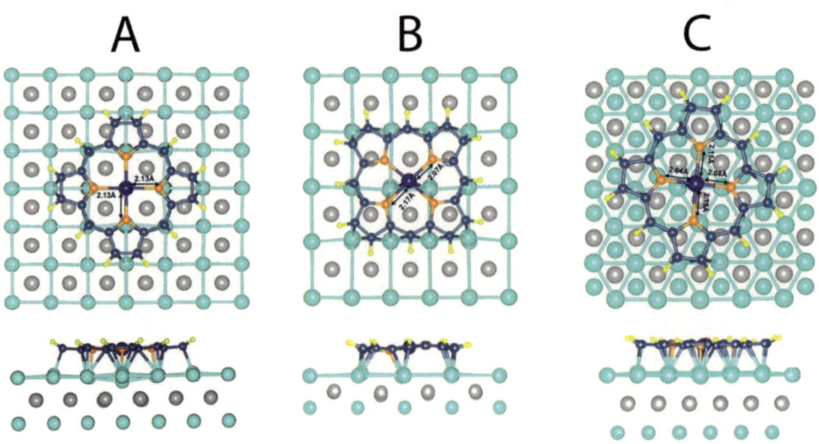

Figure 3. Top view and side view of relaxed adsorption structures for (A) FeP on Co(001), top site; (B) FeP on Ni(110), bridge site; and (C) FeP on Ni(111), hollow site. Data from Refs. [2, 3].

when adsorbed on the top site on Co(001), to values of 2.07–2.17 Å on the top site of Ni(110), to 2.04–2.15 Å on the bridge site of the Ni(111) surface. The molecular stretching corresponds a buckling of the top layer, as well as a contraction in the bond length of the underlying Co atoms in the surface layer. For example, the distance between two Co atoms beneath a FeP adsorbed on a top site on Co(001) is contracted from 2.50 to 2.44 Å.

To understand how the spin state is affected by the new strained structure of the molecule, it is necessary to look at how the electronic structure is modified by the changes in the molecular geometry. The intramolecular bonding between the N 2p levels and the Fe 3d is strictly coupled to the planar square ligand field generated by the Fe^{2+} ion, which dictates the energy separation between the 3d levels. A simple picture can illustrate how different spin states can arise in the gas phase and in the adsorbed FeP, depending on the ligand field splitting. In the gas-phase geometry, with shorter Fe-N distances, the energy splitting between the Fe 3d electrons and the N 2p electrons is such that the $d_{x^2-y^2}$ orbital remains unoccupied in both spin channels. In this case, FeP is in an intermediate spin (S = 1) configuration with four electrons in the majority and two in the minority spin channel. When the Fe-N distances are increased, the ligand field becomes weaker, leading to a smaller energy separation of the 3d states. In this case, electronic level occupation according to the Hund's rule prevails, and five orbitals in the majority spin channel and one in the minority are occupied, giving a high spin configuration (S = 2).

2.3. Magnetic coupling mechanisms between molecule and substrate

The magnetic coupling of SMM with magnetic surfaces has been studied for the iron phthalocyanine (FePc) [10, 11]. The FePc is a more stable molecule and can be easily purchased with a large variety of central atoms. Experimental studies by means of X-ray magnetic circular dichroism (XMCD) performed at HZB-BESSY II investigated the magnetic coupling between FePc and the Co(001) surfaces. The XMCD measurements were carried out at 32 and 300 K on 0.8 ML of FePc deposited first on the bare Co(001) and then on the same surface covered by an oxygen adlayer, forming a $O(\sqrt{2} \times 2\sqrt{2})R45°$ superstructure on top of the Co(001) surface. The XMCD spectra reported in **Figure 4** show how the ferromagnetic (FM) coupling of the FePc with the Co substrate is turned into antiferromagnetic (AFM) coupling in the presence of the O adlayer. The adsorption of FePc on the Co(001) surface has different characteristics with respect to the FeP.

A DFT investigation using the PBE+U method and including van der Waals corrections (D2) confirmed the FM coupling of the FePc to the Co surface. Structural relaxations starting from different high-symmetry adsorption sites, and from one nonsymmetric adsorption site, lead in all cases to chemisorption and FM coupling of the molecule on the surface. A detailed analysis of the ground state configuration (**Figure 4**) shows that the planar structure of the molecule is perturbed by a buckling, where the Fe and the N atoms directly bonded to Fe are lifted by 0.28 and 0.35 Å with respect to the metal substrate. This indicates that the C atoms in the benzene rings and the outer N atoms play a larger role in the hybridization with the surface, since they are closer to the underlying atoms. This enlightens a different adsorption mechanism in comparison to the FeP described in the previous section, where the

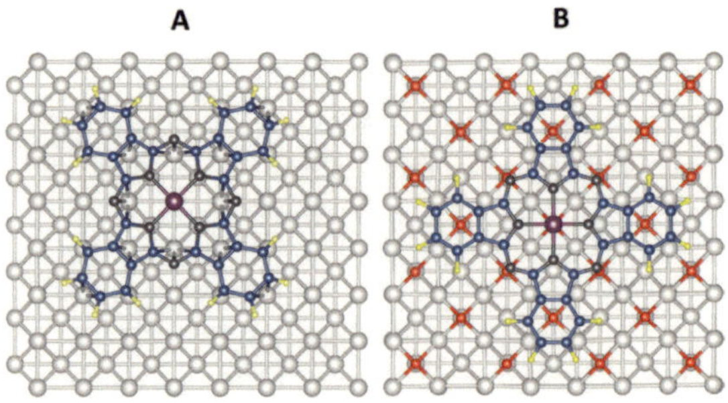

Figure 4. (A) Ground state of FePc adsorbed on Co(001) (top site). (B) Ground state of FePc adsorbed on O 2 × 2/Co(001) (top site on O atom). The dark circles indicate the O adatoms on the Co(001) surface.

hybridization with the surface mainly involved the N atoms. Moreover, the Fe-N bonding lengths are not strongly changed when the FePc are adsorbed, with calculated values of 1.97 Å in comparison to experimental values of 1.93 Å [10]. The adsorbed FePc is in the intermediate spin state (S = 1) as in the gas phase.

An important change in the magnetic structure of the hybrid interface is obtained when the Co surface is decorated by a 2 × 2 O layer. The presence of the O layer reduces the interaction of the molecules with the substrate. In fact, the magnetic coupling between the FePc and the Co substrate turns into AFM, as shown by the XMCD spectra (**Figure 5**) and also confirmed from the DFT calculations.

The lowest energy adsorption configuration obtained for the bare Co(001) and for the Oxygen plus Co is shown in **Figure 4**. It has to be observed that the relaxation of the adsorbate system formed by the molecule plus the oxidized substrate leads to different possible adsorption positions having similar energies (i.e., energies differing by less than 0.7 eV), suggesting that the molecule could in principle occupy different adsorption sites. When the O adlayer is present, the center of the FePc molecule (i.e., the Fe ion) is positioned on top of an O atom, as had been also observed for FeP on the same substrate. In this case, the distance between the Fe ion and the O atom beneath amounts to 2.19 Å, suggesting the possible formation of a chemical. The distance between the Fe and the Co atom beneath the O is 3.17 Å, indicating a weaker coupling between the Fe and the Co surfaces than was obtained for the direct adsorption on the bare Co(001). The former configuration favors an AFM superexchange coupling between the Fe(II) ion in the FePc and the underlying Co atom via the O atom in-between. An illustration of this comes from the magnetization density isosurface depicted in **Figure 6** that shows how the Fe in FePc is ferromagnetically coupled with the Co atoms while it is antiferromagnetically coupled to the same atoms in the presence of the O adlayer.

2.4. FeP on graphene/Ni(111): defect controlled magnetism

A free-standing graphene lattice with sp^2-bonded C network is spin degeneracy. The creation of specific kinds of defects breaks the spin degeneracy locally, but the correlation among

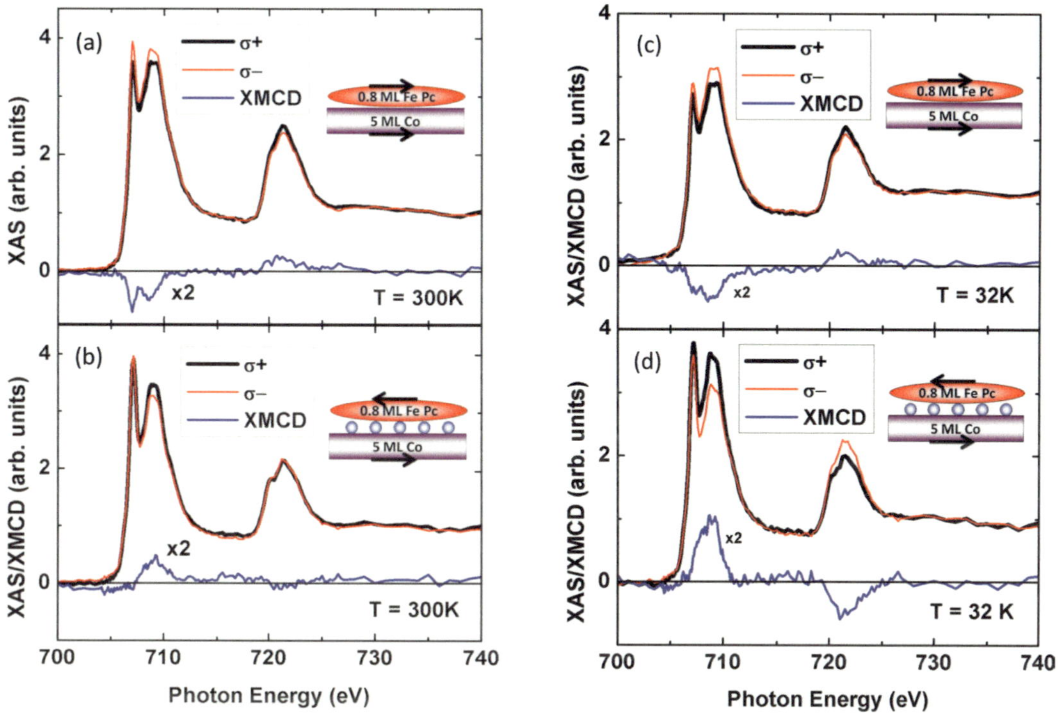

Figure 5. Fe $L_{2,3}$ XMCD spectra of 0.8 ML FePc adsorbed on (a) bare Co(001) and on (b) O/Co(001). The spectra are measured at T = 300 (left panel) and 32 K (right panel) and at a grazing angle of 70°. Data from Ref. [10].

those moments is insignificant. For the applications in the realm of molecular switch or valve, the role of graphene is inadequate albeit being extremely interesting in bringing in the spin crossover in FeP molecule. A very strong magnetic coupling and spin crossover is feasible with molecular adsorption on magnetic surface, as described in Section 2. Both the magnetic coupling and the spin state are extremely robust and are hard to manipulate. From this aspect, a composite surface, composed of a magnetic surface, such as Ni and Co and nonmagnetic "buffer" layer, fits in perfectly as it retains a magnetic coupling between the magnetic molecule and the surface but tones down the coupling strength in order to be influenced with external means. Graphene, in the presence of natural defects, appears to be an extremely potential candidate to play the crucial role of the "buffer" layer to manipulate both the spin states and magnetic coupling.

The hexagonal (111) facet of Ni surface provides an excellent platform for the deposition of the graphene layer owing to an almost perfect lattice matching. However, the unit cells of Ni(111) and graphene possess one and two atoms, respectively. In a pristine graphene layer, one of the two sub-lattice C atoms, C_A, makes direct bond with Ni atom. The bond forms out of a strong overlap between Ni-d_z and dispersive C-p_z orbitals. The other sub-lattice C atom, C_B, has much weaker overlap with Ni-d orbitals. This asymmetric hybridization has a twofold effect. Firstly, the spin degeneracy is broken and finite magnetization appears in graphene lattice. Secondly, C_A and C_B are magnetized differently.

In a pristine graphene layer on Ni, the C_A atoms acquire magnetization of 0.018 $\mu_{B,}$ each aligned antiparallel with respect to the Ni moments. The other sub-lattice atoms, $C_{B'}$ gain relatively larger moment of 0.029 μ_B coupled ferromagnetically to the Ni moments. Altogether, the graphene layer acquires the characteristics of a ferrimagnet [12] with alternating sub-lattices having unequal opposite spin polarization, as shown in **Figure 6(a)**.

The presence of defects in graphene results in an intricate magnetization profile. The absence of C_A in a monovacancy defect does not change either structure or magnetization drastically compared to those of pristine layer. As shown in **Figure 6(b)**, sub-lattice magnetization remains the same except for the missing atom. The divacancy defect, however, has a larger impact on the local structure and hence the magnetization. A Ni atom is dragged up from the surface to heal divacancy making chemical bonds with unsaturated C bonds. The effect transmits through the graphene lattice, and the total induced moment is reduced, although the sub-lattice patterning of the magnetization remains unaltered, as presented in **Figure 6(c)**. The largest deviation from pristine graphene structure and magnetization is introduced by Stone-Wales defects. A pronounced reconstruction at the defect center creates ripple in the graphene lattice, which in turn affect graphene-Ni bonding. The corresponding magnetization profile is presented in **Figure 6(d)**. As can be seen, the alternate sub-lattice magnetization patterning at the defect site is destroyed, and effective total moment in the layer is also reduced.

The magnetic moments of the Ni atoms at the topmost layer also face the consequences of modified hybridization with C atoms at the defect sites. For a pristine graphene lattice deposited on Ni, the magnetic moments in the first Ni layer have a regular periodic pattern. The absence of C

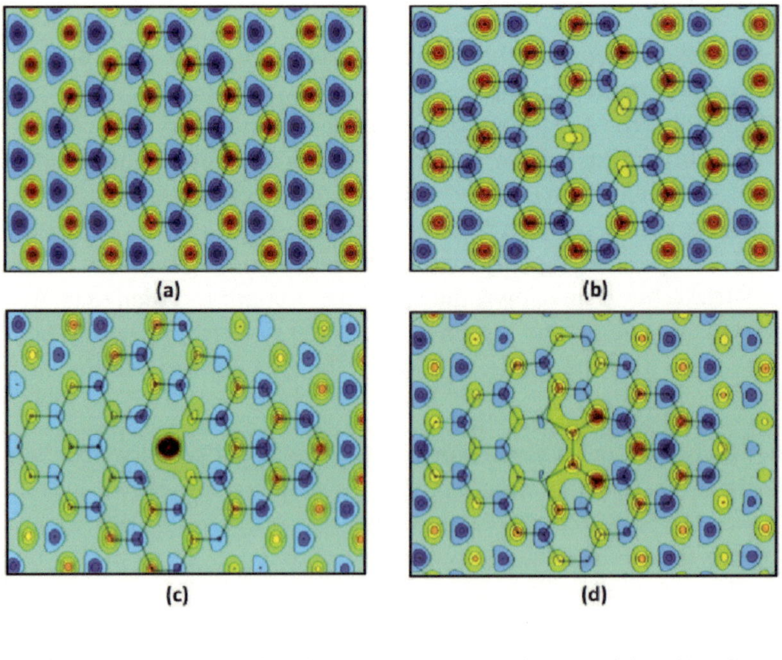

Figure 6. Magnetization densities in the graphene lattice on Ni(111) (a) without any defect, (b) with a monovacancy, (c) with a divacancy, and (d) with a SW defect. Both positive and negative densities are shown. Data from Ref. [12].

atoms in monovacancy defects reduces the coordination of Ni atoms closest to it, resulting in a larger moment. The closest Ni atom in divacancy defect comes out of Ni layer with a quenched moment. Interestingly in Stone-Wales defect, the surface Ni magnetization follows the ripple pattern. The adsorption of magnetic molecules and their properties largely depends on these combined effects at the pristine and defected sites [12].

The adsorption sites for the FeP molecule on the defect-free graphene can be identified as, Top-A, that is, on top of C_A; Top-B, that is, on top of C_B; and Hex, that is, on top of the C hexagon. The local interactions are evidently different on these three sites owing to the arrangement of molecular pyrrole ring with respect to the C atoms underneath. The induced strains in the molecule are different on these adsorption sites, and it is on the Hex position where sufficient molecular stretching occurs to realize a spin crossover. On Top-A and Top-B, molecule retains its free molecule spin state, S = 1. The average distance between the molecule and graphene is calculated to be 3.1 Å, and the graphene is 2.1 Å separated from topmost layer of Ni surface. This makes an effective separation of 5.2 Å between Fe in FeP and Ni topmost layer. On the defect-free graphene, the binding energy is highest on Hex (2.22 eV) position, but it is merely 2 and 17 meV away on Top-A and Top-B sites, respectively. The magnetic coupling is evidently much weaker than the chemisorption scenario. On the Hex site, the molecule is hardly coupled magnetically with the Ni surface. On Top-A, the exchange coupling strength is ~1.8 meV with Fe moment aligned parallel to Ni moments but antiparallel to the C_A moment (−0.018 μ_B). The alignment is quite opposite on Top-B site, where Fe moment is coupled parallel to the C_B moment (0.029 μ_B) and they both align antiparallel to the Ni moment. The coupling is also significantly strong (7 meV) compared to the other two adsorption sites.

The adsorption scenario and magnetic states of FeP are quite diverse on defect sites, depending on the local structural modifications. On the monovacancy defect, the Fe center of FeP molecule resides right in-between three C_B atoms, which have local moments aligned parallel to the Ni surface moments. The spin state of FeP resembles that of the free molecule. The molecule is coupled antiferromagnetically to the Ni surface with an exchange coupling of 7 meV. The healing of divacancy with Ni atom makes adsorption scenario strikingly different. The divacancy defect behaves like a magnetic surface, leading to chemisorption through a chemical bond formation between Fe in FeP and displaced Ni atom. The molecular pyrrole ring, however, experiences electrical repulsion form the graphene layer which imparts a convex structure to the molecule. The displacement of the Fe atom from molecular plane results in the stretching of Fe-N bond, and the molecule exhibits a high spin state (S = 2). Due to a strong orbital overlap, the magnetic coupling strength is rather high (22 meV) and is antiferromagnetic in nature. The free molecular spin state is retained on Stone-Wales defect, despite having heavily rippled graphene layer. Adsorbed FeP couples antiferromagnetically to the Ni surface. The exchange coupling is 14.5 meV which is fairly strong among all the physisorption scenarios [12].

As discussed above, the feasibility of magnetic state manipulation on this composite surface is particularly dependent on the controlled creation of specific defects. It is, hence, essential to have a quantitative estimation of the defect formation energies (DFE). In a free-standing graphene layer, the estimated DFE for the vacancy formation is 7–8 eV, while the calculated values of DFE for monovacancy and divacancy on Ni(111) surface are 2.91 and 3.83 eV, respectively. One can safely conclude that the creation of defects, either naturally of ion-beam irradiation, should be easier in the Ni-graphene composite surface which provides a large boost for the abovementioned magnetic state manipulation. Moreover, the spin state and the magnetic coupling are adsorption site dependent; the protection of it requires a sufficient energy barriers between adsorption sites. The energy barrier in moving a FeP molecule between a Hex site and a Top-A site (which are energetically comparable adsorption sites) on a pristine graphene is calculated to be 33 meV, which translates to a temperature, higher than the room temperature. On the defect sites, this energy barrier is expected to be much higher, making the abovementioned value to be the lower limit for the diffusion barrier. One may envisage controlled formation of specific types of defects and achieving either parallel or antiparallel orientation of Fe moments relative to the moments in the Ni layers [12].

3. Nonmagnetic substrates

In the previous section, TMP or TMPc molecules deposited on magnetic substrates have been discussed in view of magnetic coupling and manipulating the spin state. In these hybrid structures, the magnetic coupling between molecules and substrate plays a dominating role. However, adsorbing molecules on nonmagnetic (NM) substrates instead allow to study aspects which are hidden in the presence of a magnetic coupling between molecule and substrates as the hybridization between molecule and substrate depending on the surface orientation and reconstruction. Thinking of future spintronic devices, the magnetic properties of the molecules can be tuned by ligands attached to the molecular center. Without the dominating magnetism of the FM substrate, only the coupling between the molecule and possible ligands exists, and this can be more easily switched by thermal treatment than the large coupling between FM and molecule.

3.1. Influence of the surface texture on the electronic structure

The electronic structure of molecules adsorbed on magnetic layers can strongly differ from the one obtained for the molecule in gas phase because of the magnetic coupling, and molecules are often chemisorbed. Here we use FeP as a model system. It corresponds to the Fe OEP structure but without the outer ethyl groups. Dangling bonds are saturated by hydrogen, see **Figure 1**.

On a nonmagnetic substrate, Cu(001) in our case, the molecules are less tightly bound to the substrate. The distance between the Fe ion of FeP and the Cu(001) substrate is about 2.66 Å (PW91, vdW-D2, $U_{eff} = 3$ eV) compared to 1.78 Å on Co(001), that is, the molecules are physisorbed [13]. The adsorption position and the relative orientation are the same as on the FM substrates, which means that in this system, the adsorption position is not influenced by the magnetism of the substrate but by geometry. Even though on Cu(001) there is no magnetic interaction with the molecule, the molecules hybridize with the substrate which can be seen

http://dx.doi.org/10.5772/intechopen.68224

from the density of state (DOS) in **Figure 7(a)**. The Fe 3d states having components in z-direction (z^2 and π) are broadened compared to the DOS of the free molecule. However, the interaction is much weaker and does not lead to strong changes in the orbital occupation other than in the case of Co or Ni substrates. The Fe-N distance is basically the same as for the molecule in gas phase (2 Å) which entails that the spin state is not affected if FeP is absorbed on Cu(001), and it remains in the S = 1 state which agrees with the experimental observation from X-ray absorption spectroscopy (XAS) and X-ray magnetic circular dichroism (XMCD) [13].

The situation changes if the Cu(001) surface is covered by 0.5 layers of oxygen. This O adlayer leads to a √2×√2R45° missing row reconstruction of the surface. The ground state configuration changes from the hollow site (as on Cu(001)) to the missing row position with two O atoms next to the Fe ion. Because the O layer is basically incorporated in the surface layer of the Cu film, the distance between the Fe ion and the substrate increases only by 0.3 Å, and despite the strong surface reconstruction, the Fe-N distance increases only to 2.03 Å, that is, the spin state remains S = 1 and only minor changes in the Fe 3d DOS can be observed; see **Figure 7(b)**. Mostly the broadening of the peaks is reduced compared to the molecule on the plain Cu surface meaning that the hybridization is weaker than the Cu(001).

While the O adlayer had a huge influence on the magnetic properties when added to a ferromagnetic substrate, for example, for FePc on Co(001) or FeOEP on Co(001) [2, 3], here the spin state and the electronic structure are basically unchanged. Only the adsorption position is affected due to the surface reconstruction.

3.2. Influence of ligands

As discussed above, an O adlayer on nonmagnetic Cu(001) has no effect on the magnetic properties of the molecule, but there exist other combinations of magnetic molecules and nonmagnetic substrates where adlayers or dopants switch the spin state, for example, GaAs(001) and Vanadyl Pc. The molecule switches to the high spin state if the (Ga-rich) GaAs substrate is doped with Si. However, thinking of spintronic devices, a reliable, controllable switching

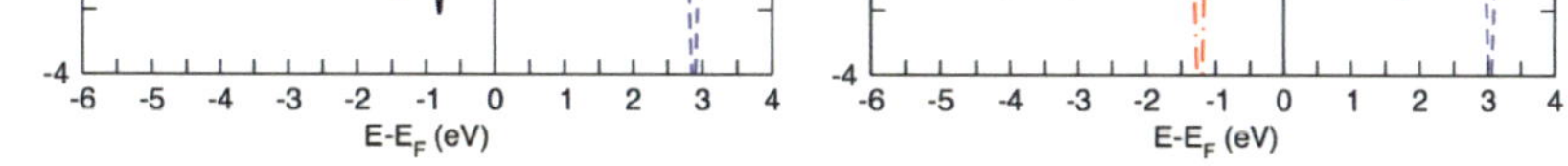

Figure 7. Calculated density of states of FeP/Cu(001) (a) and FeP/√2×v2R45°O/Cu(001) (b) for the ground state configurations with FeP on the hollow site of the Cu(001) surface with the N atoms on top of the underlying Cu atoms (a) and adsorbed on the missing row position of the reconstructed surface. Note only the Fe 3d states are shown here. Data are taken from Ref. [13].

between two configurations, for example, low spin to high spin state, is needed. This can hardly be achieved if the manipulation of the spin state arises from adlayers or doping of the substrate layer. A more realistic way is the manipulation of the spin state by ligands since ligands can be thermally attached and removed. This has been demonstrated for Co tetraethylporphyrin on Ni(001) and also works for nonmagnetic substrates. This works also on nonmagnetic substrates. Fe octaethylporphyrin (OEP) can be stabilized in air by pyridine (Py) or Cl. Depositing the two types of Fe OEP on Cu(001), they show a completely different spectroscopic signature [13], as can be seen from the X-ray absorption spectra and the X-ray magnetic circular dichroism for Fe OEP-Py and Fe OEP-Cl on Cu(001) (**Figure 8**). Especially for normal incidence, the XMCD signal is four times larger in the case of Cl ligands, whereas for grazing incidence of the photon beam, no significant differences exist; see **Figure 8**. This indicates that the magnetic and electronic properties depend on the ligand. While Py dissolves during the deposition process, that is, pure FeP remains on the substrate, in the case of Cl, about 50% of the ligand remains, which leads to an increase of the magnetic signal. Theoretical calculations for FeP (porphyrin without the outer ethyl groups) with an axial Cl ligand deposited on Cu(001) confirm the observation. The Fe-d level occupation has changed, the previously occupied d_π levels have moved above the Fermi level, and the hybridization with the substrate has decreased leading to sharper peaks compared to FeP/Cu(001). The reason is that the Fe ion also interacts strongly with the ligand.

Having seen that for a sub-monolayer coverage of Fe OEP on Cu(001) 50% of the Cl atoms which remain after deposition of the surface is sufficient to cause significant changes in the magnetic behavior, a detailed study of possible ligands and their influence on the magnetic properties has been performed for free and deposited FeP molecules [14]. In gas phase, the influence of various combinations of porphyrin or phthalocyanine and ligands (axial or peripheral) has been investigated. Here we focus on the molecule ligand complex

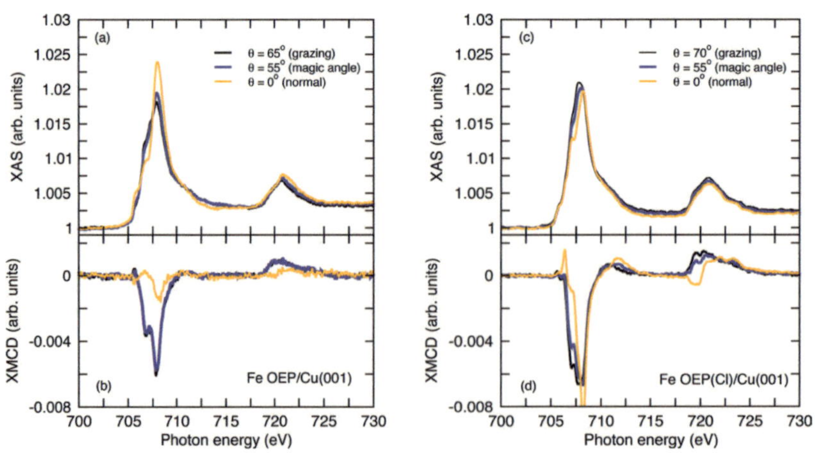

Figure 8. Measured Fe $L_{2,3}$ edge XAS (a) and XMCD (b) for 0.4 ML Fe OEP on Cu(001). Panels (c) and (d) show the analogous results for Fe OEP (Cl)/Cu(001). The angle θ denotes the angle between surface normal and photon beam. Taken from Ref. [13].

FeP+L deposited on Cu(001) with L = Cl, O,O_2. DFT calculations (VASP, PAW, PBE, and vdW-D2) show a significant dependence of the electronic and magnetic structure depending on the ligand. In all cases, the molecules have been adsorbed on the hollow site position (**Figure 9(a)**), which has been found to be the ground state for FeP/Cu(001). In agreement with the XAS/XMCD experiments described above, an axial Cl ligand enhances the magnetic moment of the FeP complex from 2 to 3 μ_B, whereby 2.69 μ_B are on the Fe ion. If the FeP-Cl complex is deposited on Cu(001), the moment becomes even larger (3.71 μ_B), and the induced moment of the Cl atom decreases from 0.2 to 0.1 μ_B. On the contrary the Fe-N distance increases from 2.05 Å for FeP/Cu(001) to 2.23 Å with Cl ligand which agrees with the observation of the transition from an intermediate spin state (S = 1) to the high spin state. The ligand has also indirect influence on the organic rings; they are driven away from the surface visible in a strong bending of the molecule; see **Figure 8**. This contrasts with FeP or FePc on magnetic substrates where also the organic ligands contribute to the interaction with the substrate.

In the case of O ligands, that is, atomic oxygen or O_2, a different magnetic behavior is observed. The spin moment on the Fe ion is reduced, while the oxygen atoms gain a moment parallel to Fe such that the spin state of the whole complex is unchanged and the total moment remains 2 μ_B. The theoretically determined spin moment of an atomic oxygen ligand is 0.58 μ_B, whereas for the oxygen dimer, the moment is evenly distributed on both O atoms (0.22 μ_B/0.21 μ_B). The latter result deviates from the gas-phase solution where the two O atoms differ in size and relative orientation. In gas phase the Fe atom has a slightly enhanced moment compensated by the antiparallel moment of the outer oxygen ligand.

Though no magnetic coupling between surface and molecule is present in the case of Cu(001), a significant interaction between substrate and molecule is observed between Fe and Cu which condenses in the change of the electronic and magnetic structure of the adsorbed molecules and is connected to severe changes in the geometrical structure. Furthermore, a deformation or buckling of the substrate next to the molecule is observed which is particularly pronounced for Cl and O ligands, while with O_2 as a ligand, the surface is much less affected. Indicating that for the dimer the interaction with the substrate is weaker than in the case of atomic Cl or O even though the adsorption distance is very similar, 2.28 Å (2.23 Å) with O_2 (O) for Cl the average distance between substrate and Fe is even larger (2.50 Å) [14].

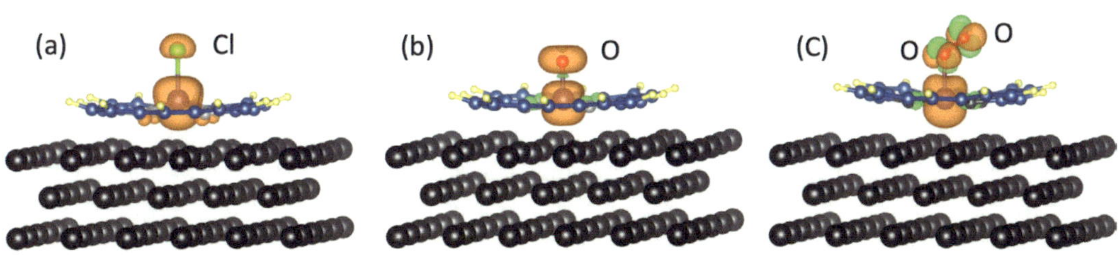

Figure 9. Calculated spin density of FeP with different axial ligands, Cl (a), atomic O (b), and O_2 (c), adsorbed on Cu (001). Both positive and negative densities are shown. Data are partially taken from Ref. [14].

In conclusion, it has been shown that the spin moment of FeP or the iron center itself can be tackled by the choice of the ligand in gas phase as well as on nonmagnetic Cu(001), whereby the changes for deposited molecules are even more expressed.

3.3. Effective spin moment and the role of the spin dipolar term

To compare the calculated spin moments with experimentally determined values, we face a problem since most experimental data are obtained from XMCD measurements. From these kinds of experiments, only orbital and effective spin moments m_{eff} are accessible. The effective spin moment differs from actual spin moment by the spin-dipole moment contribution, and depending on the symmetry, this contribution can be large [1, 10]. Comparing the XMCD signal of the Fe OEP L_3 edge (**Figure 8(c)**) for different incidence angles of the photon beam, it turns out that the intensity strongly varies with the angle. A large signal is observed for grazing incidence in much larger than for normal incidence of the photon beam. This can be caused by a large contribution of the spin-dipole term or be related to large magnetocrystalline anisotropy. Following Oguchi [16], the spin-dipole operator T is defined by

$$T = \sum_i Q^{(i)} s^{(i)} \tag{1}$$

with Q being the quadrupole tensor:

$$Q^{(i)}_{\alpha\beta} = \delta_{\alpha\beta} - 3\,\hat{r}^{(i)}_{\alpha}\,\hat{r}^{(i)}_{\beta}. \tag{2}$$

Hence, the spin-dipole moment arises from the asphericity of the spin density, that is, for transition metals where spin-orbit coupling is weak; this is related to the crystal field. If the cubic symmetry is not broken, Q vanishes, but for systems with reduced symmetry as for clusters [15, 16] or molecules adsorbed on substrates, the spin-dipole moment plays an important role [17]. To compare calculated spin moments to experimentally reported effective spin moments for low symmetry systems, the dipolar term must be included, especially to rule out other sources for the discrepancy between the effective spin moment and the total spin moment, such as a large magnetocrystalline anisotropy or not fully saturated magnetic moments. For simplicity, we focus only on the z component of the spin-dipole operator. Its expectation value $<T_z>$ is given by the trace of the density matrix multiplied by T_z. the density matrix can be obtained from DFT calculations. If for transition metals the spin-orbit coupling is negligible, the size of $<T_z>$ depends on the existence of a finite spin moment on the nonequivalent charge distribution on the orbitals. To obtain the spin-dipole moment and the effective spin moment by van der Laan [18] provided a scheme how to apply the general approach to a typical XMCD experiment such as in **Figure 10**.

The intensity which is measured in XMCD experiments as response to the photon beam hitting the surface depends on the relation between the magnetization direction **M**, the polarization of the incident photon beam **P**, and the surface normal **n** as depicted in **Figure 10**. The angular dependence of the dipole operator reads then

$$\langle 7T(n,P,M)_i \rangle = \frac{1}{4}\langle 7\,T_z \rangle (cos(\varphi) + 3cos(\varphi + 2\theta))) \tag{3}$$

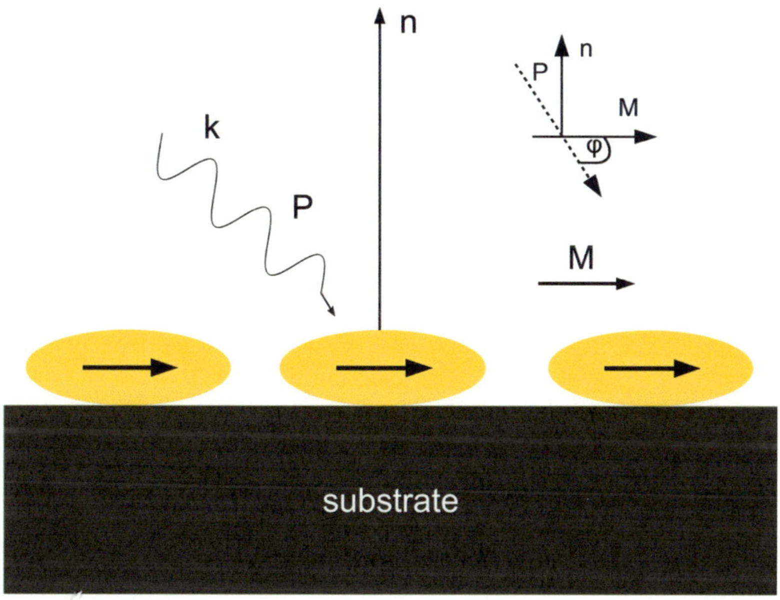

Figure 10. Sketch of a typical experimental XMCD setup. The polarization of the photon beam and the magnetization are denoted by **P** and **M**, respectively, as well as **n** denoting the direction of the surface normal.

with φ being the angle between the magnetization **M** and the polarization of the beam **P**. The deviation of the magnetization direction from the surface normal **n** is denoted by the angle θ. In the present case, that is, for nonmagnetic substrates, the alignment of the magnetic moments of the transition metal centers of the molecules is achieved by applying an external magnetic field which is usually parallel aligned to the polarization of the incident photon beam **M**∥**P**. In this special case, Eq. (3) reduces to

$$\langle 7T(\theta)_i \rangle = \frac{1}{4}\langle 7\,T_z \rangle (3\cos^2(\varphi)-1) \quad (4)$$

and the experimentally observed effective spin moment becomes

$$m_{eff} = m_s + \langle 7T(\theta) \rangle . \quad (5)$$

As can be seen from **Figure 11**, the effective moment strongly depends of the incidence angle of the photon beam. Only for measurements carried out at the *magic angle* (54.5°), the argument in the brackets on the right side of Eq. (4) vanishes, and it yields $m_{eff} = m_s$.

In an ideal case, the calculated and the measured effective moment should be identical; however, comparing the calculated m_{eff} to the experimental data (triangles) in **Figure 11(a)**, distinct deviations occur even for the magic angle where T_z vanishes. This is related to the fact the experimental sample could not be fully saturated in the magnetic field (5.9 T) [13]. For the oxidized surface, the density matrix is almost identical with the one for the plain Cu substrate; hence, also the m_{eff} and the angular dependence do not change. In this case, the theoretical and

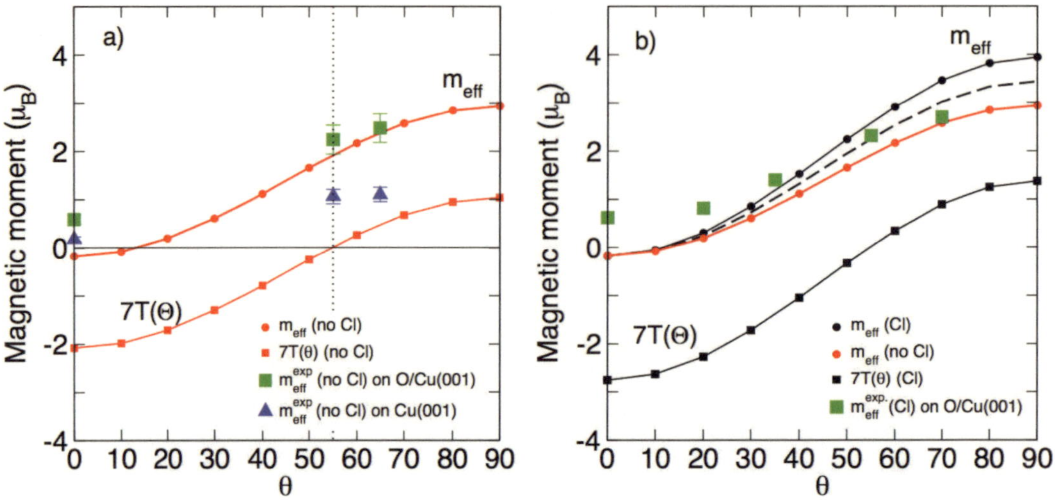

Figure 11. Dipolar term and effective spin moment for FeP on Cu(001) (a). Open (filled) symbols denote the dipolar term (effective spin moment). The data for FeP with Cl ligand are given in (b). The lighter solid line corresponds to meff without Cl, and the dashed line is the average of meff with and without Cl. Deviations between the calculated and measured m_{eff} at small incidence angles may result from limited accuracy of the determination of the dipolar term. Data are partially taken from Ref. [13].

experimental data are in good agreement, because the saturation could be reached. With a Cl ligand attached to the FeP, the dipolar term is different due to changes in the occupation of the Fe 3d levels (cf. **Figure 11**). The m_{eff} at $\theta = 90°$ would be 4 μ_B instead of 3 μ_B as without Cl. Even though the sample is basically saturated, the data deviate from the theoretically predicted spin moments. Assuming only 50% of the Cl ligands remain at the FeP molecules after deposition (dashed line in **Figure 7(b)**) improves the agreement between theory and experiment significantly. Scanning tunneling microscopy images of the Fe OEP (Cl)/Cu(001) have verified the assumption that between 40 and 60% of the ligands have been dissolved during deposition.

In summary, as shown for the example of FeP (OEP) on Cu(001), the dipolar term is an important factor to interpret and understand experimental XAS and XMCD data since effects from non-saturated samples as well as incomplete dissolved ligands can be detected.

4. Field-regulated switching of magnetic coupling

In this section, we move a step ahead to devise a practical mechanism to control magnetic bistability with external means. The composite is specifically designed with three layers of CoOEP molecules deposited on graphene-covered clean Ni(111) single crystal, as shown schematically in **Figure 12**. The focus in this composite will be on the manipulation of magnetic coupling leaving the spin crossover feasibility aside.

The CoOEP molecular layers are physisorbed on the graphene-Ni composite surface, quite resembling the FeP adsorption, discussed above. The arrangement of the molecular layers

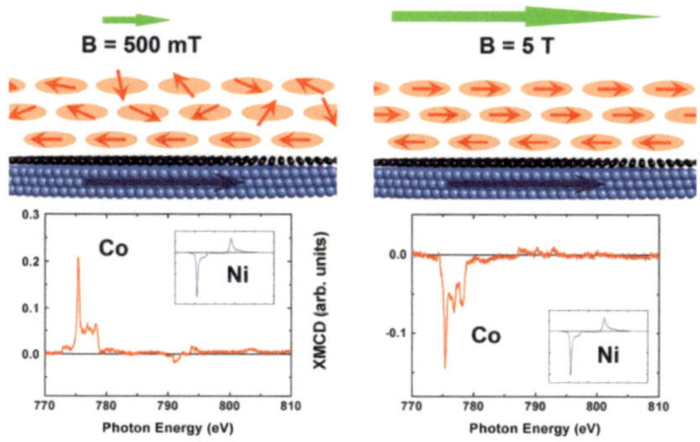

Figure 12. Top: schematic images of the sample 3ML CoOEP/graphene/Ni(111). The light and dark arrows indicate the direction of the magnetic moments of Ni and Co in a low magnetic field of 500 mT (left) and in a high magnetic field of 5 T (right). Bottom: XMCD at the Co $L_{2,3}$ edges of the CoOEP molecules in the low magnetic field (left) and in the high magnetic field (right). Insets show the XMCD at the $L_{2,3}$ edges of the saturated Ni crystal at 500 mT and at 5 T. All spectra are recorded at T = 2 K and Θ = ~70°. Data from Ref. [19].

appears in a particular fashion. The molecules in the two consecutive layers do not reside right on top of each other but slightly horizontally shifted. An indirect overlap between Co-d_z orbitals via N-p orbitals favors this specific geometry. As discussed in the previous section, the single layer graphene on Ni is spin polarized with two different kinds of sub-lattice moments. Unlike FeP, in CoP the S = 1/2 spin state is quite robust and solely arises from the singly occupied Co-d_z orbital, while other orbitals are doubly occupied giving no contribution to the molecular moment. However, in the presence of three layers, the net magnetization of the CoOEP molecules appears to be a sum of two contributions; the first layer couples antiferromagnetically with the Ni substrate, and the other two layers are magnetically decoupled but grow parallel magnetization with applied magnetic field. To resolve the strength of the magnetic coupling between the first CoOEP layer and Ni surface, we considered three adsorption sites in a defect-free graphene, Top-A, Top-B, and Hex, as described above. Unlike FeP, the Top-A site is energetically most favorable, while binding energies on Top-B and Hex sites are 14.6 and 23.5 meV lower. The magnetic coupling also varies in strength on these three sites exhibiting 4.2, 9.9, and 3.1 meV, respectively, on Top-A, Top-B, and Hex sites, while in all cases, molecular moments align antiferromagnetically with respect to Ni moments [19].

The CoP layers in a free-standing, perfectly parallel bilayer couple antiferromagnetically to each other. However, in practice the extended outermost ligands in CoOEP destroy the flat arrangement of the layers, reducing drastically the interlayer coupling. We modeled the scenario by increasing spatial separation and introducing angle between molecules, which essentially results in a sharp drop in the exchange coupling strength [19].

The system is then exposed under a magnetic field (B). The magnetization of the Ni layer is saturated for B > 200 mT, which is inadequate to magnetize the paramagnetic layers of CoOEP molecule. The antiferromagnetic coupling between the first layer and Ni surface requires extremely strong field to switch magnetization. The effective coupling between CoOEP layers and Ni remains antiferromagnetic under a sufficiently low field. However, as the strength of the applied field is increased, at about 1 T, the paramagnetic molecular layer grows sufficient magnetization to revert the net orientation of the molecular magnetization parallel to the Ni magnetization. In **Figure 12**, we represent this field-induced switching of the magnetization. With an applied magnetic field as low as 500 mT, the material-specific X-ray magnetic circular dichroism (XMCD) spectra for Ni and Co show opposite orientation, as can be seen in the left part of **Figure 12**. In a sufficiently strong field, the paramagnetic layers get magnetized and overpower the antiferromagnetic contribution from the first layer. The XMCD signals for Co and Ni at 5 T are presented in the bottom-right part of **Figure 12**, which show this parallel alignment [19].

The realization of the field-regulated switching of molecular magnets is a key advancement toward molecular spintronic concepts. The spin injection into the organic layers can be regulated by an external magnetic field which may lead to the practical realization of a spin switch or spin valve in the near future.

Author details

Heike C. Herper[1], Barbara Brena[1], Sumanta Bhandary[2] and Biplab Sanyal[1]*

*Address all correspondence to: Biplab.Sanyal@physics.uu.se

1 Department of Physics and Astronomy, Uppsala University, Uppsala, Sweden

2 Institute of Solid State Physics, TU Wien, Wien, Austria

References

[1] Bogani L, Wernsdorfer W. Molecular spintronics using single-molecule magnets. Nature Materials. 2008;**7**:179

[2] Wende H, et al. Substrate induced magnetic ordering and switching of iron porphyrin molecules. Nature Material. 2007;**6**:516

[3] Bhandary S, Brena B, Panchmatia PM, Brumboiu I, Bernien M, Weis C, Krumme B, Etz C, Kuch W, Wende H, Eriksson O, Sanyal B. Manipulation of spin state of iron porphyrin by chemisorption on magnetic substrates. Physical Review B. 2013;**88**:024401

[4] Bhandary S, et al. Correlated electron behavior of metal-organic molecules: Insights from density functional theory combined with many body effects using exact diagonalization. Physical Review B. 2016;**93**:155158

[5] Gottfried JM. Surface chemistry of porphyrins and phthalocyanines. Surface Science Reports. 2015;**70**:259

[6] Bhandary S et al. Graphene as a reversible spin manipulator of molecular magnets. Physical Review Letter. 2011;**107**:257202

[7] Geim AK. Graphene: Status and prospects. Science. 2009;**324**:1530

[8] Banhart F, Kotakoski J, Krasheninnikov AV. Structural defects in graphene. ACS Nano. 2011;**5**:26

[9] Liao M, Scheiner S. Electronic structure and bonding in metal porphyrins, metal = Fe, Co, Ni, Cu, Zn. Journal of Chemical Physics. 2002;**116**:205

[10] Klar D, Brena B, Herper HC, Bhandary S, Weis C, Krumme B, Schmitz-Antoniak C, Sanyal B, Eriksson O, Wende H. Oxygen-tuned magnetic coupling of Fe-phthalocyanine molecules to ferromagnetic Co films. Physical Review B. 2013;**88**:224424

[11] Herper HC, Bhandary S, Eriksson O, Sanyal B, Brena B. Fe phthalocyanine on Co(001): Influence of surface oxidation on structural and electronic properties. Physical Review B. 2014;**89**:085411

[12] Bhandary S, Eriksson O, Sanyal B. Defect controlled magnetism in FeP/graphene/Ni(111). Scientific Reports. 2013;**3**:3405

[13] Herper HC, Bernien M, Bhandary S, et al. Iron porphyrin molecules on Cu(001): Influence of adlayers and ligands on the magnetic properties. Physical Review B. 2013;**87**:174425

[14] Brena B, Herper HC. Influence of ligands on the electronic and magnetic properties of Fe porphyrin in gas phase and on Cu(001). JAP. 2015;**117**:17B318

[15] Stepanow S, Mugarza A, Ceballos G, Moras P, Cezar JC, Carbone C, Gambardell P. Giant spin and orbital moment anisotropies of a Cu-phthalocyanine monolayer. Physical Review B. 2010;**82**:014405

[16] Oguchi T, Shishidou T. Anisotropic property of magnetic dipole in bulk, surface, and overlayer systems. Physical Review B. 2004;**70**:024412

[17] Šipr O, Minár J, Ebert H. On the importance of the magnetic dipole term T_z in analyzing X-ray magnetic circular dichroism spectra of clusters. Europhysics Letter. 2009;**87**:67007

[18] van der Laan G. Relation between the angular dependence of magnetic x-ray dichroism and anisotropic ground-state moments. Physical Review B. 1998;**57**:5250

[19] Klar D, Bhandary S, et al. Field-regulated switching of the magnetization of Co-porphyrin on graphene. Physical Review B. 2014;**89**:144411

Chapter 5

Heterostructures Based on Porphyrin/Phthalocyanine Thin Films for Organic Device Applications

Marcela Socol, Nicoleta Preda, Anca Stanculescu,
Florin Stanculescu and Gabriel Socol

Additional information is available at the end of the chapter

http://dx.doi.org/10.5772/67702

Abstract

Multilayer or blend heterostructures based on porphyrins and phthalocyanines were obtained on different substrates using VTE and MAPLE methods. Stacked structures based on ZnPc and C60 with NTCDA were prepared by VTE on ITO/glass, their current value being increased by the deposition of the materials in an inverted configuration or by using ITO/PEDOT:PSS as a substrate. Multilayer structures comprising ZnPc and NTCDA were fabricated by MAPLE on an AZO/glass. Treating the AZO in oxygen plasma, a higher current value was obtained for the deposited heterostructures. The oxygen plasma treatment can increase the work function of the AZO resulting in a decrease of the energetic barrier from AZO/organic interface and finally improving the charge transport. Stacked layers or blend heterostructures having ZnPc, MgPc and TPyP were deposited by MAPLE on ITO/PET. In the case of those containing MgPc and TPyP, an increase in the current value (in dark) was obtained for the blend compared to the stacked layer configuration. For those with ZnPc and TPyP, under illumination, a photovoltaic effect was observed for the blend structure. All heterostructures are featured by a large absorption in the visible domain of the solar spectrum and suitable electrical properties for their use in OPV applications.

Keywords: ZnPc, TPyP, MgPc, VTE, MAPLE

1. Introduction

During the last years, the organic materials have attracted the attention of researchers because they can be used in different types of applications: organic photovoltaic (OPV) cells, organic-based light-emitting devices (OLEDs) and organic field effect transistors (OFETs) [1–6]. Heliatek reports a conversion efficiency of about 13.2% for an OPV fabricated by vacuum evaporation and having three absorbers [7]. OLEDs are already integrated in commercially available devices such as mobile phone displays, TV sets, etc.

The field of organic materials for applications in photovoltaic cells has begun in 1906 and 1913 with the observation of the anthracene photoconductivity [8, 9]. Kearns introduced, in 1958, the first organic photovoltaic cell with a film based on magnesium phthalocyanine (MgPc) [10]. In 1986, Tang makes a step forward for the OPV, fabricating a photovoltaic cell with two organic layers in configuration donor/acceptor (D/A) using copper phthalocyanine (CuPc) as a donor and perylenediimide (PDI) as an acceptor [11]. Since then the photovoltaic effect (PV) was reported in different organic compounds such as porphyrins, phthalocyanines or their derivatives [12]. The organic materials are part from the third generation of photovoltaic, after those based on inorganic materials (the first and second generation).

Comparatively with the inorganic compounds, the organic materials present the following advantages: they can be deposited at low temperature (decreasing in this way the processions costs), are compatible with plastic substrates (a good premise for the flexible electronics) and their properties can be tuned by various processing techniques which allow their deposition even on a large area. In photovoltaic cells, organic compounds present absorption coefficients greater than 10^5 cm^{-1} allowing an increased absorption of the incident light even under 100 nm [13]. The light collection efficiency is dependent on the organic active layer thickness and the absorption properties of the used materials [14].

Porphyrins and phthalocyanines are the most used organic compounds as active layers in photovoltaic cells due to their several absorption maxima in the visible part of the solar spectrum (less than 700 nm [14, 15]). Furthermore, in the porphyrins and derivatives, the range of absorption spectrum in the near infrared part can be increased due to the extended conjugation [14].

The impact of the porphyrins and phthalocyanines on the OPV domain can be evaluated, as shown in **Figure 1**, which contains the histograms with the publications number (from ISI web of science) from the last 5 years (2012 to 2016) having as subject porphyrins or phthalocyanines

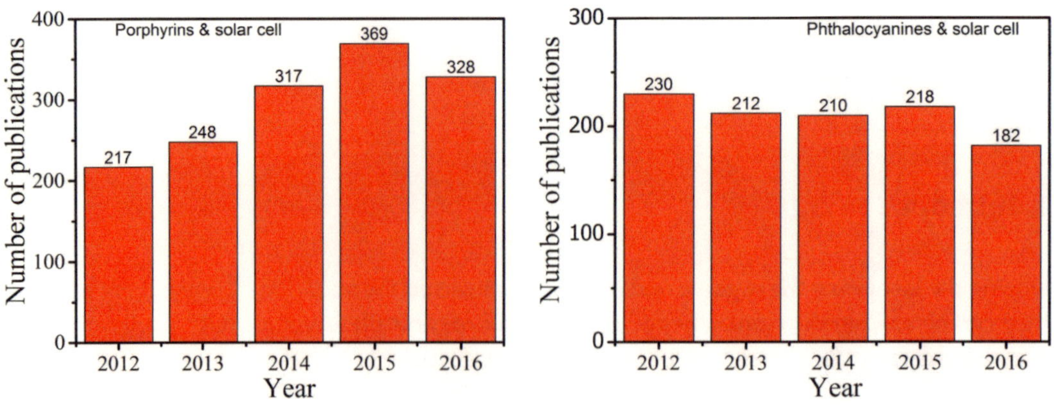

Figure 1. Number of publications per year in the last five years having as topic porphyrins or phthalocyanines and solar cell.

and solar cells. Moreover, it has to be mentioned that there is a journal entirely dedicated to these organic compounds.

Photovoltaic cells based on porphyrins with high performances were achieved. Thus, in 2011 Yella et al. reported 12.3% efficiency for a structure with a zinc porphyrin (YD2-oC8) co-sensitised with Y123 deposited on a TiO_2 [16]. Also, in 2014, a conversion efficiency of about 13% was obtained for a porphyrin dye, coded SM315 [17]. In 2015, a teoretical study made for a new porphyrin-based molecular complex shows that an open circuit voltage of about ~1.8 V can be obtained using this kind of materials [18].

Additionally, the bioinspired structures of porphyrins can be attractive in different forms (nanoparticles, nanosheets, nanorods and nanorings, nanowires, nanotubes, aggregates) as summarised by Monti et al. [19] in applications as catalysts (for O_2 reduction or H_2O oxidation [20]), sensors [21], in photodynamic therapy as photosensitizers [22], for drug delivery [23] and for the treatment of tumours [24].

One of the most important advantages of the phthalocyanines over other organic materials is their increased value of the exciton diffusion length, which is usually in the range of 10 nm [25]. Thus, for CuPc a diffusion length of about ~68 nm was reported [26]. Increased cell performances (efficiency) were also recorded for the OPV based on phthalocyanines: 3.6% for a double layer cell with CuPc and C60 [27], 4.2% for a structure with 1,4,8,11,15,18,22,25-octahexylphthalocyanine ($C6PcH_2$) and [6,6]-phenyl-C61 butyric acid methyl ester (PCBM) prepared by spin-coating [28], 5% for a cell also with CuPc:C60 [29] and the highest reported efficiency of about 5.7 % was achieved also for a structure based on CuPc and C60 [30]. And in the field of the perovskite cells was reported an increased efficiency (11.75%) for a structure containing a ZnPc thin film as a donor material [31].

Complementary, the phthalocyanines and their derivatives have wide range of applications such as OLEDs, gas sensors and optical communications [32]. These compounds are promising candidates in the non-linear optical devices due to their large third-order non-linearity [32, 33]. They are also used in therapy of cancer, infectious or neurodegenerative diseases [34] and in the xerographic photoreceptors of laser printers due to the strong Q-band absorption [35].

In this chapter, we summarised some of our results regarding the preparation and characterisation of porphyrins and metallic phthalocyanine layers for applications in OPV. These materials were obtained as thin films (in multilayer structures or blends) on solid (glass coated with indium tin oxide-ITO or aluminium-doped zinc oxide-AZO) or on a flexible substrate (polyethylene terephthalate-PET coated with ITO).

2. Deposition techniques for single and multilayer thin films: description and advantages

In the applications such as OPV or OLED, the heterostructures consist in one or more organic thin films (active material) sandwiched between two electrodes, an anode that must be transparent (in order to pass the light) and the metallic cathode [13].

The transparent electrode can be prepared by several techniques: sol-gel, magnetron sputtering, oxygen ion beam assisted deposition, pulsed laser deposition (PLD), spray pyrolysis. The widely used methods are RF magnetron sputtering and PLD because films obtained by using these methods are characterised with adequate properties [36, 37].

In the case of the organic compounds, vacuum thermal evaporation (VTE) was one of the most used deposition techniques. Other methods such as spin-coating, doctor blading or inkjet printing were also involved in the preparation of the organic layers [38–40].

The techniques used to prepare our organic heterostructures based on porphyrins and phthalocyanines are briefly described in the following section.

2.1. Transparent conductive oxide (TCO) thin films obtained by pulsed laser deposition (PLD)

PLD is a versatile deposition method frequently used for the preparation of the thin films based on TCO materials [41]. The depositions are made inside a vacuum chamber. A solid target comprising the raw materials is ablated under a pulsed laser beam. When the elements from the target reach their evaporation temperature (above a certain value of the laser intensity), they are ejected from the target and form the plasma plume and pass to the deposition support starting the nucleation process which leads to the formation of the thin layer [42]. In order to improve the properties of the layers, the deposition can be also made in inert gases such as nitrogen (N_2) or in reactive gas such as oxygen (O_2). Also, the deposition target can be rotated during the deposition process to avoid a local deterioration which can affect the uniformity of the obtained film. The deposition parameters that must be controlled are fluence of the laser beam, a number of the laser pulses, target-substrate distance and, sometimes, substrate temperature [41].

High-quality TCO materials with an increased transmittance, low electrical resistivity and a reduced roughness of layer surfaces are obtained by PLD [37, 43].

2.2. Organic thin films prepared by vacuum thermal evaporation (VTE)

VTE is a dry technique, frequently used for the deposition of the metallic layers, inorganic materials but also for the organic compounds. The method is simple and it can be applied for deposition on a large scale being used in the industry. Heliatek fabricated a cell made with three organic layers and high efficiency by VTE [7].

The solubility is another reason for choosing this deposition method which does not imply a solvent, if the organic materials are insoluble or poorly soluble. Thus, can be deposited successive organic layers, the previous deposited layer not being affected by the deposition of the next layer.

Using this method, materials can be evaporated which are vacuum compatible and chemically stable up to their evaporation temperature. In the vacuum evaporation, the material of interest is heated until its vapour pressure is greater than 10^{-2} Torr [44]. The high vacuum in the deposition chamber ensures a particles flow (atoms, molecules) from the evaporated material. The process is followed by the condensation of the formed vapours on an adequate substrate [45]. As deposition substrates can be used glass, quartz, silicon, ITO or other plate materials.

The evaporation and condensation of the materials are influenced by the following parameters: temperature of the heater (influences the evaporation rate), evaporation rate (depends by the system geometry), substrate temperature (control the surface atom mobility), heater and substrate geometry (related to film uniformity) and substrates (as smooth and clean as possible) [44]. All these parameters are very important because they affect the quality of the obtained thin film. It is known that the thin films have the tendency to copy the form of the substrate used for deposition.

The organic compounds adequate to be deposited by vacuum evaporation are those from the small molecules class, because they do not suffer stoichiometric changes during the transfer, having low melting temperatures (~300°C).

A disadvantage of this method is the time necessary until it is reached the high vacuum in the deposition chamber. But the thin films obtained are uniform, have a good adherence and have the wished geometry (shadow mask being used) [46].

2.3. Organic thin films prepared by matrix-assisted pulsed laser evaporation (MAPLE)

MAPLE is a laser technique that has been developed from the PLD method. It was developed at the end of the 1990s from the necessity to deposit soft organic thin films (unicomponent layers or blends) preserving the properties of the used raw materials. MAPLE is also useful in the deposition of the polymers when the use of VTE results in the broken of the molecular chains [47, 48]. It is also used for the deposition of thin films from small molecule compounds or oligomers [49, 50].

In order to avoid the deterioration of the organic materials during the deposition were used lower laser fluences (<500 mJ/cm²) compared to those used in the classical PLD [47] and targets (frozen in liquid nitrogen) formed from a mixture between the organic material and an adequate solvent used as a matrix [47, 48]. The solvent is chosen in order to obtain homogeneous mixture with the organic material and to be compatible with the laser wavelength. The solvent evaporation takes place at the absorption of the laser energy that is converted in thermal energy. Further, the solvent is pumped outside from the deposition chamber by the vacuum system [48]. The material of interest reaches the support where the nucleation process starts and the thin layer is formed. The concentration of the target is usually ~3%, depending on the material type.

Multilayer organic heterostructure can be fabricated by MAPLE, because the second deposited layer does not affect the first obtained layer [51].

3. Organic heterostructures with single and multilayer thin films: influence of the deposition technique type on their structural, morphological and optical properties

Different organic heterostructures were obtained either by VTE or MAPLE on a solid glass substrate (covered with ITO or AZO) and on a flexible substrate (covered with ITO). The prepared layers and heterostructures were investigated by various techniques: X-ray diffraction (XRD), atomic force microscopy (AFM), ultraviolet–visible (UV–VIS) spectroscopy, photoluminescence spectroscopy (PL) and infrared Fourier transform spectroscopy (FTIR). The used organic materials were metal phthalocyanines (ZnPc or MgPc), porphyrins (15,10,15,20-tetra(4-pyridyl)-21H,23H-porphine -TPyP) or other small molecule compounds (1,4,5,8-naphthalenetetracarboxylic dianhydride—NTCDA, fullerene-C60), and their chemical structure is presented in **Figure 2**.

3.1. Heterostructures based on ZnPc and NTCDA thin films obtained by VTE and MAPLE

Phthalocyanines are materials that are often used in OPV due to their large absorption domain in the visible part of the spectrum. These compounds are characterised by a high chemical stability having the property to form uniform layer on different solid substrates [52]. Thus, they can be easily deposited by the VTE method.

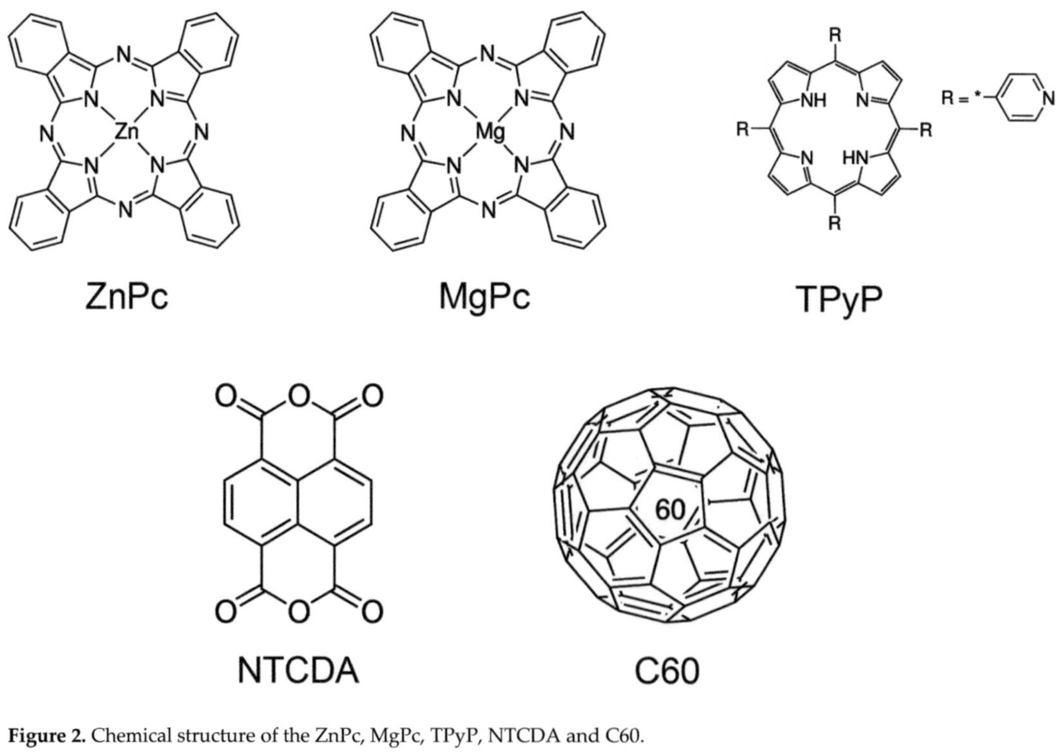

Figure 2. Chemical structure of the ZnPc, MgPc, TPyP, NTCDA and C60.

http://dx.doi.org/10.5772/67702

ZnPc, C60 and NTCDA layers were deposited by the VTE technique [53] on a ITO/glass, poly(3,4-ethylenedioxythiophene)-poly(styrenesulfonate)-PEDOT:PSS/ITO/glass, silicon and glass using the following experimental conditions: 8 × 10^{-6} mbar pressure in the chamber, at ~218°C for ZnPc, ~255°C for C60 and ~166°C for NTCDA. A PEDOT:PSS layer (20 nm) was prepared by spin-coating on ITO (15 Ω/sq) at a rotation speed of 3000 rot/min for 30 s. After that the obtained layers were supposed to a thermal treatment at 120°C for 5 min [53].

The following heterostructures were fabricated: 1-[(ITO/ZnPc(50 nm)/C60(30 nm)/NTCDA(110 nm)/Al)], 2-[(ITO/PEDOT:PSS(20 nm)/ZnPc(50 nm)/C60(20 nm)/NTCDA(120 nm)/Al)] and 3-[glass/Al/NTCDA(90)/C60(20)/ZnPc(50)/ITO]. For the second type of heterostructure (starting from glass/Al), the ITO electrode (the last material deposited) was prepared by PLD with an excimer laser source (λ = 248 nm and τ_{FWHM} ~25 ns) using the following experimental conditions: room temperature, 5 Hz repetition rate, 30,000 number of laser pulses, in oxygen atmosphere at 1.5 Pa pressure [53]. The resistivity of ITO was 3.9 × 10^{-4} Ωcm. Aluminium was used as a top metallic electrode (80 nm) being also obtained by VTE at 10^{-4} Pa pressure in the deposition chamber. The schematic representations of the prepared organic heterostructures are given in **Figure 3**.

The XRD diffractograms (**Figure 4**) indicated that the organic films obtained by VTE are not completely amorphous. The ZnPc layer presents lower diffraction peaks at 6.9, 9.6 and

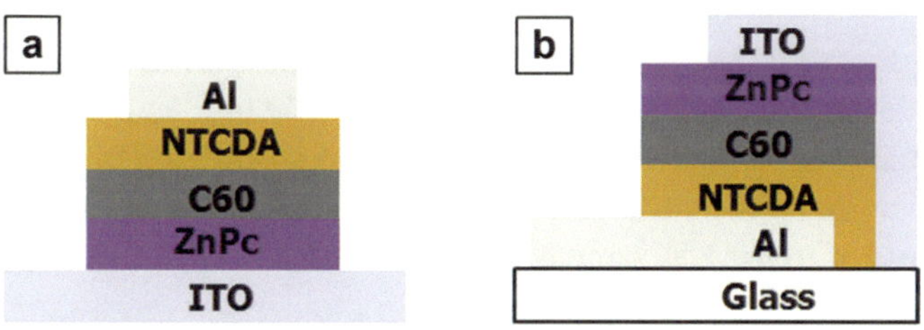

Figure 3. Schematic representation of the organic heterostructures deposited by VTE: standard structure (a) and inverted structure (b).

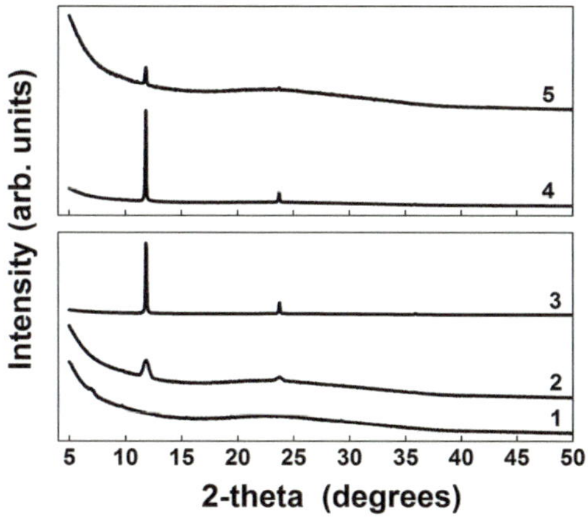

Figure 4. XRD patterns of the ZnPc film (curve 1), C60 film (curve 2), NTCDA film (curve 3), standard structure (curve 4) and inverted structure (curve 5) deposited by VTE.

29.3° [29], attributed to the powder raw material [54]. For the C60 layer, remarked lines were obtained at 11.8 and 23.7° specific to the (111) and (311) diffraction plane of this material [55]. The XRD diagram of the NTCDA film presents three peaks, including an intense one at 11.9° obtained also in the diffractogram of the powder [53]. Depending on the method used for deposition of the organic layers, in the multilayer structures are remarked only the diffractions lines originating from NTCDA, which are more intense when NTCDA is deposited on top (**Figure 4**, curve 4).

For the ZnPc and C60 films, the AFM images (**Figure 5**) show topography characteristic to these materials deposited by thermal evaporation [56, 57]. Thus, it can be observed a low roughness for the ZnPc film (root mean square, RMS = 5.1 nm) in comparison with C60 (RMS = 14.7 nm) and NTCDA (RMS = 20.9 nm). The RMS higher value of the NTCDA can be attributed to the layer thickness, this layer being thicker than ZnPc and C60 films.

For the heterostructures containing three organic layers, a reduced roughness was obtained in the inverted structure (RMS = 6.8 nm) compared to the normal structure (RMS = 8.3 nm) as it is expected, taking into account that in standard structure the top organic layer is NTCDA which is characterised by the highest roughness.

The vibrational properties of the raw materials were identified in the FTIR spectra (**Figure 6**) of ZnPc and NTCDA layers deposited by VTE, indicating that no chemical decomposition took place during the VTE transfer. The C60 film was too thin to remark some FTIR peaks on it. In the ZnPc layer, the peak from 727 cm^{-1} is specific to C-H out of plane deformation, the peaks situated at 748, 1095, 1118 and 1288 cm^{-1} appear due to the in-plane C-H bending, the peak at 1333 cm^{-1} evidenced the C-C stretching in isoindole and the peaks from 1481 to 1608 cm^{-1} are attributed to the C-C stretching in benzene [56, 58].

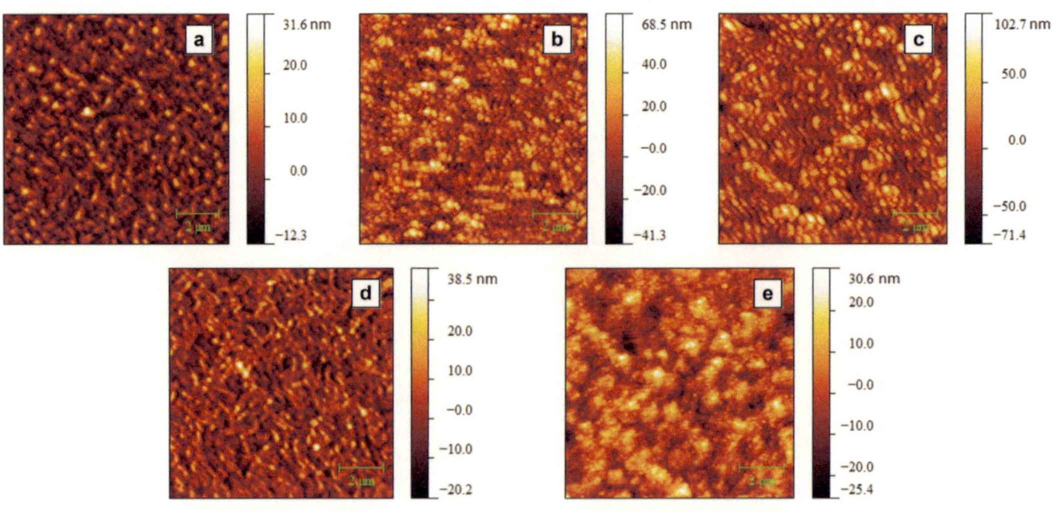

Figure 5. AFM images of the ZnPc film (a), C60 film (b), NTCDA film (b), standard structure (d) and inverted structure (e) deposited by VTE.

http://dx.doi.org/10.5772/67702

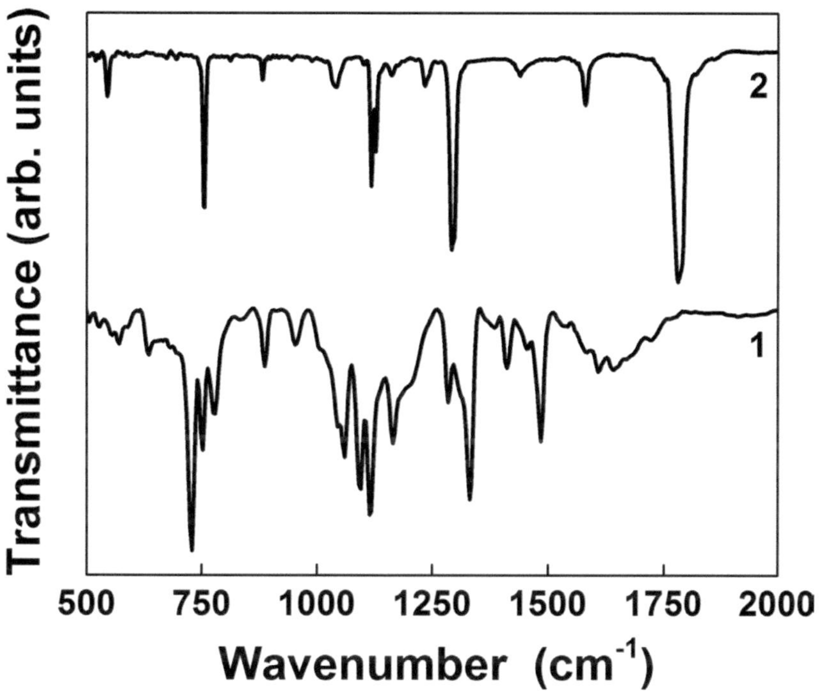

Figure 6. FTIR spectra of the ZnPc (curve 1) and NTCDA (curve 2) single layers deposited by VTE.

For the NTCDA layer, the peaks at 1780 cm^{-1} (characteristic to the dianhidrydecarbonylic group [59]), at 543, 753 and 882 cm^{-1} (specific to the C-H out-of-plane bending vibrations [60, 61]), at 698 and 754cm^{-1} (attributed to C-H bending vibration [36]), at 1044, 1120, 1161, 1234 and 1293 cm^{-1} (characteristic to the stretching vibration of C-O in the anhydride groups and the C-H in-plane bending vibration) and at 1442 and 1582 cm^{-1} due to the C-C bending [61]) were evidenced.

The UV–VIS spectra of the VTE prepared organic thin films are given in **Figure 7**. For the ZnPc layer, a high transparency (~90% at 500 nm) was emphasised, covering a broad part of the VIS region and presenting the band B (so-called Soret band) and band Q [62, 63]. Several absorption maxima are remarked for the C60 layer, at 340, 400 and 440 nm which are characteristic to this material prepared by the VTE technique [64, 65]. The NTCDA layer used as buffer in our structure reveals absorption maxima in UV (at 370 and 390 nm) attributed to the π-π^* transition [66]. The structure comprising all the organic layers (**Figure 7**, curve 4) is characterised by a high transmittance, showing the absorption maxima of all components.

ZnPc, C60 and NTCDA thin films deposited by the VTE method are polycrystalline and have morphologies specific to raw materials (ZnPc, C60 and NTCDA), being characterised by different roughness values (RMS ranged between 5.1 and 20.9 nm). The materials present adequate absorption bands in the visible region. The peaks disclosed by the FTIR spectra are assigned to each organic material, evidencing that no chemical decomposition appears in the thin-film deposition.

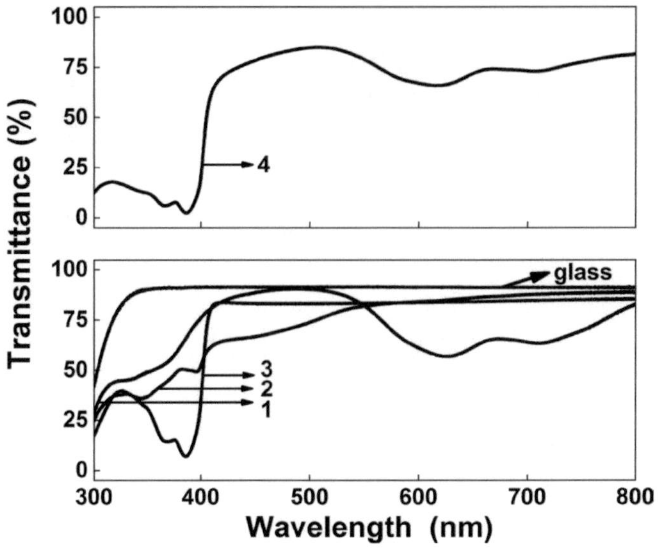

Figure 7. Transmission spectra of the ZnPc film (curve 1), C60 film (curve 2), NTCDA film (curve 3) and standard structure (curve 4) deposited by VTE.

Along the time, ITO was the most used transparent electrode, due to its high optical transmittance and reduced electrical resistivity. Because the required indium is rare and expensive, many attempts were made in order to replace the ITO in various applications, including the OPV field.

As transparent electrode we choose a large band gap semiconductor, ZnO doped with Al (AZO) because it presents adequate electrical resistivity (~10^{-4} Ωcm), a high optical transmission in the visible-NIR domain, and a higher chemical stability in comparison with ITO [67–69].

AZO (ZnO doped with 2% Al) thin films were prepared at room temperature on a glass substrate by PLD using the KrF* excimer laser in the following experimental conditions: 10 Hz repetition rate 5 cm, substrate-target distance, 3 J/cm^2 laser fluence, 32,000 laser pulses, in oxygen atmosphere at 10^{-2} mbar pressure [70]. Subsequently, the obtained AZO layers were treated in oxygen plasma at 0.6 mbar and P_{max} = 130 W (for 5 and 10 min) in order to observe how this treatment affects the properties of the formed layers. The samples were labelled as follows: AZO (untreated film), 5AZO (film treated for 5 min) and 10AZO (film treated for 10 min).

The MAPLE technique was used to process organic films from ZnPc and NTCDA on the AZO substrate. The same laser source was used to prepare thin films from a frozen target containing ZnPc or NTCDA and dimethyl sulphoxide (DMSO) as a solvent compatible with the laser wavelength. Two different laser fluences were used for the deposition of the ZnPc layer: 0.4 J/cm^2 (1ZnPc) and 0.3 J/cm^2 (2ZnPc). Organic heterostructures with two stacked layers were formed by the deposition of the NTCDA layer over ZnPc films. For the NTCDA, the deposition parameters were 0.3 J/cm^2 laser fluence, 90,000 and 100,000 laser pulses [70]. **Table 1** presents the experimental conditions for the deposition of the organic layers.

http://dx.doi.org/10.5772/67702

Sample	ϱ of AZO before/after treatment (10^{-4} Ωcm)	Laser fluence (J/cm²)	No. of laser pulses (K)	d/RMS of AZO (nm)	d/RMS of ZnPc (nm)	d/RMS of NTCDA (nm)
AZO/1ZnPc/NTCDA	3.1/3.1	0.4/0.3	65/90	1310/9.3	360/51	120/59
AZO/2ZnPc/NTCDA	2.7/2.7	0.3/0.3	65/90	1300/9.2	380/53	150/73
5AZO/1ZnPc/NTCDA	2.9/2.8	0.4/0.3	65/100	1290/6.4	550/60	140/42
5AZO/2ZnPc/NTCDA	2.9/3.1	03/0.3	65/100	1290/6.1	390/61	90/50
10AZO/1ZnPc/NTCDA	3.1/3.1	0.4/0.3	65/100	960/4.4	490/61	120/53
10AZO/2ZnPc/NTCDA	3.2/2.5	0.3/0.3	65/100	940/3.3	440/60	100/56

Table 1. Resistivity of the AZO layers before and after treatment, used laser fluences in the MAPLE deposition, the thickness (d) of the layers and the roughness value obtained from AFM.

The heterostructures (**Figure 8**) were carried out by the gold (Au) electrode of ~100 nm thickness deposited also by VTE.

The morphological investigations of the AZO substrates and of the ZnPc/NTCDA structures are represented in **Figure 9**. Only the AFM images collected for the structures with the ZnPc layer deposited at 0.4 J/cm^2 laser fluence are presented, but the roughness (RMS) values both for structures with ZnPc deposited at 0.4 and 0.3 J/cm^2 laser fluences are presented in **Table 1**.

Oxygen plasma treatment leads to a decrease in the RMS value of the AZO substrate, from 9.3 nm for the untreated film to 3.3 nm for the treated film for 10 min (**Table 1**). A similar behaviour was remarked by others authors [71]. The AFM images exhibit a topography characterised by small grains for the organic layers obtained on treated substrate compared to that formed on the untreated substrate. The RMS value increases from the single to bilayer structures prepared on the untreated AZO substrate. The RMS recorded for the ZnPc layer shows an increase when the AZO substrate is treated (**Table 1**). Probably, the ZnPc deposition is affected by the surface energy of AZO layer modified during the oxygen plasma treatment. The higher roughness of the ZnPc layer obtained on the AZO-treated substrate leads to a better arrangement of the NTCDA molecules having an effect on lowering the RMS value recorded for the bilayer heterostructures.

From the UV–VIS spectra, a transparency between 75 and 87% in the range 400–800 nm was obtained for the AZO layers (**Figure 10**, curve 1). The thickness of the AZO films was evaluated using the formula from [72] which takes into consideration successive interference maxima and minima. The obtained values (between 940 and 1310 nm) are given in **Table 1**.

The UV–VIS spectra of the AZO layers revealed a slight improvement in the transparency with the increase in duration of the applied plasma treatment (**Figure 10**, curves 1′ and 1′′). This can be attributed either to a reduction of the defects number inside the AZO layer (these can act as scattering centres), due to decrease in the AZO layer thickness (**Table 1**) or to the reduction in scattering at the surface in wavelength domain (>750 nm).

The thickness of the organic films was also estimated from the UV–VIS spectra, using the absorption coefficients at $\lambda = 355$ nm reported in the literature, $\alpha_{ZnPc} = 3.5 \times 10^4$ cm^{-1} [73] and $\alpha_{NTCDA} = 2.1 \times 10^5$ cm^{-1} [74]. The thickness varied between 360 and 550 nm for the ZnPc layer and between 90 and 150 nm for NTCDA (**Table 2**).

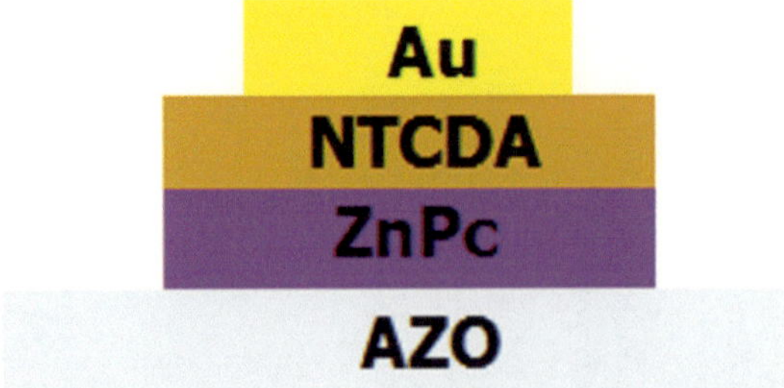

Figure 8. Schematic representation of the organic heterostructure deposited by MAPLE on AZO substrate.

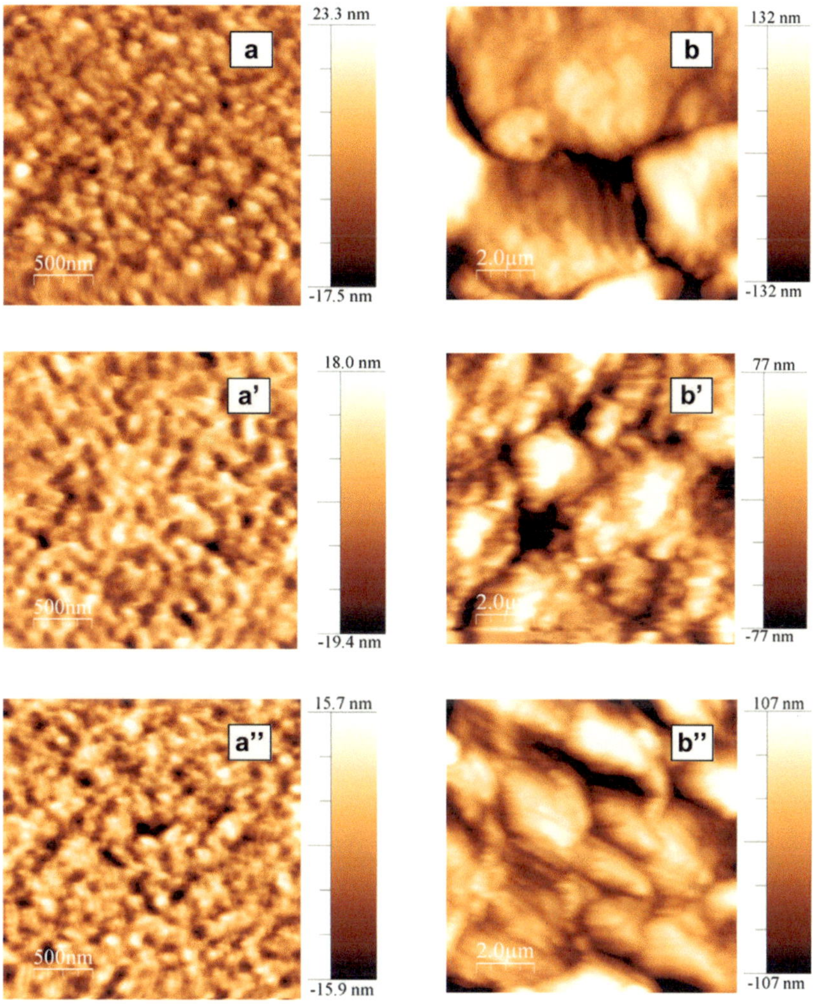

Figure 9. AFM images of the glass/AZO substrates (a, a' and a'') and ZnPc/NTCDA layers deposited by MAPLE—0.4 J/cm² fluence for ZnPc and 0.3 J/cm² fluence for NTCDA (b, b' and b''): untreated (a, b), treated in oxygen plasma for 5 minutes (a', b') and treated in oxygen plasma for 10 minutes (a'', b'').

The ZnPc layers (**Figure 10**, curves 2 and 3) present a structured absorption in the range of 550–750 nm, this large absorption domain being useful in generation of the charge carriers. As mentioned above, the oxygen plasma treatment can modify the surface energy of the AZO layer, can change the way in which the organic molecules are arranged on the substrate and as consequence the optical properties of these organic layers. Comparison of the 1ZnPc and 2ZnPc samples was found that for the second film the absorption is smaller. Additionally, the NTCDA layer does not affect the shape of the transmission spectrum (**Figure 10**, curve 3).

The emission properties of the samples under excitation with λ_{exc} = 335 nm were also investigated (**Figure 11**). The AZO layer is characterised by an intense emission band with maximum at ~430 nm and a shoulder at ~480 nm, linked to point defects as Zn^{2+} interstitial [75, 76].

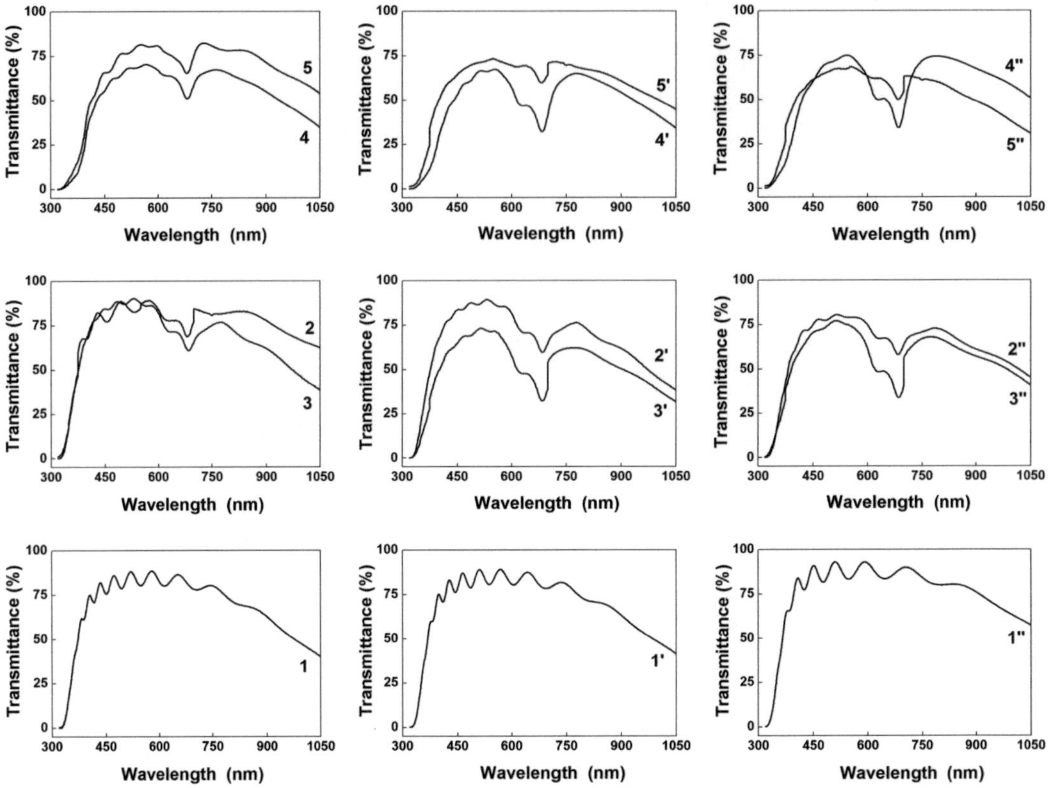

Figure 10. Transmission spectra of the organic films deposited by MAPLE on a glass/AZO substrate untreated (1–5), treated in oxygen plasma for 5 min (1′-5′) and treated in oxygen plasma for 10 min (1″-5″): glass/AZO substrate (curves 1, 1′ and 1″), 1ZnPc film (curve 2, 2′ and 2″), 2ZnPc film (curve 3, 3′ and 3″), 1ZnPc/NTCDA layers (curve 4, 4′ and 4″) and 2ZnPc/NTCDA layers (curve 5, 5′ and 5″).

Sample	Laser pulses	Thickness (nm)	RMS (nm)
ZnPc/ITO	100k	570	36
MgPc/ITO	85k	470	35
TPyP/ITO	74k	440	34
TPyP/ZnPc/ITO	30k/30k	430	25
TPyP:ZnPc/ITO	60k	380	32
TPyP/MgPc/ITO	30k/30k	530	49
MgPc:TPyP/ITO	60k	440	57

Table 2. MAPLE conditions used for the deposition of organic films and structures on ITO/PET, layer thickness and RMS values interpolated from AFM.

The oxygen radicals from the plasma can lower the number of the Zn^{2+} interstitials due to the reduction of the defects which appear near to the film surface [77]. The thickness of the samples has a decisive role in the intensity of the emission. In the AZO and 5AZO thicker layers, the

http://dx.doi.org/10.5772/67702

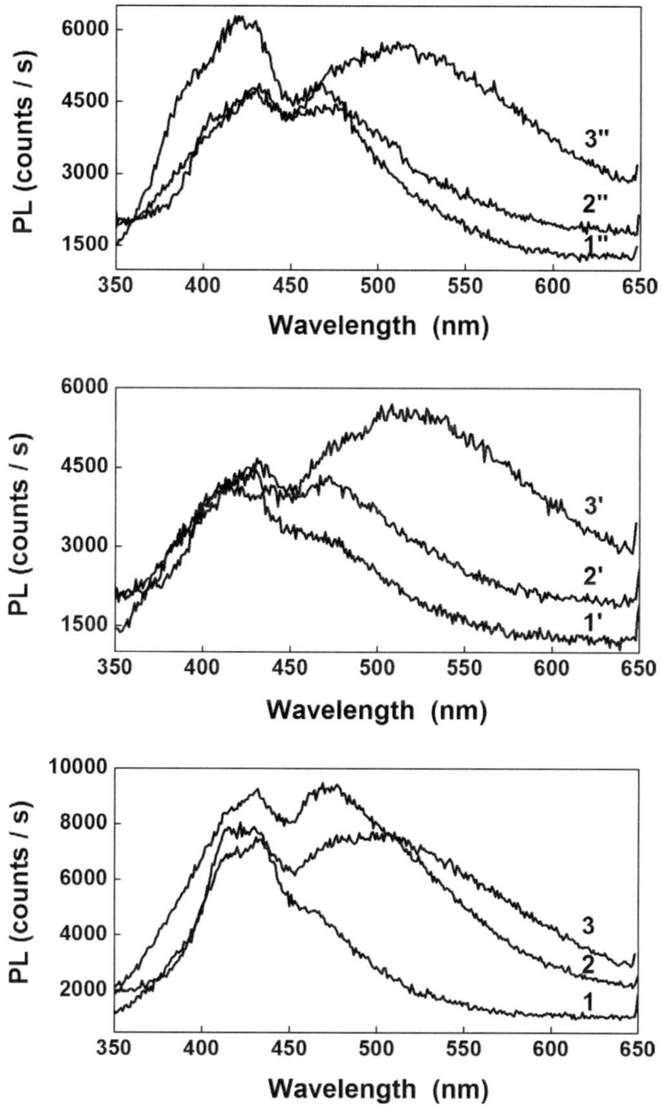

Figure 11. Photoluminescence spectra of the organic films deposited by MAPLE on a glass/AZO substrate untreated (1–3), treated in oxygen plasma for 5 min (1′-3′) and treated in oxygen plasma for 10 min (1″-3″): glass/AZO substrate (curves 1, 1′ and 1″), 1ZnPc film (curve 2, 2′ and 2″) and 1ZnPc/NTCDA layers (curve 3, 3′ and 3″).

emission band attributed to the deep level point defects (2.6 eV) is lower while in the thinner 10AZO layer the emission increases (**Figure 11**, curves 1).

The emission band situated at 430 nm in the AZO spectrum can be remarked also in the structures prepared with ZnPc and ZnPc/NTCDA (**Figure 11**, curves 2 and 3). The ZnPc layer discloses also a peak in the range of 400–450 nm [78]. The shoulder situated at 480 nm from AZO became a well-structured band in the structures containing ZnPc. No supplementary maxima were observed by adding NTCDA, probably because the emission bands specific to this material, one situated at ~430 nm and other situated in 475–575 nm range are masked by the emission of the AZO and ZnPc layers, respectively [79]. The intensity of the emission band with the maximum at 480 nm from the AZO substrate decreases in the structures prepared with one or two organic layers.

AZO layers were successfully transferred by PLD, in order to be further used to prepare organic heterostructure by MAPLE. The oxygen plasma treatment influences the roughness of the AZO layers. A decrease of the films roughness is obtained with the increase duration of the applied treatment. For this TCO, a high transmittance and emission with maxima at about 430 and 480 nm were evidenced. The organic heterostructures formed on the AZO substrate present also a high transmittance in the visible domain. The surface topography of the organic heterostructures is characterised by grains, smaller grains being remarked for the AZO/1ZnPc/NTCDA structure made on the treated AZO substrate.

3.2. Heterostructures based on metal phthalocyanines (ZnPc or MgPc) and TPyP thin films prepared by MAPLE

Another type of organic heterostructure has bulk active layer. The bulk heterojunction concept [80] was introduced to overpass the mismatch between the energy bands of the constituent organic materials used to form an organic cell with different layers. A bulk heterojunction can be obtained using a wet method for the deposition of the organic materials, these being mixed in a solution with an adequate solvent from which are subsequently deposited films. The organic p-n materials form an interpenetrating network. In this way, the interface between them is enlarged, having effect on the exciton dissociation and the charge transport [81].

The MAPLE method described above was used for obtaining structures with metallic phthalocyanines (ZnPc or Mg Pc) and a non-metallic porphyrin, 5,10,15,20-tetra(4-pyrydil)21H,23H-porphyne (TPyP) as a bulk active layer or as a stacked layer, to investigate the effect of the cell architecture on the properties. In these structures, the phthalocyanines are the p-type material and the TPyP is the n-type material.

A flexible ITO/PET substrate (14 Ω/sq resistivity) was used as a TCO electrode. For the MAPLE deposition, the same above presented laser was involved, keeping the constant experimental conditions: 2.5% concentration of the organic material in DMSO, 300 mJ/cm^2 laser fluence, 5 Hz laser frequency and 5 cm target-substrate distance. In order to obtain layer with appropriate thickness, the number of the laser pulses was varied (**Table 2**). Besides ITO/PET, substrates as a glass and silicon were used. In the structures containing blends, the materials were used in the weight ratio of 1:1 and in those having two stacked layers: the first deposited layer was the metallic phthalocyanine [50]. A schematic representation of the transferred MAPLE layers is presented in **Figure 12**.

The layers prepared were analysed from structural point of view, the diffractograms of ZnPc, MgPc, TPyP and their structures are presented in **Figure 13**. The single layers and the heterostructure based on MgPc are amorphous. In the case of the diffractograms of heterostructures based on ZnPc, some lines characteristic to this material are observed (6.8, 9.1 and 13.8°) [54, 63], meaning that ZnPc presents some degree of crystallinity. The amorphous

http://dx.doi.org/10.5772/67702

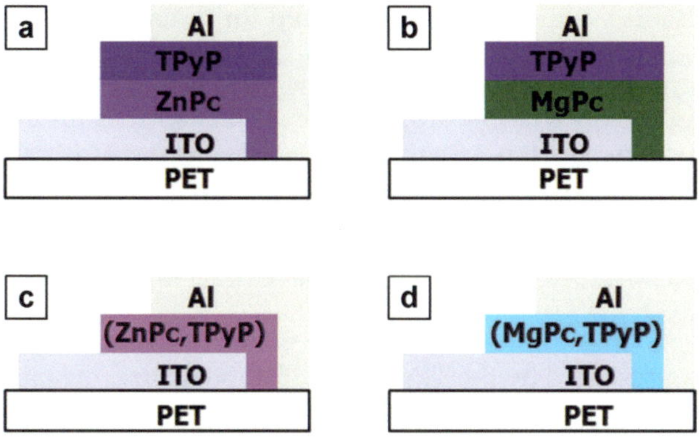

Figure 12. Schematic representation of the organic heterostructures deposited by MAPLE with stacked films (a, b) and mixed layers (c, d).

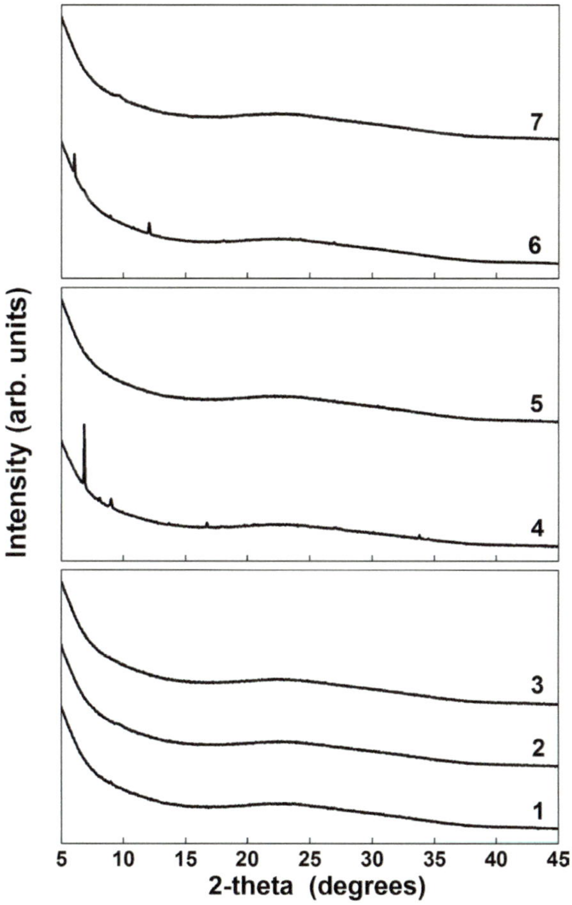

Figure 13. XRD patterns of the organic layers deposited by MAPLE: single layers—ZnPc (curve 1), MgPc (curve 2), TPyP (curve 3), heterostructures containing stacked films—ZnPc/TPyP (curves 4), MgPc/TPyP (curve 5) and heterostructures with mixed layers—ZnPc:TPyP (curves 6), MgPc:TPyP (curve 7).

behaviour of the phthalocyanines was also reported for films prepared by VTE [82]. So, this behaviour is independent of a deposition technique.

The AFM images (**Figure 14**) were recorded on thin films and on the structures. The granular morphology showed by the deposited films was also reported in other papers, this morphology being characteristic to the MAPLE prepared films but also to the phthalocyanines [56, 62, 83]. Small and large grains were disclosed by the AFM images performed on mixed layers (ZnPc:TPyP and MgPc:TPyP). The RMS values extracted from AFM are between 25.0 and 56.8 nm (**Table 2**).

A small RMS value is presented by the TPyP/ZnPc/ITO structure, meaning that in the stacked structure appears a better accommodation of the TPyP molecules on the rough ZnPc film (35.7 nm). The highest RMS value was obtained for the MgPc:TPyP/ITO structure. Probably, the MgPc:TPyP blend is less homogenous in DMSO, the obtained films being characterised by bigger grains comparable with ZnPc:TPyP blend.

The optical properties of the phthalocyanines and porphyrins films were also analysed. The FTIR spectra (**Figure 15**) were recorded in order to observe if some changes appear in the structure of the materials deposited by MAPLE. Thus, the FTIR spectra of the films are shown in comparison with those of the raw powders, the IR bands from the powders appearing also in the thin films, with lower intensity (due to the film thickness). In the phthalocyanines, films

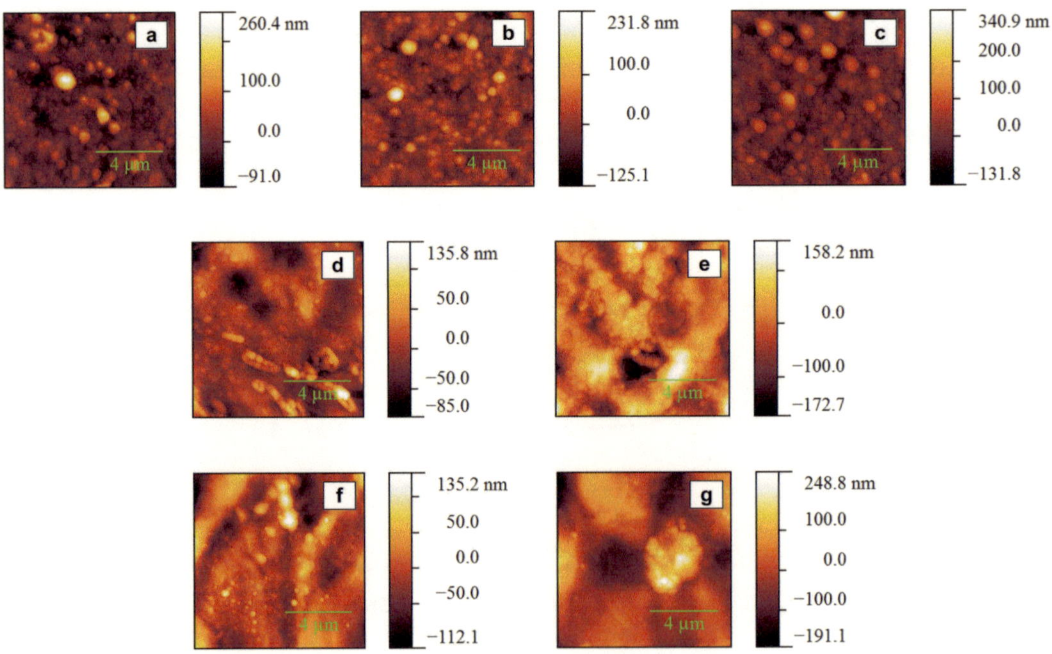

Figure 14. AFM images of the organic layers deposited by MAPLE: single layers—ZnPc (a), MgPc (b), TPyP (c), heterostructures containing stacked films—ZnPc/TPyP (d), MgPc/TPyP (e) and heterostructures with mixed layers—ZnPc:TPyP (f), MgPc:TPyP (g).

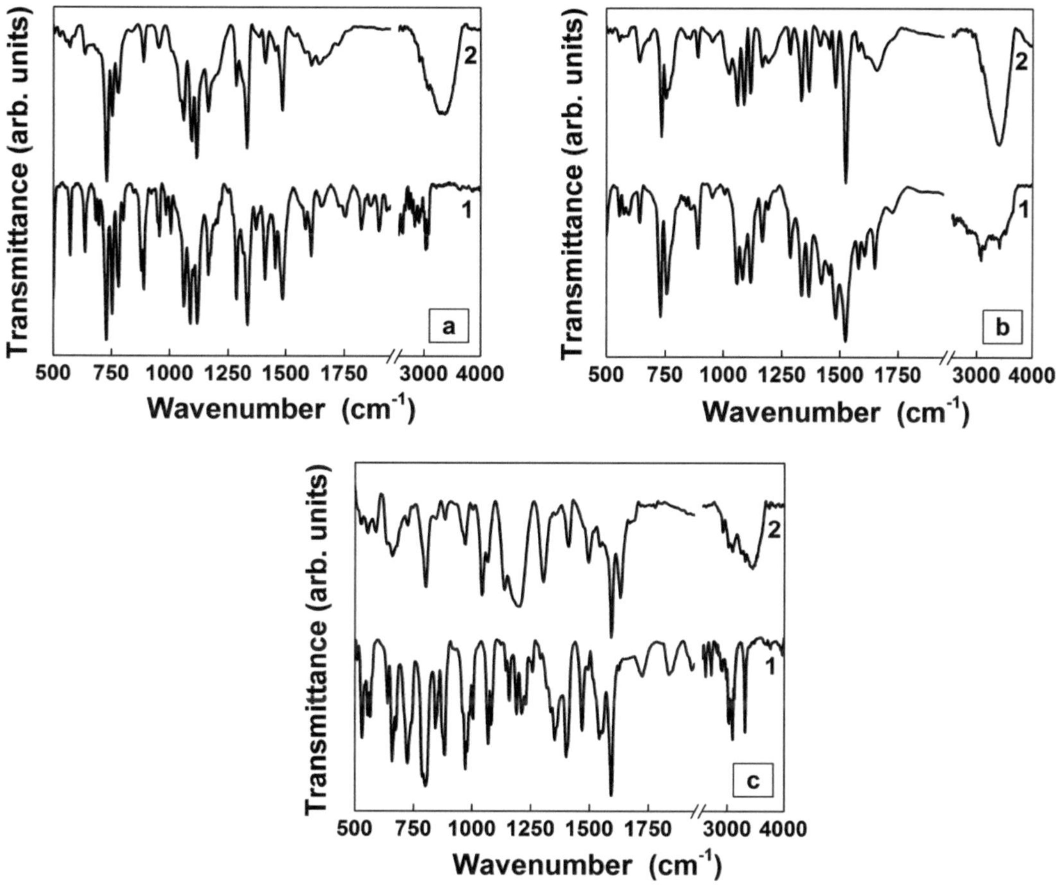

Figure 15. FTIR spectra of the ZnPc (a), MgPc (b) and TPyP (c) as powders (curves 1) and single layers deposited by MAPLE (curves 2).

were identified the vibrations attributed to the C-H out of plane deformation at 725 cm^{-1}, in-plane C-H bend at 754, 1088, 1114 and 1285 cm^{-1}, the C-C stretching in isoindole at 1333 cm^{-1}, the C-C stretching in benzene at 1482 and 1606 cm^{-1}, the C-H bending in aryl at 1490 cm^{-1} [56, 58]. For the TPyP, the following vibrations were attributed: 798 cm^{-1} to thc C-H bond in pyrrole, 1500 and 1590 cm^{-1} to the C-C stretching in the pyridyl aromatic ring, 970and 3306 cm^{-1} to porphyrin free-base signature [15].

Based on these results, it can be concluded that no modification appears at the MAPLE transfer of the organic materials.

The UV–VIS spectra (**Figure 16**) of the organic thin films deposited on flexible substrates have identified the absorption maxima typical to the used compounds (**Figure 16**). It can be evidenced the presence of the B and Q bands (between 550 and 750 nm) characteristic to ZnPc and MgPc [62, 63]. Two submaxima are remarked in the Q band due to the π-π^* transition, this band being localised on the phthalocyanine ring [84, 85]. The π-π^* absorption is emphasised

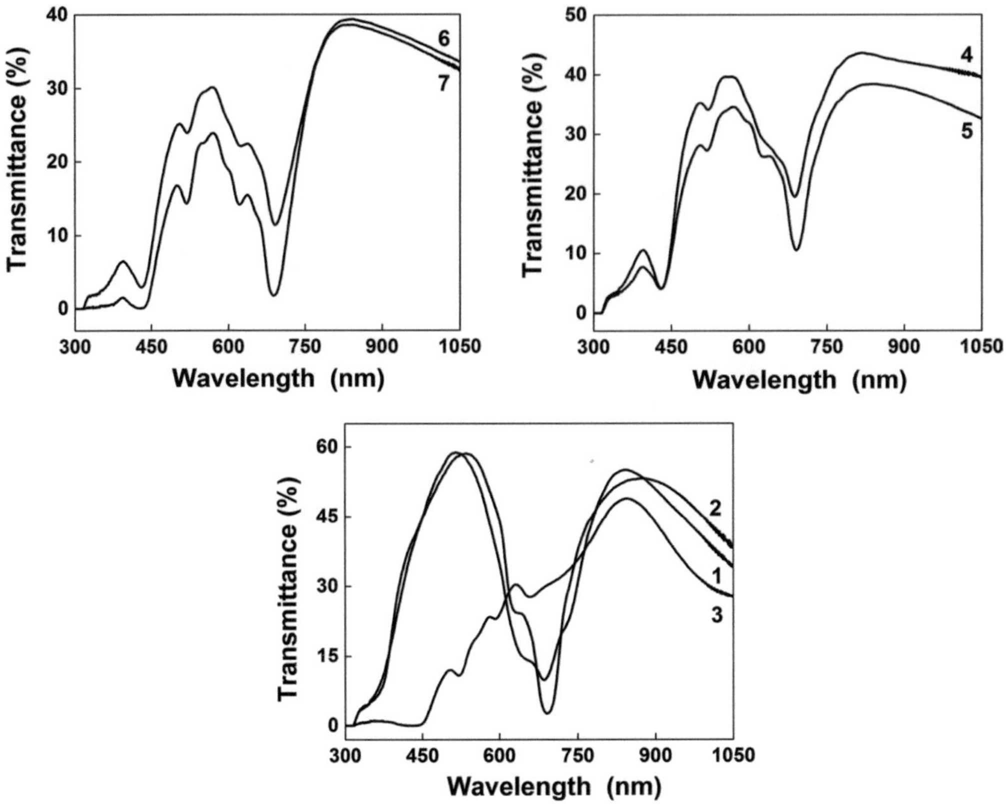

Figure 16. Transmission spectra of the organic layers deposited by MAPLE: single layers—ZnPc (curve 1), MgPc (curve 2), TPyP (curve 3), heterostructures containing stacked films—ZnPc/TPyP (curves 4), MgPc/TPyP (curve 5) and heterostructures with mixed layers—ZnPc:TPyP (curves 6), MgPc:TPyP (curve 7).

also in the TPyP film, being specific to the free-base ethio-type porphyrin, with the B (428 nm) and Q (with maxima at 520, 590 and 660 nm) bands [86].

Investigating the emission properties (at 435 nm excitation wavelength) of the samples based on phthalocyanines and porphyrins was noted that those containing ZnPc present a large emission band (**Figure 17**) with a maximum at 690 nm and those with MgPc show a broader band having the maximum at ~800 nm, associated with the Davidov coupling in the phthalocyanine solid films [84].

An emission band with two maxima at 660 and 713 nm was observed in the TPyP film, these peaks being characteristic to TPyP free base [86]. In the heterostructures prepared with stacked layers, the TPyP emission bands were also evidenced (**Figure 17** curves 4 and 5). For the heterostructures made with blends, a decrease in the emission intensity attributed to TPyP (leading even to its quenching) was observed (**Figure 17**, curves 6 and 7).

Layers based on ZnPc, MgPc and TPyP were successfully transferred by MAPLE on ITO flexible substrates. Only the ZnPc presents a certain crystallinity degree when is deposited both in stacked and blend forms with TPyP, all the others organic films being amorphous.

http://dx.doi.org/10.5772/67702

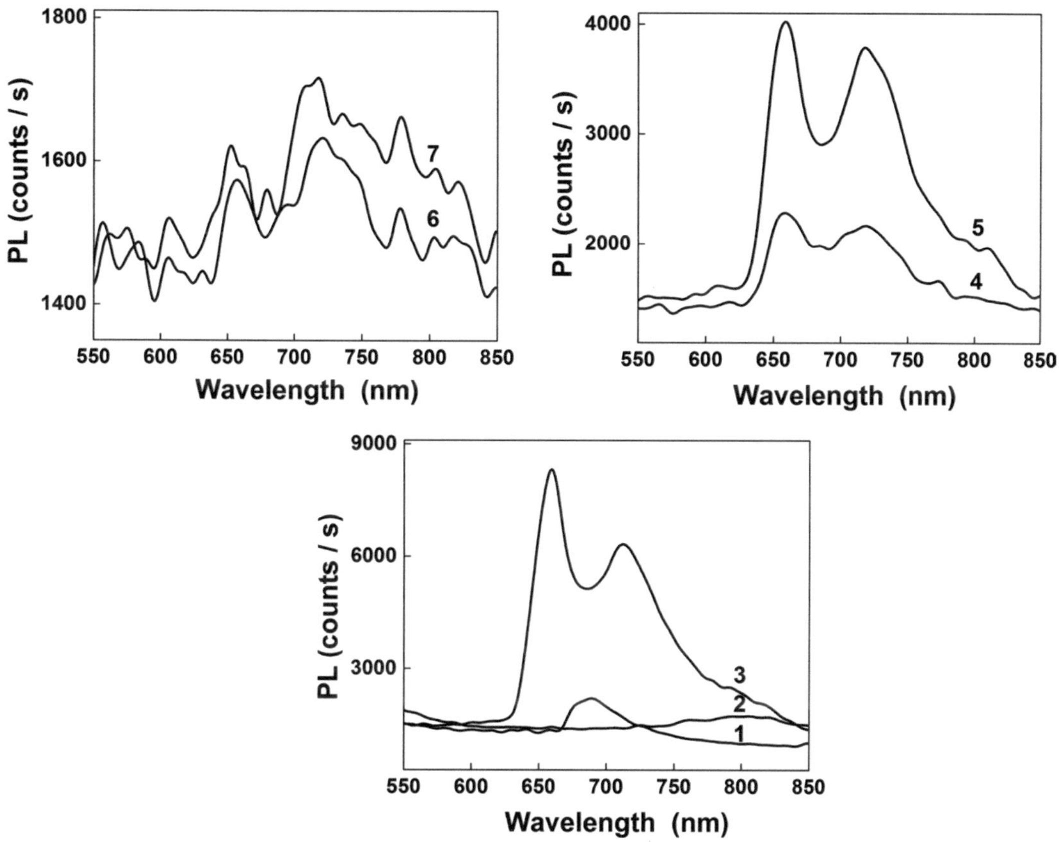

Figure 17. Photoluminescence spectra of the organic layers deposited by MAPLE: single layers—ZnPc (curve 1), MgPc (curve 2), TPyP (curve 3), heterostructures containing stacked films—ZnPc/TPyP (curves 4), MgPc/TPyP (curve 5) and heterostructures with mixed layers—ZnPc:TPyP (curves 6), MgPc:TPyP (curve 7).

The morphology with grains can be attributed also to the phthalocyanines but also to the MAPLE deposition method. The FTIR spectra confirm that the film deposited by MAPLE preserves the vibrational properties of the raw materials. The optical properties of the films evidenced a large absorption domain in the visible range. A quenching of the photoluminescence in the bulk heterostructures was observed.

4. Organic heterostructures based on single and multilayer thin films: electrical properties for device applications

The *I-V* characteristics (**Figure 18**) for the heterostructures prepared by VTE were recorded under dark conditions, in the range of 0V–1V. By using an additional layer of PEDOT:PSS, the current value in the standard structures increased from 1.6×10^{-5} A (glass/ITO/ZnPc/C60/NTCDA/Al) at 6×10^{-4}A (glass/ITO/PEDOT:PSS/ZnPc/C60/NTCDA/Al). This supplementary layer favours the hole injection from the ITO electrode in the first organic film [52].

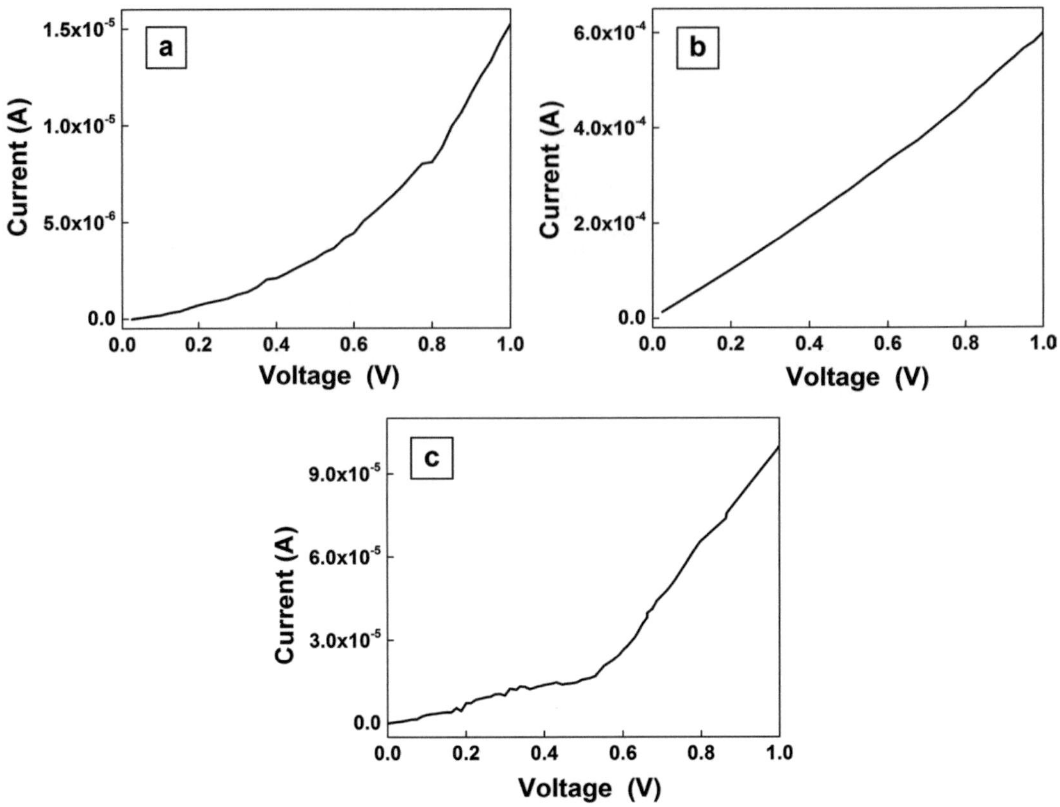

Figure 18. Current-voltage characteristics of the organic heterostructures deposited by VTE: ITO/ZnPc/C60/NTCDA/Al (a), ITO/PEDOT:PSS/ZnPc/C60/NTCDA/Al (b) and glass/Al/NTCDA/C60/ZnPc/ITO (c) structures.

By the deposition of these materials in the inversed order from Al to ITO, an improvement in the current value was also obtained, from 1.6×10^{-5} (glass/ITO/ZnPc/C60/NTCDA/Al) A at 1×10^{-4} A (glass/Al/NTCDA/C60/ZnPc/ITO). Preparing the heterostructure in this way was avoided the interaction of the hot Al atoms with the organic layer which can determine the appearance of some recombination centres at the interface [87].

Thus, the high current values obtained for these heterostructures can be useful for the OPV applications. It was remarked that the current value can be increased either using a supplementary PEDOT:PSS layer or by preparing the heterostructure in the inverted way.

The resistivity for the AZO layers prepared by PLD was determined using a four-point probe method, the values being between 2.7×10^{-4} and 3.2×10^{-4} Ω cm in the case of untreated layers and between 2.5×10^{-4} and 3.1×10^{-4} Ωcm in the case of the treated layers in oxygen plasma (**Table 1**).

For analysis of the NTCDA/ZnPc/AZO structures from electrical point of view, an injection contact behaviour was evidenced for both structures deposited on untreated AZO and treated AZO films (**Figure 19**). 5AZO and 10AZO films were characterised by a lower resistivity compared to that of the untreated AZO layer, and are chosen to facilitate the charge carrier injection. It was observed that the *I-V* characteristics became asymmetric for the heterostructures prepared on

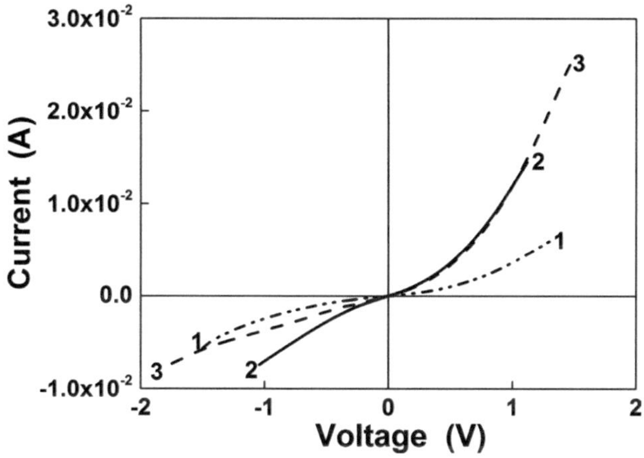

Figure 19. Current-voltage characteristics of the organic heterostructures deposited by MAPLE: AZO/ZnPc/NTCDA/Au structure on substrates untreated (curve 1), treated in oxygen plasma for 5 min (curve 2) and treated in oxygen plasma for 10 min (curve 3).

treated AZO substrates. At low voltages (under 0.4 V), the characteristics are linear and at higher voltages the effect of the space charge limited currents (SCLC) becomes dominant. At direct polarisation, at 1V, the current value increases from 3×10^{-3} A in the structure with AZO at 1.5×10^{-2} A in the structures with 5AZO and 10AZO (**Figure 19** Quadrant 1). An increase in the work function of the AZO electrode was induced by the oxygen plasma treatment [88], having as effect a decrease in the energetic barrier at the AZO/organic interface which improves the injection of the charge carriers from AZO in the organic layer.

Taking into consideration the properties of the AZO, this TCO can be integrated in organic heterostructure, instead of the ITO electrode. The heterostructures prepared on AZO are characterised by current values suitable for the photovoltaic applications. Moreover, treating in oxygen plasma the AZO substrate can be increased the current value in the heterostructures based on ZnPc and NTCDA.

The electrical properties of stacked and blend layers deposited by MAPLE on flexible substrate were also investigated. Regarding the stacked structures, they were energetically favourable, taking into account the ionisation potential (IP) and electron affinity (EA) levels in ZnPc ($E_{IP;ZnPc}$ = 5.28 eV and $E_{EA;ZnPc}$ = 3.28 eV [89]), MgPc ($E_{IP;MgPc}$ = 5.4eV and $E_{EA;MgPc}$ = 3.9 eV [90]) and TPyP ($E_{IP;TPyP}$ = 6.8 eV, $E_{EA;TPyP}$ = 4.1 eV [86]).

The *I-V* characteristics recorded under dark and under illumination conditions, in 0V–1V domain, are near linear (**Figure 20**). In the dark, the higher value of the current (~10^{-6} A) was obtained for the Al/MgPc:TPyP/ITO structure, with ~ 3 orders of magnitude higher than the value presented by the Al/MgPc/TPyP/ITO structure. As was remarked in the AFM images (**Figure 10** g), MgPc:TPyP layer seems to be characterised by a larger roughness which can lead to the formations of some dipoles which reduce the energetic barrier at interfaces favouring the charge transport [83].

An increase in the current value and a photovoltaic effect was also evidenced in the Al/ZnPc:TPyP/ITO structure after exposure to light (**Figure 21**). The solar cell parameters are:

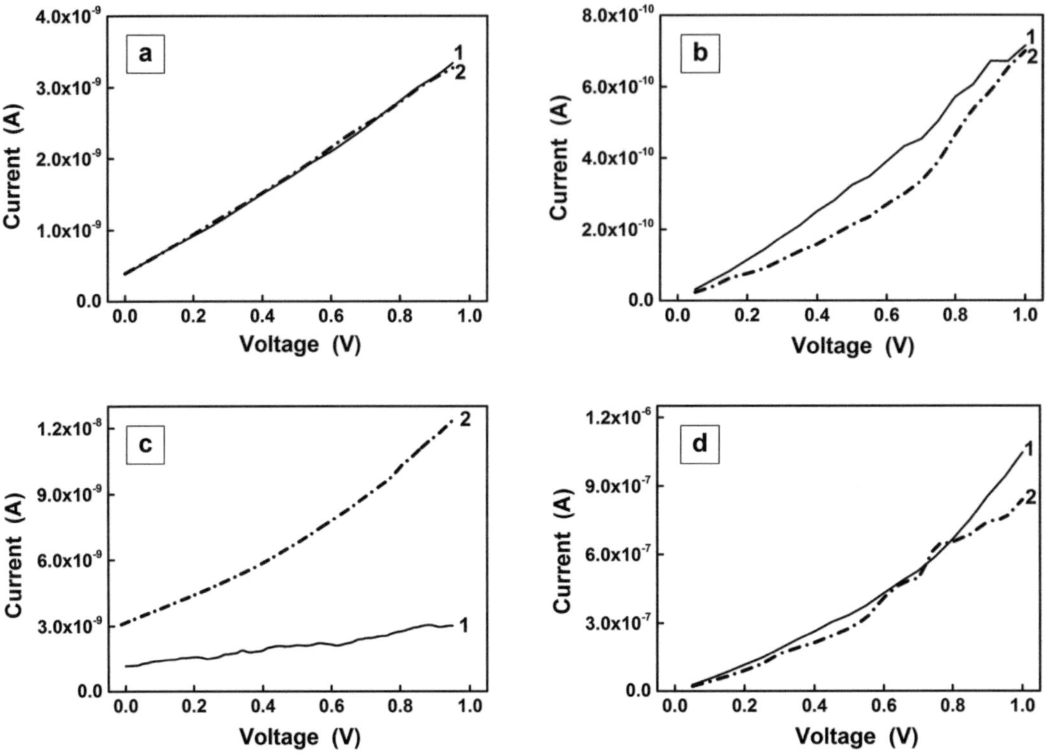

Figure 20. Current-voltage characteristics (in dark conditions—curves 1 and under illumination—curves 2) of the organic heterostructures deposited by MAPLE: PET/ITO/ZnPc/TPyP/Al (a), PET/ITO/MgPc/TPyP/Al (b), PET/ITO/ZnPc:TPyP/Al (c) and PET/ITO/MgPc:TPyP/Al (d) structures.

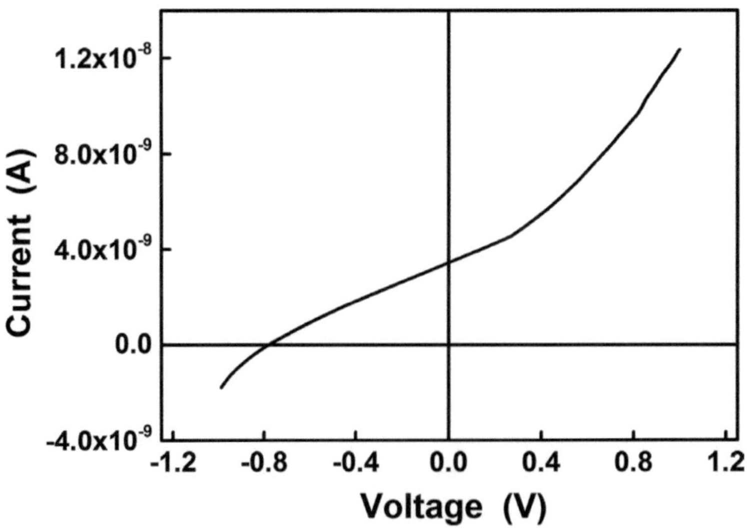

Figure 21. Current-voltage characteristic (under light, −1 V–1 V domain) of the PET/ITO/ZnPc:TPyP/Al heterostructure deposited by MAPLE.

$I_{SC} = 3.4 \times 10^{-9}$ A; $U_{OC} = 0.77$ V and FF = 0.28. Even if in the PL spectra of the MgPc:TPyP structure (**Figure 17**) was remarked a quenching of the photoluminescence, in the *I-V* characteristic recorded under illumination between −1 and 1V, the photovoltaic effect was not evidenced. This means that for this case the collection of the charge carrier at the electrodes has not occurred.

In analysis of the structures prepared with stacked layer, an increased current value was obtained in Al/TPyP/ZnPc/ITO structure comparing to that based on MgPc. This can be explained, considering the position of the HOMO and LUMO levels in these materials, a better hole injection from the ITO electrode in the ZnPc layer being ensured by the lower barrier at the ITO/ZnPc interface ($WF_{ITO} = 4.6$ eV [91] and $\Delta E_{ZnPc\text{-}ITO} = 0.68$ eV).

Organic heterostructures based on ZnPc, MgPc and TPyP are suitable to be used in OPV due their electrical properties, especially in bulk forms instead of stacked layers.

5. Conclusions

Heterostructures based on ZnPc were prepared using two different deposition techniques. In a first step, ZnPc was deposited into a multilayer structure in combination with C60 and NTCDA by VTE, the most accessed method for the deposition of organic materials.

The heterostructure were fabricated starting from glass/ITO or glass/ITO covered by a thin film of PEDOT:PSS on which were deposited ZnPc, C60, NTCDA and Al electrode or starting from glass/Al, followed by the deposition of NTCDA, C60, ZnPc and ITO electrode. The structures present the absorption maxima characteristic to the used materials. It was evidenced that the way in which the layers are deposited influenced the properties. *I-V* characteristics revealed that the value of the current is increased in normal configuration glass/ITO/ZnPc/C60/NTCDA/Al when an additional PEDOT:PSS layer is used. An increase in the current value was also achieved depositing the layers in inverted order (glass/Al/NTCDA/C60/ZnPc/ITO).

A p-n heterostructure based on ZnPc and NTCDA layers was also fabricated by a laser technique. Moreover, the structures are obtained on the AZO substrate, and a TCO used to replace the ITO, the material most used as a transparent conductor electrode. AZO layers with adequate optical and electrical properties were prepared by PLD. The influence of an oxygen plasma treatment of AZO on the properties of the organic structures deposited on this TCO was analysed. The UV–VIS spectra show features typical to the used materials, covering a large region of the visible domain. The PL emission bands attributed to the ZnPc and NTCDA were overlapped by the emission bands showed by the AZO substrate. AFM images evidenced a decrease in the size of the grains of the organic heterostructures with the increase in duration of the applied plasma treatment of the AZO substrate. The *I-V* characteristics of the heterostructures revealed an injector contact behaviour and the appearance of the space charge limited currents (characteristic in organic materials) at voltages higher than 0.4V. AZO substrates treated in oxygen plasma (for 5 and 10 min) can favour the injection of the charge carrier in the organic layer, probably as a result of the increasing of AZO work function (leading to a decrease in the energetic barrier at the interface), and determining a current higher with 1 order of magnitude in the heterostructures prepared on the treated substrate.

Organic heterostructures based on metal phthalocyanines and a porphyrin (ZnPc orMgPc and TPyP) were deposited by MAPLE on flexible substrate PET/ITO (in stacked or mixed form). The films preserve their IR absorption properties indicating that no decomposition appears at the laser transfer. The *I-V* characteristics of the heterostructures measured in dark conditions show an increased current value with 3 orders of magnitude higher for the structure with MgPc:TPyP compared to the structure formed with stacked films based on the same compounds. The appearance of the photovoltaic effect was remarked in the heterostructures with ZnPc:TPyP when the structure was exposed to the light.

In conclusion, thin films based on porphyrins and/or phthalocyanines can be deposited in multilayers or blend configurations on various substrates (ITO/glass, AZO/glass, Al/glass or ITO/PET) by different deposition techniques, including laser techniques. The obtained results are promising and very useful for further applications in the photovoltaic field.

Acknowledgements

The work has been funded by the Romanian National Authority for Scientific Research, CNCS-UEFISCDI, projects TE 188/2014, PN-II-RU-TE-2014-4-1590 and the National Authority for Research and Innovation in the frame of Core Program - contract 4N/2016 and contract PN16-480102.

Author details

Marcela Socol[1*], Nicoleta Preda[1], Anca Stanculescu[1], Florin Stanculescu[2] and Gabriel Socol[3]

*Address all correspondence to: marcela.socol@infim.ro

1 National Institute of Material Physics, Bucharest-Magurele, Romania

2 Faculty of Physics, University of Bucharest, Bucharest-Magurele, Romania

3 National Institute for Lasers, Plasma and Radiation Physics, Bucharest-Magurele, Romania

References

[1] Yu J.S., Yin X.X., Xu Z.S., Deng P., Han Y.B., Zhou B.J., Tang W.H. Bisalkylthio side chain manipulation on two-dimensional benzo[1,2-b:4,5-b′]dithiophene copolymers with deep HOMO levels for efficient organic photovoltaics. Dyes and Pigments. 2017;**136**:312–320. DOI: 10.1016/j.dyepig.2016.08.057

[2] Park Y., Berger J., Tang Z., Muller-Meskamp, L., Lasagni A. F., Vandewal K., Leo K. Flexible, light trapping substrates for organic photovoltaics. Applied Physics Letters. 2016;**109**(9): Article Number: 093301. DOI: 10.1063/1.4962206

[3] Lee S., Kim B., Jung H., Shin H., Lee H., Lee J., Park J. Synthesis and electroluminescence properties of new blue dual-core OLED emitters using bulky side chromophores. Dyes and Pigments. 2017;**136**:255–261. DOI: 10.1016/j.dyepig.2016.08.010

[4] Zhou H., Cheong H. G., Park J. W. Charge carrier transport through the interface between hybrid electrodes and organic materials in flexible organic light emitting diodes. Journal of Nanoscience and Nanotechnology. 2016;**16**(5):5179–5185. DOI: 10.1166/jnn.2016.12265

[5] Xiang L. Y., Ying J., Wang W., Xie W. F. High mobility n-channel organic field-effect transistor based a tetratetracontane interfacial layer on gate dielectrics. IEEE Electron Device Letters. 2016;**37**(12):1632–1635. DOI: 10.1109/LED.2016.2616517

[6] Kheradmand-Boroujeni B., Schmidt G.C., Hoft D., Haase K., Bellmann M., Ishida K., Shabanpour R., Meister T., Carta C., Hubler A. C., Ellinger F. Small-signal characteristics of fully-printed high-current flexible all-polymer three-layer-dielectric transistors. Organic Electronics. 2016;**34**: 275–283. DOI: 10.1016/j.orgel.2016.04.037

[7] Heliatek. Heliatek sets new organic photovoltaic world record efficiency of 13.2% [Internet]. 2016. Available from: http://www.heliatek.com/en/press/press-releases/details/heliatek-sets-new-organic-photovoltaic-world-record-efficiency-of-13-2

[8] Pochettino A. The development of organic conductors, including semiconductors, metals and superconductors cont. Academy Lincei Rendus. 1906;**15**:355.

[9] Volmer M. Different Photoelectric Phenomena in Anthracene, their Relation to one another, to Fluorescence and to the Formation of Dianthracene. Annalen der Physik. 1913;**345**(4):775–796. DOI:10.1002/andp.19133450411

[10] Kearns D., Calvin M. Photovoltaic effect and photoconductivity in laminated organic systems. The Journal of Chemical Physics. 1958;29(4):950–951. dx.doi.org/10.1063/1.1744619

[11] Tang C. W. Two-layer organic photovoltaic cell. Applied Physics Letters. 1986;**48**:183–185. DOI: http://dx.doi.org/10.1063/1.96937

[12] Spanggaard H., Krebs F. C. A brief history of the development of organic and polymeric photovoltaics. Solar Energy Materials & Solar Cells. 2004;**83**:125–146. DOI:10.1016/j.solmat.2004.02.021

[13] Hoppe H., Sariciftci N. S. Organic solar cells: An overview. Journal of Material Research. 2004;**19**(7):1924–1945. DOI: 10.1557/JMR.2004.0252

[14] Walter M. G., Rudine A. B., Wamser C. C. Porphyrins and phthalocyanines in solar photovoltaic cells. Journal of Porphyrins and Phthalocyanines, 2010;**14**:759–792. DOI: 10.1142/S1088424610002689

[15] Fagadar-Cosma E., Enache C., Armeanu I., Fagadar-Cosma G. Comparative investigations of the absorption and fluorescence spectra of tetrapyridylporphyrine and Zn(ii) tetrapyridylporphyrine. Digest Journal of Nanomaterials and Biostructures. 2007;**2**(1):175–183

[16] Yella A., Lee H. W., Tsao H. N., Yi C., Chandiran A. K., Nazeeruddin M. K., Diau E. W., Yeh C. Y., Zakeeruddin S. M., Gra¨tzel M. Porphyrin-sensitized solar cells with cobalt (II/III)-based redox electrolyte exceed 12 percent efficiency. Science. 2011;**334**:629–634. http://dx.doi.org/10.1126/science.1209688

[17] Mathew S., Yella A., Gao P., Humphry-Baker R., Curchod B. F. E., Ashari-Astani N., Tavernelli I., Rothlisberger U., Nazeeruddin Md. K., Grätzel M. Dye-sensitized solar cells with 13% efficiency achieved through the molecular engineering of porphyrin sensitizers. Nature Chemistry. 2014;**6**:242–247.doi:10.1038/nchem

[18] Orns O K. B., Garcia-Lastra J. M., De La Torre G., Himpsel F. J., Rubio d A., Thygesen K. S. Design of two-photon molecular tandem architectures for solar cells by ab initio theory. Chemical Science. 2015;**6**:3018–302. DOI: 10.1039/c4sc03835e

[19] Monti D., Nardis S., Stefanelli M., Paolesse R., Di Natale C., D'Amico A. Porphyrin-based nanostructures for sensing applications. Journal of Sensors. 2009;2009: 1-10. http://dx.doi.org/10.1155/2009/856053

[20] Baran J. D., Grönbeck H., Hellman A. Analysis of porphyrines as catalysts for electrochemical reduction of O_2 and oxidation of H_2O. Journal of American Chemical Society. 2014;**136**(4):1320–1326.DOI: 10.1021/ja4060299

[21] Ishihara S., LabutaJ., Van Rossom W., Ishikawa D., Minami K., Hill J. P., Ariga K. Porphyrin-based sensor nanoarchitectonics in diverse physical detection modes. Physical Chemistry Chemical Physics. 2014;**16**(21):9713–9746. DOI: 10.1039/C3CP55431G

[22] Dougherty T. J., Gomer C. J., Henderson B. W., Jori G., Kessel D., Korbelik M., Moan J., Peng Q. Photodynamic therapy. Journal of the National Cancer Institute. 1998;**90**:889–905. PMC4592754

[23] Huynh E., Leung B. Y. C., Helfield B. L., Shakiba M., Gandier J., Jin C. S., Master E.R., Wilson B. C., Goertz D. E., Zheng G. In situ conversion of porphyrin microbubbles to nanoparticles for multimodality imaging. Nature Nanotechnology. 2015;**10**:325–332. DOI:10.1038/nnano.2015.25

[24] Aviezer D., Cotton S., David M., Segev A., Khaselev N., Galili N., Gross Z., Yayon A. Porphyrin analogues as novel antagonists of fibroblast growth factor and vascular endothelial growth factor receptor binding that inhibit endothelial cell proliferation, tumor progression, and metastasis.Cancer Research.2000;**60**:2973–2980. PubMed ID:**10850445**

[25] Stanculescu A., Socol G., Vacareanu L., Socol M., Rasoga O., Breazu C., Girtan M., Stanculescu F. MAPLE preparation and characterization of mixed arylenevinylene based oligomers: C60 layers. Applied Surface Science. 2016;**374**:278–289. http://dx.doi.org/10.1016/j.apsusc.2015.11.250

[26] Stubinger T., Brutting W. Exciton diffusion and optical interference in organic donor-acceptor photovoltaic cells. Journal of Applied Physics. 2001;**90**:3632–3641.

[27] Peumans P., Forrest S.R. Very-high-efficiency double-heterostructure copper phthalocyanine/C-60 photovoltaic cells. Applied Physics Letters. 2001;79:126–128. DOI: http://dx.doi.org/10.1063/1.1384001

[28] Dao Q.D., Hori T., Fukumura K., Masuda T., Kamikado T., Fujii A., et al. Effects of processing additives on nanoscale phase separation, crystallization and photovoltaic performance of solar cells based on mesogenic phthalocyanine. Organic Electronics. 2016;**14**:2628–2634. http://dx.doi.org/10.1016/j.orgel.2013.05.041

[29] Xue J., Rand B.P., Uchida S., Forrest S.R. A hybrid planar-mixed molecular heterojunction photovoltaic cell. Advanced Materials. 2005;**17**:66–71. DOI: 10.1002/adma.200400617

[30] Xue J., Uchida S., Rand B.P., Forrest S.R. Asymmetric tandem organic photovoltaic cells with hybrid planar-mixed molecular heterojunctions. Applied Physics Letters. 2004;**85**:5757–5759. DOI: http://dx.doi.org/10.1063/1.1829776

[31] Cho K. T., Rakstys K., Cavazzini M., Orlandi S., Pozzi G., Nazeeruddi M. K. Perovskite solar cells employing molecularly engineered Zn(II) phthalocyanines as hole-transporting materials. Nano Energy. 2016;**30**:853–857. http://dx.doi.org/10.1016/j.nanoen.2016.09.008.

[32] Leznoff C. C., Lever A. B. P. (eds.). Phthalocyanines: Properties and Applications : Volume 2, New York: VCH Publishers, 1993, 355p.

[33] Rao S. V., Rao D. N. Excited state dynamics in phthalocyanines studied using degenerate four wave mixing with incoherent light. Journal of Porphyrins Phthalocyanines. 2002;**6** 233–237. DOI: http://dx.doi.org/10.1142/S1088424602000270

[34] Mali S. S., Dalavi D. S., Bhosale P. N., Betty C. A., Chauhancand A. K., Patil P. S. Electro-optical properties of copper phthalocyanines (CuPc) vacuum deposited thin films. RSC Advances. 2012;**2**:2100–2104.DOI: 10.1039/C2RA00670G

[35] Law K. Y. Organic photoconductive materials: recent trends and developments. Chemical Reviews. 1993;**93**(1): 449–486. DOI: 10.1021/cr00017a020

[36] Ghorannevis Z., Akbarnejad E., Elahi A. Salar, Ghoranneviss M. Application of RF magnetron sputtering for growth of AZO on glass substrate. Journal of Crystal Growth. 2016;**447**:62–66. http://dx.doi.org/10.1016/j.jcrysgro.2016.04.062

[37] Socol G., Socol M., Stefan N., Axente E., Popescu-Pelin G., Craciun D., et al. Pulsed laser deposition of transparent conductive oxide thin films on flexible substrate. Applied Surface Science. 2012;**260**:42–46. DOI:10.1016/j.apsusc.2012.02.14

[38] Kymakis E., Stylianakis M. M., Spyropoulos G. D., Stratakis E., Koudoumas E., Fotakis C. Spin coated carbon nanotubes as the hole transport layer in organic photovoltaics. Solar Energy Materials and Solar Cells. 2012;**96**:298–301: http://dx.doi.org/10.1016/j.solmat.2011.09.046

[39] Xiong K., HouL., Wu M., Huo Y., Mo W., Yuan Y., Sun S.,Xu W., Wang E. From spin coating to doctor blading: A systematic study on the photovoltaic performance of an isoin-

digo-based polymer. Solar Energy Materials and Solar Cells. 2015;**132**:252–259. http://dx.doi.org/10.1016/j.solmat.2014.08.039

[40] Haldar A., Liao K.-S., Curran S. A. Fabrication of inkjet printed organic photovoltaics on flexible Ag electrode with additives. Solar Energy Materials and Solar Cells. 2014;**125**:283–290. http://dx.doi.org/10.1016/j.solmat.2014.03.013

[41] Kim H. Transparent Conducting Oxide Films. In: Easton R., editor. Pulsed Laser Deposition of Thin Films: Applications-Led Growth of Functional Materials. 1st ed. New York, United States: John Wiley & Sons Inc.; 2006, pp. 240–255. DOI: 10.1002/9780470052129.ch11

[42] Norton D. P. Pulsed laser deposition of complex materials: progress toward applications. In: Eason R., editor. Pulsed Laser Deposition of Thin Films: Applications-Led Growth of Functional Materials. 1st ed. New York, United States: Wiley & Sons; 2006, pp. 3–32. DOI: 10.1002/9780470052129.ch1

[43] Socol G., Craciun D., Mihailescu I.N., Stefan N., Besleaga C., Ion L., et al. High quality amorphous indium zinc oxide thin films synthesized by pulsed laser deposition. Thin Solid Films. 2011;**520**:1274–1277. DOI:10.1016/j.tsf.2011.04.196

[44] Thomas M. T. Vacuum deposition techniques. In: G.L. Weissler and Carlson R.W., editors. Methods in Experimental Physics. Volume 14, Vacuum Physics and Technology. New York, Elsevier. 1980, pp.521–575.http://dx.doi.org/10.1016/S0076-695X(08)60385-3

[45] Graper E. B. Resistance evaporation. In: Glocker D. A., Shah S. I., editors. Handbook of Thin Film Process Technology (Inst. Phys. Publ., Bristol 1995) A 1.1: pp.A1.1:1–A1.1:7.

[46] Stanculescu A., Stanculescu F., Tugulea L., Socol M. Optical properties of 3,4,9,10-perylenetetracarboxylic dianhydride and 8-hydroxyquinoline aluminium salt films prepared by vacuum deposition. Material Science Forum. 2006;**514–516**:956–960

[47] Luches A., Caricato A. P. Fundamentals and applications of MAPLE. In: Miotello A., Ossi P. M., editors. Laser-Surface Interactions for New Materials Production. 1st ed. Berlin: Springer Berlin Heidelberg; 2010, pp. 203–233. DOI 10.1007/978-3-642-03307-0_9

[48] Caricato A. P. Lasers in materials science. MAPLE and MALDI: theory and experiments. In: M. Castillejo, M. O., Paolo,L. Zhigilei, editors. Lasers in Materials Science. 1st ed. Switzerland: Springer International Publishing; 2014, pp. 295–323. DOI: 10.1007/978-3-319-02898-9_12

[49] Socol M., Preda N., Vacareanu L., Grigoras M., Socol G., Mihailescu I.N., et al. Organic heterostructures based on arylenevinylene oligomers deposited by MAPLE. Applied Surface Science. 2014;**302**:216–222. DOI: 10.1016/j.apsusc.2013.12.091

[50] Socol M., Preda N., Rasoga O., Breazu C., Stavarache I., Stanculescu F., et al. Flexible heterostructures based on metal phthalocyanines thin films obtained by MAPLE. Applied Surface Science.2016;**374**:403–410. DOI 10.1016/j.apsusc.2015.10.166

[51] Caricato A. P., Cesaria M., Gigli G., Loiudice A., Luches A., Martino M., et al. Poly-(3-hexylthiophene)/6,6 -phenyl-C-61-butyric-acid-methyl-ester bilayer deposition by matrix-assisted pulsed laser evaporation for organic photovoltaic applications. Applied Physics Letters. 2012;**100**:073306. DOI:10.1063/1.3685702

[52] Lin C.-F., Zhang M., Liu S.-W., ChiuT.-L., Lee J.-H. High photoelectric conversion efficiency of metal phthalocyanine/fullerene heterojunction photovoltaic device. International Journal of Molecular Sciences. 2011;**12**: 476–505.doi:10.3390/ijms12010476

[53] Socol M., Rasoga O., Breazu C., Socol G., Preda N., Pasuk I., et al. Heterostructures based on small molecules organic compounds. Digest Journal of Nanomaterials and Biostructures. 2015;**10**(4):1383–1392

[54] Senthilarasu S., Velumani S., Sathyamoorthy R., Subbarayan A., Ascencio J.A., Canizal G., et al. Characterization of zinc phthalocyanine (ZnPc) for photovoltaic applications. Applied Physics A-Materials Science & Proces. 2003;**77**(3–4):383–389. DOI: 10.1007/s00339-003-2184-7

[55] Ginzburg B. M., Tulchiev Sh., Tabarov S. K., Shepelevski A. A., Shibaev L. A. X-ray diffraction analysis of C[sub 60] fullerene powder and fullerene soot. Technical Physics. 2005;**50**(11):1458. DOI 10.1134/1.2131953

[56] Gaffo L., Cordeiro M.R., Freitas A.R., Moreira W.C., Girotto E. M., Zucolotto V. The effects of temperature on the molecular orientation of zinc phthalocyanine films. Journal of Material Science. 2010;**45**:1366–1370. DOI:10.1007/s10853-009-4094-3

[57] Zhang H., Wu C., Liang L., Chen Y., He Y., Zhu Y., et al. Structural, morphological and optical properties of C_{60} cluster thin films produced by thermal evaporation under argon gas. Journal of Physics: Condensed Matter. 2001;**13**(13):2883–2889

[58] Seoudi R., El-Bahy G.S., El Sayed Z.A. FTIR, TGA and DC electrical conductivity studies of phthalocyanine and its complexes. Journal of Molecular Structure. 2005;**753**(1):119–126. http://dx.doi.org/10.1016/j.molstruc.2005.06.003

[59] Triboni E. R., da Silva M. F. P., Finco A. T., Rodrigues M. A., Demets G. J.-F., Dyszy F. H., et al. Synthesis and properties of new paramagnetic hybrid bayerite from Al(0)/naphthalene dianhydride reaction. Materials Research. 2010;**13**(4):505–511.

[60] Braatz C. R., Ohl G., Jakob P. Vibrational properties of the compressed and the relaxed 1,4,5,8-naphthalene-tetracarboxylic dianhydride monolayer on Ag(111). The Journal of Chemical Physics. 2012;**136**(13):134706. DOI: http://dx.doi.org/10.1063/1.3699030.

[61] Torsi L., Dodabalapur A., Cioffi N., Sabbatini L., Zambonin P.G. NTCDA organic thin-film-transistor as humidity sensor: weaknesses and strengths. Sensors and Actuators B. 2001;**77**(1–2):7–11. http://dx.doi.org/10.1016/S0925-4005(01)00664-5

[62] Sathyamoorthy R., Senthilarasu S. Influence of RMS strain on optical band gap of Zinc Phthalocyanine (ZnPc) thin films. Solar Energy. 2006;**80**:201–208. http://dx.doi.org/10.1016/j.solener.2005.06.005

[63] Senthilarasu S., Hahn Y.B., Lee S.H. Structural analysis of zinc phthalocyanine (ZnPc) thin films: X-ray diffraction study. Journal of Applied Physics. 2007;**102**(4):043512. DOI: http://dx.doi.org/10.1063/1.2771046

[64] Jayatissa A. H., Nadarajah A., Dutta A.K. Investigation of C_{60} films for surface finishing applications. Proceeding SPIE. 2005;**6002**:60021A-1. doi:10.1117/12.631051

[65] Chu C.-W., Shao Y., Shrotriya V., Yang Y. Efficient photovoltaic energy conversion in tetracene-C60 based heterojunctions. Applied Physics Letters. 2005;**86**,243506. DOI: 10.1063/1.1946184

[66] Tachikawa H., Kawabata H., Miyamoto R., Nakayama K., Yokoyama M. Experimental and theoretical studies on the organic–inorganic hybrid compound: aluminum-NTCDA co-deposited film. The Journal of Physical Chemistry B. 2005;**109**(8):3139–3145. DOI: 10.1021/jp046168e

[67] Dong B.-Z., Fang G.-J., Wang J.-F., Guan W.-J., Zhao X.-Z. Effect of thickness on structural, electrical, and optical properties of ZnO: Al films deposited by pulsed laser deposition. Journal of Applied Physics. 2007;**101**(3):033713. DOI: http://dx.doi.org/10.1063/1.2437572

[68] Kim H., Gilmore C. M., Horwitz J. S., Piqué A., Murata H., Kushto G. P., et al. Transparent conducting aluminum-doped zinc oxide thin films for organic light-emitting devices. Applied Physics Letters. 2000;**76**(3):259. DOI: http://dx.doi.org/10.1063/1.125740

[69] Wu H.-W., Yang R.-Y., Hsiung C.-M., Chu C.-M. Characterization of aluminum-doped zinc oxide thin films by RF magnetron sputtering at different substrate temperature and sputtering power. Journal of Materials Science: Materials in Electronics. 2013;**24**(1):166–171. DOI: 10.1007/s10854-012-0769-7

[70] Stanculescu A., Socol M., Rasoga O., Mihailescu I. N., Socol G., Preda N., et al. Laser prepared organic hetrostructures on glass/AZO substrates. Applied Surface Science. 2014;**302**:169–176. http://dx.doi.org/10.1016/j.apsusc.2014.01.181

[71] Martins R., Fortunato E., Nunes P., Ferreira I., Marques A., Bender M. Zinc oxide as an ozone sensor. Journal of Applied Physics. 2004;**96**: 1398–1408. DOI: http://dx.doi.org/10.1063/1.1765864

[72] Chou Y.-H., Yan J.-T., Lee H.-Y., Lee C.-T. AZO films with Al nano-particles to improve the light extraction efficiency of GaN-based light-emitting diodes Proceeding SPIE. 2008;**6894**:68941C.1–68941C.6. DOI: 10.1117/12.765633

[73] Wojdyla M., Derkowska B., Lukasiak Z., W. Bala. Absorption and photoreflectance spectroscopy of zinc phthalocyanine (ZnPc) thin films grown by thermal evaporation. Materials Letters. 2006;**60**(29–30):3441–3446. http://dx.doi.org/10.1016/j.matlet.2006.03.029

[74] Forrest S.R., Holmes R. Organic Polariton Laser Patents. US 20050195874 A1. http://www.google.de/patents/US20050195874.

[75] Ding J., Chen H., Zhao X., Ma S. Effect of substrate and annealing on the structural and optical properties of ZnO: Al films. Journal of Physics and Chemistry of Solids. 2010;**71**(3):346–350. DOI: 10.1016/j.jpcs.2009.12.088

[76] Ding J.J., Ma S.Y., Chen H.X., Shi X.F., Zhou T.T., Mao L.M. Influence of Al-doping on the structure and optical properties of ZnO films. Physica B: Condensed Matter. 2009;**404**(16): 2439–2443. DOI: 10.1016/j.physb.2009.05.006

[77] Ahn K., Jeong Y.S., Lee H.U., Jeong S.Y., Ahn H.S., Kim H.S., et al. Physical properties of hydrogenated Al-doped ZnO thin layer treated by atmospheric plasma with oxygen gas. Thin Solid Films. 2010;**518**(14):4066–4070. DOI: 10.1016/j.tsf.2010.02.028

[78] Wojdyla M., Bala W., Derkowska B., Lukasiak Z., Czaplicki R., Sofiani Z., et al. Photoluminescence and third harmonic generation in ZnPc thin films. Nonlinear Optics Quantum Optics. 2006;**35**:103–119.

[79] Ng A.M.C., Djurisic A.B., Chan W.K. Organic luminescent nanowires: fabrication and characterization. Proceeding SPIE. 2007;**6828**:682807.1–682807.10. DOI:10.1117/12.760844

[80] Shoheen S. E., Brabec C. J., Sariciftci N. S., Padinger F., Fromherz J., Hummelen J. C. 2.5% efficient organic plastic solar cells. Applied Physics Letters. 2001;**78**(6):841–843. DOI: http://dx.doi.org/10.1063/1.1345834

[81] Zhang C., Hu Y., Tang A., Deng Z., Teng F. Investigating the reduction in the absorption intensity of P3HTin polymer/fullerene "bilayers" coated using orthogonal solvents. Journal of Applied Polymer Science. 2015;**132**(14):41757.DOI: 10.1002/app.41757

[82] El-Nahass M. M., Atta A. A., El-Sayed H. E. A., El-Zaidia E. F. M. Structural and optical properties of thermal evaporated magnesium phthalocyanine (MgPc) thin films. Applied Surface Science. 2008;**254**(8):2458–2465. http://dx.doi.org/10.1016/j.apsusc.2007.09.064

[83] Stanculescu A., Socol M., Socol G., Mihailescu I. N., Girtan M., Stanculescu F. Maple prepared organic heterostructures for photovoltaic application. Applied Physics A: Materials Science & Processing. 2011;**104** (3):921–928. DOI: 10.1007/s00339-011-6440-y

[84] Bala W., Wojdyla M., Rebarz M., Szybowic M., Drozdowski M., Grodzicki A., et al. Influence of central metal atom in MPc (M = Cu, Zn, Mg, Co) on Raman, FT-IR, absorbance, reflectance, and photoluminescence spectra. Journal of Optoelectronic and Advanced Materials. 2009;**11**(3):264–269.

[85] Gu D., Chen Q., Shu J., Tang X., Gan F., Shen S., et al. Optical recording performance of thin films of phthalocyanine compounds. Thin Solid Films. 1995;**257**(1):88–93. http://dx.doi.org/10.1016/0040-6090(94)06327-3

[86] Socol M., Rasoga O., Stanculescu F., Girtan M., Stanculescu A. Effect of the morphology on the optical and electrical properties of TPyP thin films deposited by vacuum evaporation. Optoelectronic Advanced Materials. 2010;**4**(12):2032–2038.

[87] Vogel M., Doka S., Breyer Ch., Lux-Steiner M. Ch., Fostiropoulos K. On the function of a bath ocuproine buffer layer in organic photovoltaic cells. Applied Physics Letters. 2006;**89**:163501–3. DOI: 10.1063/1.2362624

[88] Park Y.S., Seo M., Yi J., Lim D., Lee J. Characteristics of aluminum-doped zinc oxide films with oxygen plasma treatment for solar cell applications. Thin Solid Films. 2013;**547**:47–51. http://dx.doi.org/10.1016/j.tsf.2013.05.065

[89] Stanculescu A., Stanculescu F., Socol M., Grigorescu O. Electrical transport in crystalline perylene derivatives films for electronic devices. Solid State Sciences. 2008;**10**(12):1762–1767. http://dx.doi.org/10.1016/j.solidstatesciences.2008.03.023

[90] Breeze A. J., Salomon A., Ginley D. S., Gregg B. A., Tillmann H., Hörhold H.-H. Polymer-perylenediimide heterojunction solar cells. Applied Physics Letters. 2002;**81**(16):3085–3087. DOI: http://dx.doi.org/10.1063/1.1515362

[91] Cui J., Wang A., Edleman N.L., Ni J., Lee P., Armstrong N.R., et al. Indium tin oxide alternatives-high work function transparent conducting oxides as anodes for organic light-emitting diodes. Advanced Materials. 2001;**13**(19):1476–1480. DOI: 10.1002/1521-4095(200110)13:19<1476::AID-ADMA1476>3.0.CO;2-Y

Chapter 6

Porphyrin-Based Organophotocatalysts

Yingzhi Chen, Zheng-Hong Huang and Lu-Ning Wang

Additional information is available at the end of the chapter

http://dx.doi.org/10.5772/intechopen.68223

Abstract

The planar geometric structure and the rich absorption feature endow porphyrins with interesting optoelectronic properties and also make it promising building blocks for supramolecular assembly. Recent advances in the photocatalytic applications of porphyrins, including homogeneous, heterogeneous photocatalysis, and photoelectrochemical solar cells are highlighted. Porphyrin photocatalysts are involved in the form of molecules, supported molecules, nanostructures, and thin film. Related rational design strategies are provided for each form with an aim to enhance the light conversion efficiency. Finally, the ongoing directions and challenges for the future development of porphyrin semiconductors in high-quality optoelectronic devices are also proposed.

Keywords: porphyrin, homogeneous photocatalysis, heterogeneous photocatalysis, nanostructure, photoelectrochemical solar cell

1. Introduction

Environment and energy issues have been presented as the biggest challenges facing humanity nowadays. Among the various solutions, photocatalysis is a promising approach both for photochemical energy conversion and for photochemical decontamination, hence to fulfill the sustainable energy supply and environment remediation by use of the abundant, natural sunlight [1–5]. To achieve efficient solar energy conversion, the photocatalysts are required to possess excellent light-harvesting capability, charge transfer efficiency (factors including exciton lifetime, mobility, etc.), as well as surface activity (specific surface area, ionic adsorption, etc.) [6–8]. Most research studies in photocatalysis have been concentrated on the use of inorganic semiconductors, such as TiO_2 [9–11], Fe_2O_3 [12–14], ZnO [15–17], and Cu_2O [17–19] which mostly suffer from inefficient light absorption and hardness. Strategies to enhance the efficiency of these catalysts correlating with band engineering [20, 21], texture modification [16, 22], or configuration organization [23, 24] always involve complicated fabrication processes. All these limit their practical affordable applications. Still, much effort is needed to find other photoactive materials as alternatives for facile preparation and economical applications.

During the last decades, increasing attention has been paid to the field of semiconducting organic materials for optoelectronic applications [25–27]. One of the most important advantages concerning these organic materials is that their molecular structure and functions can be easily modulated via molecular design and tailoring. Additionally, integration of them into lightweight, large-area devices can be simply realized through solution processing at low cost. In addition, organic semiconductors, also referred to as π-conjugated molecules, are characterized by a delocalized π-electron system that makes them ideal building blocks for the fabrication of advanced functional nanomaterials and nanodevices [28–30]. As a typical representative of π-conjugated molecules, porphyrins are of particular interest due to some key aspects, such as their excellent light-harvesting property, p-type semiconducting behavior, ease of chemical modification, good supramolecular assembly, and film-forming features by means of either solution-based or thermal-based techniques [31–33]. Coupled with their chemical stability and flexibility, the use of porphyrin in optoelectronics has become a fast-growing research focus, and great development has been made in the field of organic solar cells (OSCs) [34, 35], organic field-effect transistors (OFETs) [36, 37], organic light-emitting diodes (OLEDs) [38, 39], even in flexible organic semiconductor devices.

As a photocatalyst, porphyrins were first used in homogenous photocatalysis [40]. The problem with it is the limited stability of porphyrin molecules and the recovery of them for successive use. Fortunately, this could be circumvented by mobilizing porphyrin molecules on solid supports or assembling them into robust nanostructures [41, 42]. Recently, more efforts have been made on the development of a semiconductor-based photoelectrochemical (PEC) water splitting device [43, 44], and thus organic photoelectrodes have aroused special attention. Relating progresses are dealt with in detail in separate sections. Before that we have a brief introduction of the relation between porphyrin molecular structure and optoelectronic properties. The use of molecular porphyrin as modification of inorganic semiconductors to achieve absorption of visible light is not covered in this chapter.

2. Porphyrins: structures and optoelectronic properties

In nature, porphyrin-related molecules are important photosynthetic pigments that perform the light-harvesting and charge/energy transfer functions in biological photosynthesis [45–48]. The role of porphyrins in photocatalysis is mainly related to their optical feature. As shown from the basic porphyrin ring (**Figure 1**), porphyrins are tetrapyrrole derivatives which are composed of four pyrrole subunits interconnected via —CH= bridges. The inner 16-membered ring with 18 π electrons constitutes its electronic "heart," which is responsible for the optical spectra. Many authors have investigated its optoelectronic properties because of simplicity. Again, molecular engineering is easily attainable by various chemical modifications to this basic ring, leading to proper tuning of the optoelectronic properties [49–51].

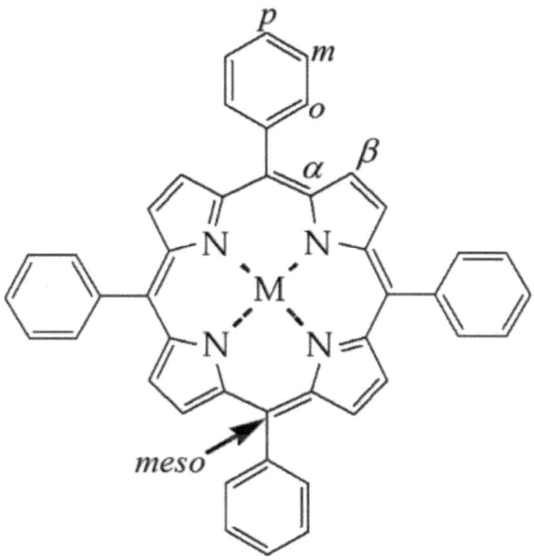

Figure 1. An illustration of a representative tetraarylporphyrin.

First, central substituent of porphyrin ring has a major effect on the optical spectra. Depending on the atom or group that occupies the center, porphyrins can be basically divided into freebase type (two hydrogens in the center) and metal-type [52, 53], or the so-called metalloporphyrin that is formed by exchange of the two protons in freebase porphyrin by a metal ion. Considerable varieties in the optoelectronic properties just arise from such center difference. Particularly, freebase porphyrin has a four-banded visible spectrum notably different from the two-banded spectrum exhibited by metal complex [54]. This spectral difference is attributed to the fact that the two freebase hydrogens in the center greatly reduce the symmetry from square to rectangular. In the case of metalloporphyrin [55], the change of metal in some cases can strongly influence absorption spectra. It is now known that the central metal perturbs the absorption spectra mainly through the interaction of the metal electrons with those of the ring, and sometimes the coordination type can also affect the spectra.

In addition to central substituents, peripheral substituents at various locations around the ring, including four *meso* and eight β-positions, can also impart different properties to a greater or lesser extent to the molecule [56–58]. Xie et al. have introduced various numbers of triphenylamine and trimethoxyphenyl groups to the *meso*-positions as electron donors, in an attempt to systematically tune the highest occupied molecular orbital-lowest unoccupied molecular orbital (HOMO-LUMO) energy levels [59]. As a photocatalyst, HOMO-LUMO bandgap determines the absorption wavelength for light-harvesting efficiency, and the suitable HOMO and LUMO levels ensure an efficient electron injection and dye regeneration process. With regard to porphyrins, the modulation of the HOMO-LUMO levels, along with the corresponding optoelectronic properties, can be simply realized through proper choice of an anchoring group to the ring. In another work, Sharma and coworkers reviewed the importance of various anchoring groups linked to either *meso* or β-positions in improving the light collection efficiency of dye-sensitized solar cells (DSSCs) [58]. As the most widely used anchoring group, the position of carboxylic acid (COOH) was found to vary the performance of solar cells. Increased photocurrent was generated when the position of COOH changed from the *para* position to the *meta* position. Also in some cases, porphyrin is functionalized with donor and acceptor moieties. Upon photoexcitation, the generated exciton diffuses to the donor-acceptor interface, affording enhanced charge transfer character. Meanwhile, the enlarged electron conjugation leads to a narrowing of the optical bandgap, giving rise to broad light-absorbing dye.

3. Porphyrin-based homogeneous photocatalysts

Increasing emphasis has been placed on photocatalysts as an environmentally friendly process to decompose organic pollutants in contaminated water and air. It is well documented that porphyrins and metalloporphyrins have contributed a lot to photooxidation catalysis in homogeneous media.

3.1. Reaction mechanisms

As for highly effective triplet-state porphyrins, two possible mechanistic pathways are involved in a photocatalytic process: energy transfer and electron transfer from the triplet excited state [60, 61]. Singlet oxygen species ($^1O_2$) is commonly involved during the energy transfer process, whereas other active oxygen species such as a superoxide radical anion ($O_2^{\bullet-}$) or hydroxyl radical ($^{\bullet}OH$) are essentially involved in the case of the electron transfer process. Time-resolved spectroscopic methods thus provide a powerful tool to detect the transient species derived from the photocatalyst for the study of fast reaction kinetics.

Homogeneous porphyrins are well known to generate $^1O_2$ [40, 62–64]. For example, the triplet quantum yield of *meso*-tetra (2,6-dichloro-phenyl) porphyrin (TDCPP) was reported to be 0.995, with its corresponding singlet oxygen quantum yield around 0.98 [65]. As highly recognized, the photochemically generated singlet oxygen acts as a primary oxidant in photodegrading organic pollutants and viruses in natural water. In the case of *meso*-tetra(2,6-dichloro-3-sulfophenyl) porphyrin (TDCPPS) [66] or its iron complex (FeTDCPPS) [67] when oxidizing phenols, the main photodegradation pathway involved reaction with singlet oxygen, as suggested by the following observations: the triplet state of the porphyrins was efficiently quenched by molecular oxygen; singlet oxygen phosphorescence was detected by time-resolved measurements.

3.2. Homogeneous photocatalysis

As the catalyst is dissolved, it is easy to get access to all active sites, resulting in high catalytic activities. For instance, water-soluble TDCPPS and its metal complexes were successfully used in the photodegradation of 4-chlorophenol, giving rise to the main photoproducts such as *p*-benzoquinone, whereas 2,6-dimethylphenol was transformed into 2,6-dimethylbenzoquinone [60]. The same product was obtained when 4-chlorophenol was treated with water-soluble FeTDCPPS [68] and sodium *meso*-tetra (4-sulphonatophenyl)porphyrin (NaTPPS). Photodegradation of atrazine and ametryn by *meso*-tetra(4-sulphonatophenyl) porphyrin (TPPS) or TDCPPS resulted in a mixture of photoproducts [69]. Further examples are the photooxidation of 2,4,6-trinitrotoluene (TNT) with TPPS and its iron complex (FeTPPS) to give trinitrobenzoic acid and trinitrobenzene [70]. In most cases given above, water-soluble porphyrin derivatives are adopted, which are more suitable for practical wastewater treatment.

As another case, hydrogen production is a typical photocatalytic reaction that occurs under light irradiation. Photoinduced hydrogen production from water is regarded as an efficient and cost-effective method for the conversion and storage of solar energy. This process is usually accomplished by a system containing a photosensitizer, electron carrier, electron donor, and a catalyst. Chlorophyll and ferredoxin are the natural photosensitizer and electron carrier, while porphyrins often act as an artificial photosensitizer [71]. An example of water-soluble zinc *meso*-tetra(1-methylpyridinium-4-yl)porphyrin chloride $[ZnTMPyP^{4+}]Cl_4$ as a photosensitizer, viologens as an electron carrier, ethylenediaminetetraacetic acid (EDTA) as an electron donor, and hydrogenase (H_2ase) as a catalyst was provided by Qian et al. Lazarides et al. reported the use of the same $[ZnTMPyP^{4+}]Cl_4$ as a photosensitizer, but with cobaloxime complex as a catalyst [72]. Using this system, the photocatalytic activity maintained for 20 h producing in total about 280 TON of hydrogen. In some other studies, Co, Fe, and Rh porphyrins were shown to be active as the hydrogen evolution catalysts via photoinitiation using other sensitizers [73]. Scandola and coworkers reported the efficient photochemical hydrogen evolution from 1 M pH 7 phosphate buffer by using water-soluble cationic cobalt (II) porphyrin as the catalyst, ascorbic acid as the electron donor, and $[Ru(bpy)_3]^{2+}$ (bpy = 2,2′-bipyridine) as the photosensitizer, in achievement of TON up to 725 [74]. Kinetic studies revealed a rapid electron transfer process from $[Ru(bpy)_3]^{2+}$ to cobalt (II) porphyrin with a calculated rate constant of 2.3×10^9 M^{-1} s^{-1}.

4. Porphyrin-based heterogeneous photocatalysts

Despite the feasibility, the homogeneous porphyrin cannot be commercialized. It has a tendency toward deactivation due to photobleaching or solvolysis by the solvent, and recovery of it from the reaction media usually constitutes another difficulty. In this case, heterogenization of porphyrins seems necessary. One possible solution is to immobilize them on organic or inorganic solid support for improved stability and easy recovery. On the other hand, nanoassembly provides another way to stabilize porphyrin molecules. The two aspects are addressed in the following sections.

4.1. Supported porphyrin photocatalysts

Grafting of porphyrin molecules onto a solid matrix is a useful and practical approach to carry out the reaction, since solid photocatalysts can be easily separated from the reaction medium and reused.

4.1.1. Inorganic support

Easily available silica has been extensively employed as a host for porphyrins [66]. Immobilized porphyrins have been obtained by covalent link to aminoalkylated silica particles for elimination of model pollutants. Through a modified Stöber technique and thiolene polymerization, porphyrin functionalized silica microspheres have been constructed and repeatedly used for the photooxidation of 1,5-dihydroxynaphthalene (DHN) [75]. Likewise, sulfonated porphyrin has been attached to silica to photooxidize 1,5-dihydroxynaphthalene (DHN) in water [67]. Results show that this solid photocatalyst is stable and can be recycled five times without significant loss of activity. Other supports such as clay, zeolites, or layered materials are also involved to host porphyrins [76, 77]. Due to immobilization, the stability of the catalysts is thereby enhanced and the efficiency loss is reduced.

4.1.2. Organic polymers as supports

To obtain a high retention, porphyrins have been included inside solid polymer backbones, which are found to impart interesting modifications on their photochemical properties [61]. For instance, anionic TPPS and its metal complex have been covalently anchored onto cationic polystyrene [78]. The resulting polymer-porphyrin system with rich ionic sites showed high activity in photokilling *Escherichiacoli* cells. In another work, porphyrin-polythiophene complex has been synthesized via electrostatic interactions, leading to a high-energy transfer process [79]. Thus, singlet oxygen could be effectively generated to kill the bacteria under white light. Biodegradable chitosan can also serve as a scaffold, as in the case of chitosan-supported metallotetraphenylporphyrin complexes [80]. Porphyrin that was covalently attached to nylon fabric was also found to be very effective against *Staphylococcus aureus* [81].

Resins are alternative supports for porphyrins due to the ease of preparation. NaTPPS has been ionically bounded at polymeric ion-exchange resin (Amberlite) toward oxidizing phenols [82]. Diffuse reflectance spectra revealed that the grafted porphyrin had an absorption feature close to the homogeneous one. Different resins such as Amberlite or 16–50 mesh have been employed to support a series of porphyrins to evaluate their activity and stability with phenols as model pollutants [83]. In all cases, the preparation is simply carried out by stirring the mixture of ion exchangers and the catalysts in an appropriate medium.

4.1.3. Carbon materials as supports

Compared to energy transfer, photoinduced electron transfer (PET) by transformation of excitation energy into chemical potentials in the form of long-lived carriers is at the heart of photoenergy conversion. On a molecular level, a large number of porphyrin-based dyads or triads (porphyrin-fullerene, porphyrin-quinone) have been intentionally designed to initiate the PET process [84, 85]. For heterogenization, more consideration has been given to choice of π-conjugated carbon materials such as carbon nanotubes (CNTs), graphene, or C_3N_4 as scaffolds. Their electron-accepting nature thus offers an opportunity to facilitate electron transfer and enhance the photoconversion efficiency [86–88]. An additional advantage is their ability to form flexible macroscopic scaffolds through different techniques such as filtration or layer-by-layer assembly.

Much efforts have been made to organize porphyrins on the semiconducting CNTs because of their unique optoelectronic properties, stability, and high surface area. Steady-state fluorescence (FL) reveals that covalently connected porphyrins function as energy-harvesting and electron-transferring antennae, while the CNTs function as electron acceptors [86, 89]. FL quenching is commonly referred to as a useful fingerprint to probe the PET process. As in the case of single-wall carbon tube-zinc porphyrin (SWCNT-ZnP) hybrids (**Figure 2A**), steady-state and time-resolved FL studies (**Figure 2B**) revealed efficient FL quenching of the singlet-excited state of zinc porphyrin with the rate constants of charge separation in the range of (3–6) × 10^9 s^{-1} [90]. In addition, because of increased PET, a uniform film made of CNTs-porphyrin conjugates by simple filtration displayed high light-activated antimicrobial activity toward *S. aureus* with easy recovery [91].

Graphene is known to provide high-quality two-dimensional (2D) support to increase the loading content of the photocatalysts. Regarding its large surface area, special surface activities, and layered structure, much efforts have been devoted for the preparation of nano-structured graphenes, in the form of sheets, films, or quantum dots (QDs), to promote the separation and transfer of photoinduced charge carriers [92–94]. So far, a series of papers have appeared dealing with covalent and noncovalent attachment of porphyrin to graphene. Noncovalent methods include electrostatic interactions, π-π interactions, and axial coordination. For instance, graphene/zinc tetraphenylporphyrin (GR/ZnTPP) composite was facilely prepared via noncovalent interaction [95]. Note that 71% FL quenching seen with ZnTPP in the GR/ZnTPP composite clearly implied a PET process from ZnTPP to GR. With the aid of GR, improved photocurrent response was found in the GR/ZnTPP composite. Likewise, multiple-bilayered graphene oxide (GO)-porphyrin film was prepared by taking advantage of the π-π and electrostatic interactions between GO sheets and porphyrin molecules, and then underwent subsequent vapor reduction to give a reduced graphene oxide (rGO)-porphyrin film [96]. The as-obtained film also showed enhanced photocurrent generation following the

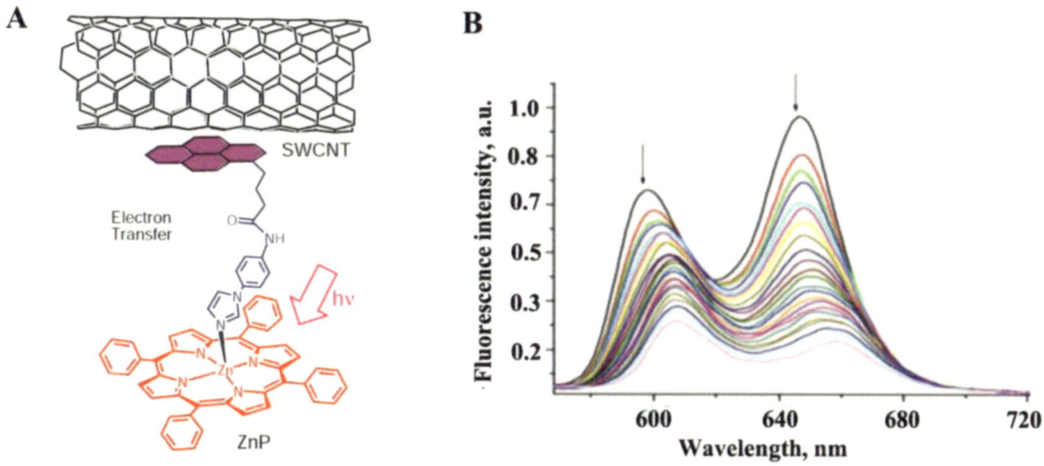

Figure 2. (A) Supramolecular structure of SWCNT-ZnP; (B) FL spectral changes in the visible region of ZnP during the titration of increasing addition of SWCNT, excitation wavelength λ_{ex} = 550 nm. Adapted from Ref. [89].

PET. Graphene quantum dots (GQDs), have also been used to bind to zinc porphyrin by π-π stacking, to give excellent photocatalytic performance toward degrading methylene blue (MB) under visible light irradiation [97].

Covalent attachment of porphyrin to graphene usually refers to the formation of covalent bonds between the different functional groups (COOH, NH_2, etc.) in the periphery of the porphyrin ring and the oxygen groups of GO. By comparison with noncovalent methods, the covalent bond is stable and well defined, and moreover, the number or type of functional groups can be controlled by fine-tuning the functionalization. It is also widely accepted that the covalent band can form channels to prompt the PET between porphyrin and graphene. For instance, amine-functionalized prophyrin (TPP-NH_2) and GO bound together via an amide bond (TPP-NHCO-Gr) [98]. In a different report (**Figure 3**), azide-terminated zinc

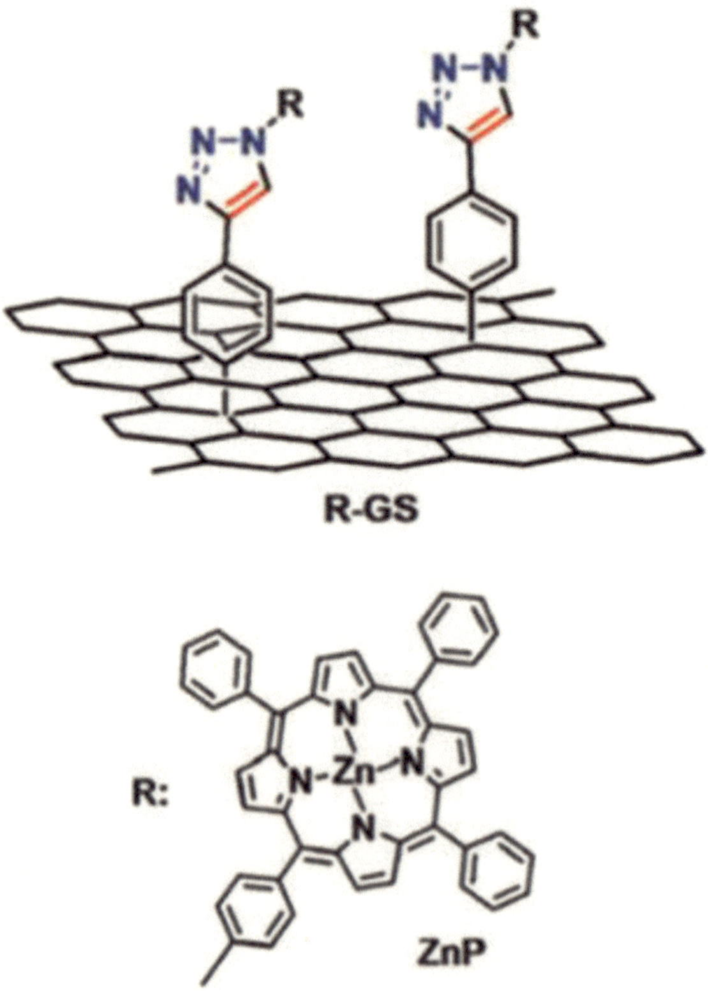

Figure 3. Schematic representation of the covalently linked ZnP-GS. Adapted from Ref. [99].

porphyrin (ZnP) and 4-(trimethylsilyl)ethynylaniline modified graphene sheets (GSs) were covalently linked to give ZnP-GS with the formation of the triazole bond [99]. Occurrence of PET was indicated in this covalently linked ZnP-GS composite based on its higher photocurrent response. Further examples are the use of different metal ions (such as K^+, Ca^{2+}, Zn^{2+}, Cu^{2+}, and Co^{2+}) as interfacial linkers to construct a series of composites between GO and 5,15-diphenyl-10,20-di(4-pyridyl)porphyrin (DPyP) [100]. The resulting strong interaction between metal ions and DPyP/GO thus facilitates the spatial separation of photogenerated charges, thereby leading to higher photocatalytic activity for hydrogen production.

Graphitic-like C_3N_4 (g-C_3N_4), as another 2D framework, is highly identified as a visible light-active polymeric semiconductor with a bandgap of ~2.7 eV and appropriate energy levels that can extract hydrogen from water [101–104]. It is expected that g-C_3N_4 can bind to porphyrin through π-π stacking, electrostatic interaction, or covalent bonding, which resembles graphene. An extra merit lies in the well-matched band structures between porphyrin and C_3N_4 that allow a good combination of them for increased PET. For instance, Cu (II) *meso*-tetra(4-carboxyphenyl)porphyrin (CuTCPP) was easily assembled on g-C_3N_4 to form CuTCPP/g-C_3N_4 composites through π-π stacking interaction [88]. Given that the LUMO band of CuTCPP lies below that of g-C_3N_4, the photoinduced electrons from excited CuTCPP can be directly transferred to g-C_3N_4. This thereby reduced the probability of charge recombination, resulting in higher photocatalytic activity for phenol degradation than individual component. Natural metalloporphyrin, hemin, has been coupled with imidazole-functionalized g-C_3N_4 through an axial coordination. The strong coordinate bond endowed the g-C_3N_4-hemin catalyst with enhanced stability. The obtained g-C_3N_4-hemin thus displayed higher and sustained photocatalytic oxidation activity for the degradation of 4-chlorophenol over a wide pH range [105]. In a different report, Co (II) porphyrin has been covalently linked to g-C_3N_4 for photocatalytic reduction of CO_2 [106]. The as-obtained hybrid possessed thirteen-fold higher photocatalytic activity (17 μmol g^{-1} h^{-1}) than pure g-C_3N_4 (1.4 μmol g^{-1} h^{-1}). Such enhancement was attributed to the increased charge separation and prolonged lifetime of the excited state due to the electron trap by Co (II) sites.

4.2. Nanostructured porphyrin photocatalysts

Nanostructured porphyrins are expected to have chemical activities and stability quite different from those free or supported porphyrins [107, 108]. Particularly, the synthetic versatility enables the controllable organization of porphyrins into well-defined nanostructures via rational assembly [109–111]. For the past decade, reports began to appear on the synthesis of porphyrin nanomaterials or nanocomposites for enhanced photocatalytic performance [112–114].

4.2.1. Supramolecular assembly

Supramolecular assembly is defined as large aggregation of molecules held together by non-covalent bonds, such as hydrogen bonds, metal coordination, van der Waals, and π-π interaction. By carefully adjusting these intermolecular interactions, aggregates with diversity and complexity can be formed [115]. For instance, π-π interaction is thought to drive the formation of nanorods of 5,15-bis(3,5-di-tert-butylphenyl)porphyrin (H_2DBuPP) [113]. The much broader absorption in the visible and near infrared regions suggested the strong supramolecular π-π interaction. As a consequence, the organized rod-crystals exhibited a broad photoresponse in the visible region (an incident photon to current conversion efficiency (IPCE): ~5.5% at 460 nm), which paralleled their corresponding absorption features. As is also shown, change of the substituents at *meso*-positions to control the intermolecular interaction finally led to some difference in the length of the rod crystals.

In a different report, highly crystalline rectangular nanotubes of *meso*-tetra(4-pyridyl)porphyrin (H_2TPyP) were synthesized [111]. The key driving forces included: hydrogen-bonding interactions along the *c*-axis, hydrogen-bonding and π-π interactions along the *a*-axis, and hydrogen-bonding and hydrogen-π intermolecular interactions along the *b*-axis, respectively. In addition, metal coordination interactions are always involved in the design of well-defined porous metal-organic frameworks (MOFs). An example is the mediation of Zn^{2+} in the assembly of *meso*-tetra(4-carboxyphenyl)porphyrin (TCPP) in **Figure 4** [116]. The X-ray diffraction (XRD) patterns of Zn^{2+}-meditated TCPP nanocubes, nanorods, and microrods are totally different from TCPP starting materials, which means that coordination bonds play a significant role in the structural change of TCPP. In another study, the involvement of Cu^{2+} has led to the assembly TCPP into 2D crystalline nanofilms [117].

All suggest that we are able to manipulate the intermolecular interactions by properly modifying the peripheral position or ring cavity of the porphyrins, thereby allowing fine-tuning of the molecular packing mode. Such a strategy enables us to construct a rich variety of nanostructures with different sizes and shapes. To date, comprehensive studies have been

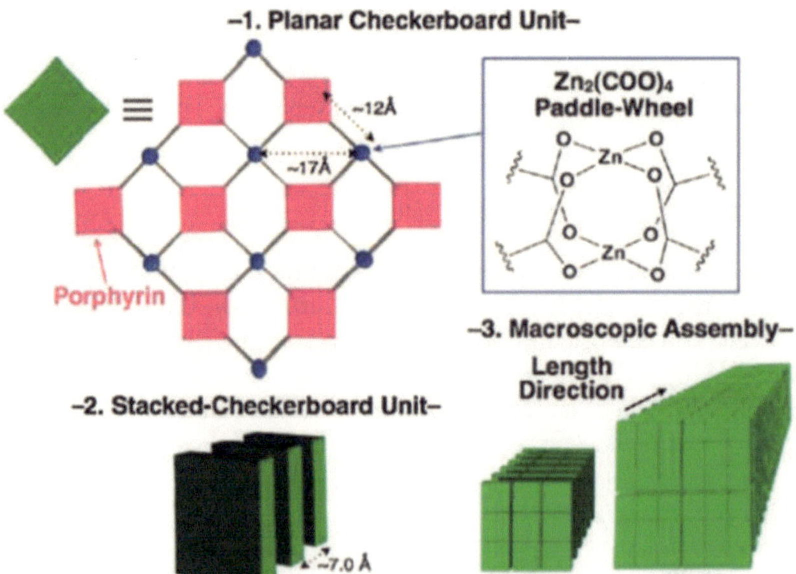

Figure 4. A schematic illustration of the proposed structures of TCPP architectures via paddle-wheel complexes in this study. Adapted from Ref. [116].

conducted on the preparation of porphyrin nanostructures by different synthetic methods, with morphologies varying from nanoparticles to nanowires, nanofibers, nanobelts, nanotubes, nanowheels, or films.

4.2.2. Preparation and applicability

Reprecipitation is the most widely used method for supramolecular assembly. It is performed by injection of a small amount of concentrated solution of porphyrin in good solvent into a pool of poor solvent. Sometimes, addition of surfactants affords a better control of the assembly process. The aforementioned H_2DBuPP nanorods were prepared by mixing of the toluene solution of H_2DBuPP with nine times volume of acetonitrile [113]. The same procedure was followed to synthesize zinc *meso*-tetra(4-pyridyl)porphyrin (ZnTPyP) nanostructures [118], with the assistance of cetyltrimethylammonium bromide (CTAB) as a surfactant to control the growth. Morphologies varied from nanoparticles to nanofibers by adjusting the concentration of the surfactant or the aging time. Compared to ZnTPyP nanoparticles, the fiber-crystals demonstrated a higher photocatalytic activity toward degrading rhodamine B (RhB) pollutants. More importantly, ZnTPyP nanofibers retained the photocatalytic efficiency after eight repeated runs because of the geometric constraint. By changing the poor/good solvents, ZnTPyP hexagonal nanocylinders are obtained [119]. The internal cavity thus enabled the encapsulation of Pt-colloids-deposited TiO_2 nanoparticles (Pt/TiO_2) to produce the final Pt/TiO_2-ZnTPyP nanorods. The as-obtained nanorods showed the photocatalytic hydrogen evolution efficiency of 2 orders magnitude greater than the simple mixture of Pt/TiO_2 + ZnTPyP/Pt.

Different morphologies of ZnTPyP can also be synthesized through acid-base neutralization. In one study, the acidified ZnTPyP (ZnTPyP-H_4^{4+}) aqueous solution was mixed with the basic surfactant solution under vigorous stirring [120]. By increasing the surfactant concentration, a series of morphologies from amorphous nanoparticles to crystalline nanodisks, tetragonal nanorods, and hexagonal nanorods were synthesized with controlled size and dimension (**Figure 5A–H**). The largest pore surface area of about 457 m^2 g^{-1} made porous nanodisks the most efficient catalyst in photodegrading methyl orange (MO) (**Figure 5I** and **J**). Moreover, the efficiency loss was greatly reduced during the repeated use due to the crystalline nature of nanodisks (**Figure 5K**). In a different report, ZnTPyP nanooctahedra were synthesized by metallization of H_2TPyP [121]. In detail, Zn^{2+} was first mixed into H_2TPyP acidic aqueous solution, and the mixture was then injected into the basic solution with CTAB. Metalation of H_2TPyP to ZnTPyP just took place during the acid-base neutralization, and it was observed that the morphology of ZnTPyP transformed from nanooctahedra to nanowires with increasing the pH value of the basic solution. ZnTPyP nanowires were found to have the best catalytic activity in photodegradation of MO and showed no sign of corrosion in the structure after 15 cycles. Acid-base neutralization was also used to synthesize various isolated TCPP aggregated structures, including spheres, rods, flakes, and flowers for photodegradation of RhB [122]. Graphene-supported TCPP nanorods have been synthesized to eliminate RhB by adding the basic suspension of TCPP-adsorbed graphene into acid aqueous solution of CTAB [123].

Ionic self-assembly is an attractive synthetic method that is managed by electrostatic interactions of two oppositely charged building blocks. The cooperative interactions between the

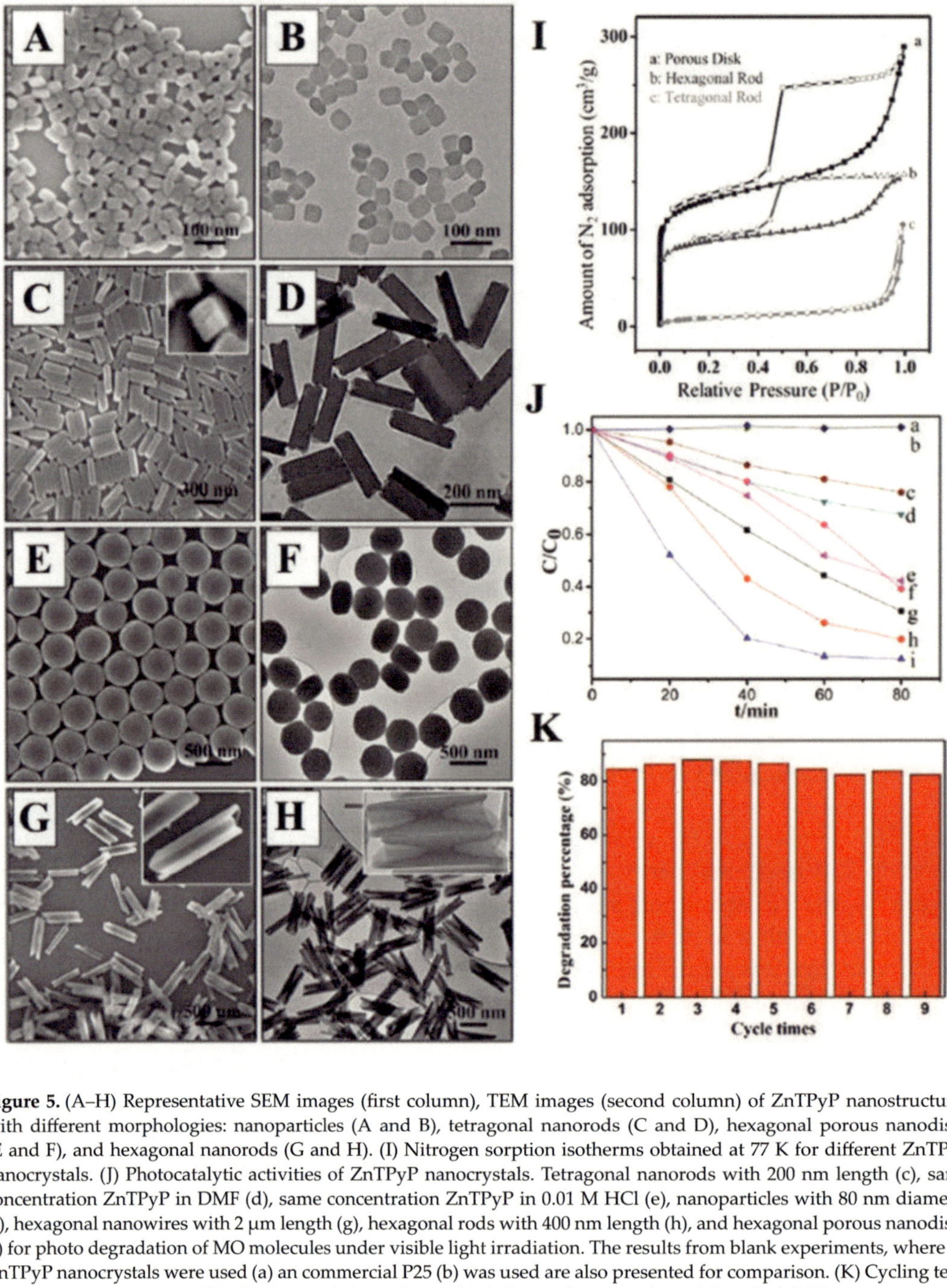

Figure 5. (A–H) Representative SEM images (first column), TEM images (second column) of ZnTPyP nanostructures with different morphologies: nanoparticles (A and B), tetragonal nanorods (C and D), hexagonal porous nanodisks (E and F), and hexagonal nanorods (G and H). (I) Nitrogen sorption isotherms obtained at 77 K for different ZnTPyP nanocrystals. (J) Photocatalytic activities of ZnTPyP nanocrystals. Tetragonal nanorods with 200 nm length (c), same concentration ZnTPyP in DMF (d), same concentration ZnTPyP in 0.01 M HCl (e), nanoparticles with 80 nm diameter (f), hexagonal nanowires with 2 μm length (g), hexagonal rods with 400 nm length (h), and hexagonal porous nanodisks (i) for photo degradation of MO molecules under visible light irradiation. The results from blank experiments, where no ZnTPyP nanocrystals were used (a) an commercial P25 (b) was used are also presented for comparison. (K) Cycling tests of photocatalytic activity of ZnTPyP nanodisks under visible light irradiation. Adapted from Ref. [120].

functional subunits may afford new interesting collective properties. For instance, four-leaf clover-like morphologies have been constructed by ionic self-assembly of Zn(II) *meso*-tetra(4-sulfonatophenyl)porphyrin (ZnTPPS) and Sn (IV) *meso*-tetra(N-2-hydroxyethyl-4-pyridinium) porphyrin (SnT(N-EtOH-4-Py)P) [124]. With Pt as the cocatalyst for hydrogen evolution, the clovers demonstrated a photoactivity far better than the sum of their individual effects. In another study, ionic assembly between tin (IV) porphyrin cation and phosphomolybdate anion led to the formation of new porphyrin-polyoxometalate hybrid nanoparticles [125]. Due to the broadened absorption and efficient electron transfer, the formed hybrid exhibited a strikingly high activity in photocatalytic hydrogen production.

Besides all that vaporization-condensation-recrystallization (VCR) organization is a commonly used way to synthesized single crystals, as in the case of H_2TPyP rectangular nanotubes [111] and the tetraphenylporphyrin (H_2TPP) nanoplates [126].

4.2.3. *Photocatalytic efficiency*

A photocatalysis event usually follows three steps: exciton formation by light absorption, charge separation, and carrier conduction. Achievement of high photoenergy conversion just relies on improving the efficiency of each step. First, aggregate formation has a strong effect on the light-harvesting efficiency. As mentioned above, the aggregation of porphyrins often arises from a long-range noncovalent interaction, while the molecular arrangement into J-aggregation (relative to H-aggregation) is crucial to the light-harvesting efficiency [127]. J-aggregates are formed with a large number of molecular building blocks arranged in one dimension. The strong intermolecular π electronic coupling within the long axes results in a coherent excitation at red-shifted wavelengths of increasing sharpness (higher absorption coefficient). For instance, the time-dependent UV-vis spectra in monitoring the growth of one-dimensional (1D) ZnTPyP hollow hexagonal nanoprisms showed that the high-energy Soret band at 424 nm, associated with monomeric ZnTPyP, gradually decreased, with an increasing high-energy The Soret band at 460 nm is associated with J-aggregated ZnTPyP [108]. The same spectral change was seen with ZnTPyP nanofibers, making the 1D nanofibers more efficient light-harvesting antenna than zero-dimensional (0D) nanoparticles [118]. On the other hand, J-aggregates are promising building blocks to direct electron transport, thereby to retard the charge recombination by stabilizing the electron transfer products. Insight into dynamics and mobility of excitons has been obtained from J-aggregates of perylene bisimides (PBIs) by transient absorption spectra [128]. The findings indicated that exciton mobility in the J-aggregates of PBIs was restricted to one dimension and exciton diffusion length was about 10 times larger than in disordered polymers. The 1D mobility thus allows for electron migration along a preferential direction without trapping effects. As proved in the case of TCPP series, J-aggregated rods exhibited more photocatalytic efficiency than the flakes and flowers [122]. As a step forward, it is proposed that alignment of the 1D nanostructures into highly ordered arrays may produce collective behavior to achieve high performance.

It is highly acknowledged that organic materials suffer from notoriously low charge carrier mobility, and photoexcited charge carriers may easily recombine before being exploited. To address it, single-crystalline organic nanostructures have been sought because the low-defect structure can largely impede the recombination of the excitons and accelerate effective charge transport. Previous reports have studied the effect of impurities on the mobility of organic pentacene, and concluded that a mobility of μ = 35 $cm^2\ V^{-1}\ s^{-1}$ at room temperature was increased to μ = 58 $cm^2\ V^{-1}\ s^{-1}$ at 225 K for pentacene single crystals [129]. Moreover, purified rubrene single crystals have showed a maximum transistor mobility of $\mu = 18\ cm^2\ V^{-1}\ s^{-1}$ [130]. Motivated by these, growing emphasis has been placed on the synthesis of photocatalytic crystals. For instance, hierarchical structured nanocrystals of Sn (IV) *meso*-tetraphenylporphyrin dichloride (SnTPPCl) were synthesized and displayed high photocatalytic activities in the reduction of platinum nanoparticles and in photodegradation of MO [131]. The XRD patterns of SnTPPCl octahedra were indexed as a tetragonal space group, while the photocatalytic ZnTPyP rectangular nanorods were indexed as a monoclinic space group [120]. In view of the perfect molecular alignment, the photocatalytic crystals are more efficient and stable compared to the amorphous nanostructures.

Interfacial heterostructuring is another strategy to effectively reduce the recombination of photoexcited electron-hole pairs. The built-in energy level offset within the heterojunction can drive the exciton to dissociate for ready charge transfer. We have succeeded in the fabrication of 1D organic single-crystal p/n nanoheterojunctions made of p-type H_2TPP and n-type N,N-(dicyclohexyl) perylene-3,4,9,10-tetracarboxylic diimide (CH-PTCDI), as shown in **Figure 6A** and **B** [126]. The large donor-acceptor interface provides a strong driving force to separate the spatial charges, and meanwhile the 1D structure facilitates the directed charge transport along the long-range axes, thereby leading to enhanced charge separation efficiency. An efficient PET process was evidenced by the significant FL quench of CH-PTCDI when coupled with H_2TPP (**Figure 6C**). As a consequence, the H_2TPP/CH-PTCDI junction showed a remarkably high photoactivity in photodegrading MB (**Figure 6D**). It is also known that porphyrin nanostructures are able to photocatalytically reduce metal ions, which encourages the preparation of serial porphyrin/metal nanohybrids. In particular, when photocatalytic hydrogen evolution is discussed, self-platinized porphyrins are preferred, with Pt nanoparticles as cocatalysts to converted water into hydrogen gas. Examples are the self-platinized porphyrin nanotubes, nanosheets, nanofibers, and clovers that have been successively synthesized for efficient hydrogen production [124, 132]. Additionally, a few reports have presented the combination of porphyrins with inorganic semiconductors as an exciting alternative. Such heterojunctions can take advantage of the two different material classes by allowing for a good combination of the wide absorption spectrum of porphyrins and the high mobility of the inorganic semiconductors. In one work, porphyrin-TiO_2 core-shell nanoparticles have been prepared and exhibit better MB photodegradation than molecular porphyrin sensitized TiO_2 [133]. We also brought together the two materials to give a new configuration of TiO_2 nanotube/H_2TPP nanoparticle hybrids for PEC water splitting [134]. The resulting hybrid displayed an intensive and broad absorption spectrum across 350–660 nm. Upon photoexcitation of H_2TPP, ultrafast charge injection from the excited H_2TPP into TiO_2 took place, and the transferred electrons were then transported away via the unidirectional electron channels of TiO_2 nanotube arrays. The increased charge separation was well proved by the largely reduced photocharge transfer resistance. Further example was provided by the aforementioned three-component Pt/TiO_2-ZnTPyP nanorods [119]. The formed electron transfer cascade from excited ZnTPyP to the conduction band of TiO_2, then to the surface of Pt nanoparticles, resulted in an enhanced hydrogen evolution efficiency. As mentioned above, the π-conjugated carbon materials have been established as an ideal scaffold to anchor

http://dx.doi.org/10.5772/intechopen.68223

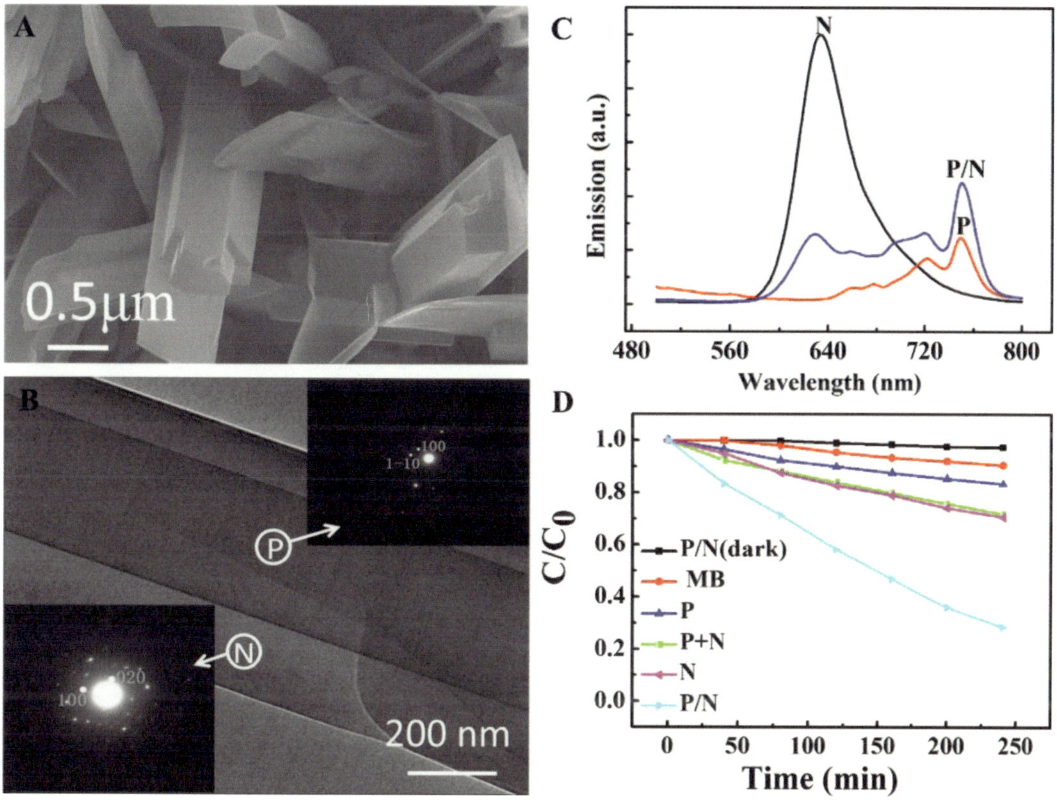

Figure 6. (A and B) SEM and TEM images of H_2TPP/CH-PTCDI nanoheterojunctions; (C) FL spectra of H_2TPP (P), CH-PTCDI (N), and H_2TPP/CH-PTCDI (p/n) nanostructures; (D) photocatalytic degradation of MB with different samples under visible light irradiation ($\lambda > 400$ nm). Adapted from Ref. [125].

molecular porphyrins for stability and recyclability. Inclusion of porphyrin nanostructures into these scaffolds may seem as a logic step forward by further increasing the loading content and stability. Certainly, such combination benefits much from the increased lifetime of the charge carriers since these π-conjugated carbon nanostructures can serve as an excellent electron acceptor and an electron transporter as well. We have integrated H_2TPP nanoparticles into graphene for the formation of free-standing H_2TPP/rGO nanohybrid film, as shown in **Figure 7A** [135]. By coupling, the average lifetime of H_2TPP emission was lengthened from ca. 362 to 473 ps (**Figure 7B**), while the photocharge transfer was reduced from 176.2 to 46.7 Ω (**Figure 7C**). The increased electron transfer thus accounted for the improved photocatalytic performance (**Figure 7D**). For the same purpose, well-dispersed TCPP nanorods [123] or ZnTPyP nanoassemblies have been successfully fabricated on the surface of graphene nanoplates [136]. The same role was also addressed in g-C_3N_4. In this respect, a combination of g-C_3N_4 and m-oxo dimeric iron (III) porphyrin ($(FeTPP)_2O$) was sketched to form g-C_3N_4/$(FeTPP)_2O$ nanocomposites, which showed dramatically improved photocatalytic hydrogen production [137].

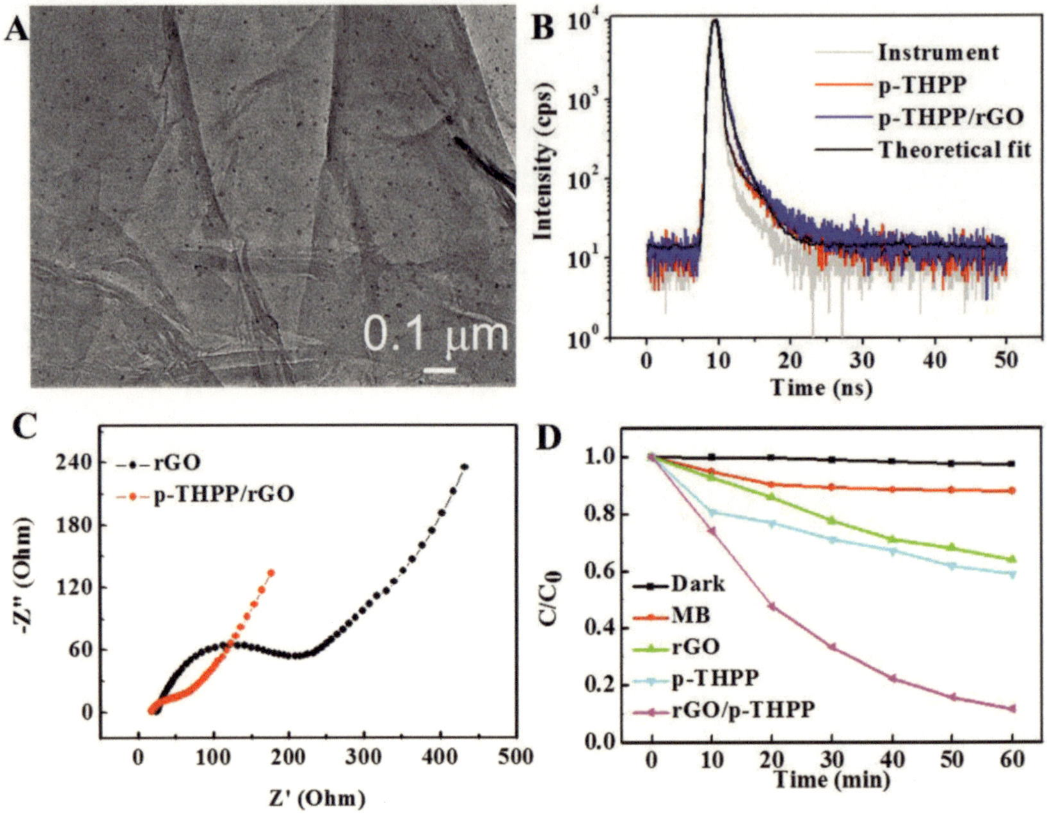

Figure 7. (A) TEM image of H_2TPP/rGO nanohybrids; (B) the FL decay profiles of H_2TPP and H_2TPP/rGO nanohybrids in H_2O (λ_{ex} = 405 nm); (C) Nyquist plots collected by electrical impedance spectroscopy (EIS) of free-standing rGO and H_2TPP/rGO films; (D) photocatalytic degradation of MB with different samples under visible light irradiation ($\lambda_{irradiation}$ > 400 nm). Adapted from Ref. [135].

5. Porphyrin-based PEC water splitting devices

PEC water splitting is attractive because of its ease with which an electric field can be created at the semiconductor/liquid junction to manipulate the charge transfer reaction. In water splitting, oxidation of water into O_2 occurs at the photoanode, and H+ is reduced to H_2 at the photocathode. Ideally, a single semiconductor must absorb light with photon energies larger than 1.23 eV, and have a conduction band energy (E_{cb}) and valence band energy (E_{vb}) that straddle the electrochemical potentials $E°$ (H^+/H_2) and $E°$(O_2/H_2O), so as to drive water splitting under illumination. Porphyrins with a band energy gap of 1.5–3.1 eV and the matched HOMO and LUMO positions are in principle able to perform this PEC reaction. As p-type semiconductors, porphyrins are usually coupled with n-type conductors to effectively motivate the water-splitting reaction. This part focuses on thin-film photoelectrodes based on porphyrin-containing systems for PEC applications.

5.1. Film deposition

Currently, the search of cheap and fast processing of large-area photoelectrodes is of intense research. In view of the low melting point, solubility, and flexibility, organic materials have the advantages of low-cost fabrication, and making flexible and lightweight devices. Generally, thin-film organic devices are fabricated either by vapor deposition or solution processing.

5.1.1. Solution-processed deposition

The solubility of organic semiconductors is a desirable characteristic in making low-cost electronic devices. As one of them, porphyrins are or can be derivatized to soluble. Therefore, thin-film porphyrins can be fabricated by solution processing near room temperature, mainly including dip coating, spin coating, or printing techniques [138, 139]. Printing methods, such as screen printing and ink jet printing, also enable fabrication and patterning of the active materials in a single step, and now the most use of them is made in fabricating organic field-effect transistors (OFETs), and organic light-emitting diodes (OLEDs) [138, 139].

Dip coating and spin coating are the most widely used methods when it comes to small-scale processing, because it is easy to handle and fairly cheap to acquire the film. In a dip-coating process, a substrate is immersed into the solution of the active materials and then exposed until dried. For instance, multilayered rGO-porphyrin films have been fabricated by alternately dipping the substrate into a GO suspension and porphyrin solutions, followed by exposure to a hydrazine vapor [96]. The resulting film showed promising applications in PEC cells.

Spin coating is often used to deposit uniform thin films to a flat substrate, simply by spreading the solution of the active material over the substrate at a high rotating speed, but this process is not quite applicable to large area, and too much material is wasted. In an example, a mixture of zinc *meso*-tetra(4-carboxyphenyl)porphyrin (ZnTCPP) and fullerene (C_{60}) was spin-coated onto the working electrodes, and exhibited efficient photocurrent generation (IPCE value up to $47 \pm 5\%$) [140].

5.1.2. Thermal vapor deposition

Vapor deposition generally includes vacuum thermal evaporation and organic vapor phase deposition (OVPD). For organic molecules that have a poor solubility, vacuum thermal evaporation is an ideal deposition method and is more suitable for multilayered fabrication or cofabrication. In a number of examples, vacuum deposition is involved to manufacture organic materials. Typically, source materials are heated under a vacuum environment, and the deposited film is usually in high uniformity. For instance, organic bilayers made of p-type porphyrin analogies (phthalocyanine) and n-type C_{60} or PBIs have been extensively deposited by this method for PEC water splitting [141, 142]. Estimation of the film thickness and the refractive index is often conducted by ellipsometry.

An alternative to vacuum deposition is organic vapor phase deposition (OVPD). It is characterized by a process in which heating of the source materials under a stream of hot inert gas. Then the sublimated molecules are transported to a low-temperature zone by gas flow, where molecules condensed into nanocrystals. The OVPD method offers good control over deposition efficiency and film morphology by adjusting the amount of the source materials, gas flow rate, or the collecting substrate. In our work, photocatalytic H_2TPP/CH-PTCDI junctions were codeposited via the OVPD method [126]. The two source materials were located into the upstream area with certain spacing distance because of their different sublimation temperature, and silicon wafers were put at the downstream cooled zone to collect samples with nitrogen as the carrier gas.

5.2. Hard porphyrin photoanodes

Appropriate choice of deposition methods makes it possible to assemble molecular, nanostructured, or thin-film porphyrins onto different electrodes. Metal substrates, such as Au, Al, or Pt, have always been involved to assemble the monolayer of porphyrin-based molecules by dip coating [143–145]. These monolayers ranged from single porphyrins to porphyrin-C_{60} dyads, or ferrocene-porphyrin-C_{60} triads, with an aim to increase intermolecular electron transfer. Furthermore, well-organized molecular assemblies can be achieved by covalent attachment of functional molecules to the chemically modified metals, as in the case of porphyrin alkane-thiolate with short alkanethiols on the gold nanoclusters [146], which resulted in an increased photocurrent density.

Due to the band match, other hard substrates such as ITO or FTO glasses, nanostructured TiO_2, SnO_2, or ZnO have been employed to couple with porphyrin nanoassemblies. Of them, ITO glass is the mostly used semiconducting substrate, and a number of photoactive organic materials have been deposited on it either by spin coating or vapor deposition. As described above, different combinations of organic bilayers made of porphyrin analogies and n-type semiconductors have been fabricated on ITO by vapor deposition. For instance, organic p/n bilayer of C_{60} and 29H, 31H-phthalocyanine (H_2Pc) was prepared by vapor deposition, and ITO glass was used as the collecting electrode [142]. As shown in **Figure 8A**, the photoanode comprised H_2Pc layer coated on ITO, and C_{60} coated on top of the H_2Pc layer (denoted as ITO/H_2Pc/C_{60}). PEC splitting of water into H_2 was confirmed across the entire visible light energy region ($\lambda < 750$ nm) in **Figure 8B**, with the faradaic efficiency for the H_2 evolution around 90%. In another work, whiskered PBIs/H_2Pc bilayer was fabricated on ITO as photoanodes (ITO/PBIs/H_2PC) [147]. Formation of the whiskered H_2Pc by proper thermal control resulted in an enhancement of the p/n interface. Therefore, the magnitude of the oxidation kinetics at the whiskered H_2Pc/water interface was demonstrated to be 2.5 times higher than the flat interface. We have adopted the dip-coating method to coat H_2TPP nanoparticles onto TiO_2 nanotube arrays as photoanodes, leading to enhanced photocurrent generation [134]. In another study, C_{60}-ZnTPyP nanorods have been deposited onto nanostructured SnO_2 films, and exhibited a power conversion efficiency of 0.63% and an IPCE of 35% [148].

5.3. Flexible porphyrin photoanodes

The low-temperature processing and low cost make organic devices one of the most important semiconductor devices for flexible optoelectronic device applications. Therefore, growing

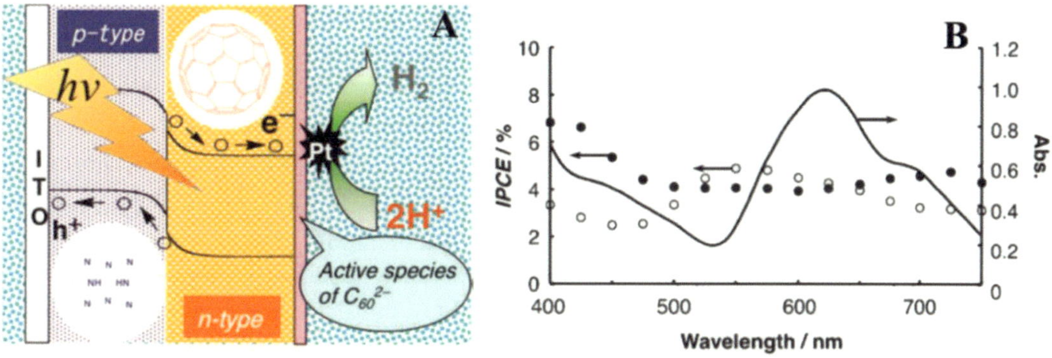

Figure 8. (A) An illustration of the ITO/H_2Pc/C_{60} photoanode; (B) action spectra of the photocurrent generated at ITO/H_2Pc/C_{60}-Pt (irradiation direction: ITO side (•) and Pt-coated C_{60} side (o)) and absorption spectrum of employed bilayer (–). Adapted from Ref. [142].

development has been made on flexible organic devices, including organic solar cells, OLEDs, OFETs, sensors, and memories. As in one case of pentacene-based OFETs, 125-μm-thick polyethylene naphthalate (PEN) film was used as a flexible substrate and 30-nm-thick pentacene was thermally deposited to create the organic active layer [149]. One early study was dealing with fully flexible OLEDs [150]. The device was built on a poly(ethylene terephthalate) substrate, with soluble polyaniline as the hole-injecting electrode, substituted poly(1,4-phenylene-vinylene) as the electroluminescent layer. Moreover, a first example of all-organic flexible photoanodes was presented to remove a gaseous pollutant (trimethylamine, TMA) [141]. Instead of ITO, H_2Pc (50 nm)/PBIs (50 nm) bilayer was coated on a self-standing fluorocarbon polymer (Nafion 112) to act as a photoanode. This all-organic catalyst is responsive to full-spectrum visible light (<780 nm), which holds promise in future use inside buildings when only interior light is present.

6. Conclusion and perspective

To sum up, this chapter presents the recent advances on the porphyrin-based organophotocatalysts. Porphyrins possess many light conversion functionalities such as light harvesting and energy/electron transfer, and are thus acknowledged to be a promising tool in homogeneous, heterogeneous photocatalysis, or even PEC solar cells. Molecular porphyrins have easy access to other reactive species and thus render a high photocatalytic efficiency, but their limitations arise from the stability and reuse. Inclusion of porphyrin molecules onto solid supports provides a robust material that can be easily recovered for successive use. In the series of solid supports, the electron-accepting and electron-conducting carbon materials are preferred in initiating an increased PET mechanism. Their large surface area also allows for high-loading content. Nanoassembly makes heterogenization of porphyrins a further step forward, opening a way to control the light conversion functionality in the aggregated state. Ordered molecular alignment is attainable by proper control over the supramolecular assembly, thus resulting in enhanced light-harvesting and charge-transfer efficiencies. Promoted charge transfer can also be fulfilled by a combination of nanostructured porphyrins with other acceptor materials, such as metals, inorganic semiconductors, and organic electron acceptors. Moreover, porphyrins can be fabricated onto hard or flexible/stretchable electrodes by low-temperature solution processing or vapor deposition, which finds wide applications in PEC solar cells.

Despite the current knowledge of the organic/inorganic hybrid, an in-depth insight into the interface geometry is essential in determining and understanding the properties and functions of the two different material classes. Supplementary information can be provided by rational calculation and simulation, especially in light of the electronic properties, excitation dynamics, and charge transport of the hybrid materials. Further efforts are required to align these molecular assemblies onto any desired substrate at the macroscopic level. The resulting collective effect will finally lead to remarkable improvement on light conversion efficiency for practical applications in optoelectronic devices.

Acknowledgements

Authors are grateful for the National Natural Science Foundation of China (Grant nos. 51503014 and 51501008).

Author details

Yingzhi Chen[1], Zheng-Hong Huang[2*] and Lu-Ning Wang[1]

*Address all correspondence to: zhhuang@mail.tsinghua.edu.cn

1 School of Materials Science and Engineering, University of Science and Technology Beijing, Beijing, China

2 Key Laboratory of Advanced Materials (MOE), School of Materials Science and Engineering, Tsinghua University, Beijing, China

References

[1] Chen C, Ma W, Zhao J. Semiconductor-mediated photodegradation of pollutants under visible-light irradiation. Chemical Society Reviews. 2010;**39**:4206-4219. DOI: 10.1039/b921692h

[2] Maeda K, Domen K. Photocatalytic water splitting: Recent progress and future challenges. The Journal of Physical Chemistry Letters. 2010;**1**:2655-2661. DOI: 10.1021/jz1007966

[3] Maeda K, Teramura K, Lu D, Takata T, Saito N, Inoue Y, et al. Photocatalyst releasing hydrogen from water. Nature. 2006;**440**:295. DOI: 10.1038/440295a

[4] Esswein AJ, Nocera DG. Hydrogen production by molecular photocatalysis. Chemical Reviews. 2007;**107**:4022-4047. DOI: 10.1021/cr050193e

[5] Chen X, Shen S, Guo L, Mao SS. Semiconductor-based photocatalytic hydrogen generation. Chemical Reviews. 2010;**110**:6503-6570. DOI: 10.1021/cr1001645

[6] Wang X, Liow C, Bisht A, Liu X, Sum TC, Chen X, et al. Engineering interfacial photoinduced charge transfer based on nanobamboo array architecture for efficient solar-to-chemical energy conversion. Advanced Materials. 2015;**27**:2207-2214. DOI: 10.1002/adma.201405674

[7] Wang X, Liow C, Qi D, Zhu B, Leow WR, Wang H, et al. Programmable photo-electrochemical hydrogen evolution based on multi-segmented CdS-Au nanorod arrays. Advanced Materials. 2014;**26**:3506-3512. DOI: 10.1002/adma.201306201

[8] Chen Y, Li A, Yue X, Wang L-N, Huang Z-H, Kang F, et al. Facile fabrication of organic/inorganic nanotube heterojunction arrays for enhanced photoelectrochemical water splitting. Nanoscale. 2016;**8**:13228-13235. DOI: 10.1039/c5nr07893h

[9] Schneider J, Matsuoka M, Takeuchi M, Zhang J, Horiuchi Y, Anpo M, et al. Understanding TiO_2 photocatalysis: Mechanisms and materials. Chemical Reviews. 2014;**114**:9919-9986. DOI: 10.1021/cr5001892

[10] Lu D, Fang P, Wu W, Ding J, Jiang L, Zhao X, et al. Solvothermal-assisted synthesis for self-assembling TiO_2 nanorods on large graphitic carbon nitride sheets with their anti-recombination in photocatalytic removal of Cr (VI) and rhodamin B under visible light irradiation. Nanoscale. 2017;**9**:3231-3245. DOI: 10.1039/C6NR09137G

[11] Cushing SK, Meng F, Zhang J, Ding B, Chen CK, Chen C-J, et al. Effects of defects on photocatalytic activity of hydrogen-treated titanium oxide nanobelts. ACS Catalysis. 2017;**7**:1742-1748. DOI: 10.1021/acscatal.6b02177

[12] Wang J, Waters JL, Kung P, Kim SM, Kelly JT, McNamara LE, et al. A facile electrochemical reduction method for improving photocatalytic performance of α-Fe_2O_3 photoanode for solar water splitting. ACS Applied Materials & Interfaces. 2017;**9**:381-390. DOI: 10.1021/acsami.6b11057

[13] Bhoi Y, Mishra B. Single step combustion synthesis, characterization and photocatalytic application of α-Fe_2O_3-Bi_2S_3 heterojunctions for efficient and selective reduction of structurally diverse nitroarenes. Chemical Engineering Journal. 2017;**316**:70-81. DOI: 10.1016/j.cej.2017.01.075

[14] Yang J, Wu Q, Yang X, He S, Khan J, Meng Y, et al. Chestnut-like TiO_2@ α-Fe_2O_3 core-shell nanostructures with abundant interfaces for efficient and ultralong life lithium ion storage. ACS Applied Materials & Interfaces. 2017;**9**:354-361. DOI: 10.1021/acsami.6b12150

[15] Barhoum A, Melcher J, Van Assche G, Rahier H, Bechelany M, Fleisch M, et al. Synthesis, growth mechanism, and photocatalytic activity of zinc oxide nanostructures: Porous microparticles versus nonporous nanoparticles. Journal of Materials Science. 2017;**52**:2746-2762. DOI: 10.1007/s10853-016-0567-3

[16] Gupta R, Eswar NK, Modak JM, Madras G. Effect of morphology of zinc oxide in ZnO-CdS-Ag ternary nanocomposite towards photocatalytic inactivation of E. coli under UV and visible light. Chemical Engineering Journal. 2017;**307**:966-980. DOI: 10.1016/j.cej.2016.08.142

[17] Wu SC, Tan CS, Huang MH. Strong facet effects on interfacial charge transfer revealed through the examination of photocatalytic activities of various Cu_2O–ZnO heterostructures. Advanced Functional Materials. 2017;**27**:1604635. DOI: 10.1002/adfm.201604635

[18] Zhang Y, Zhou X, Zhao Y, Liu Z, Ma D, Chen S, et al. One-step solvothermal synthesis of interlaced nanoflake-assembled flower-like hierarchical Ag/Cu_2O composite microspheres with enhanced visible light photocatalytic properties. RSC Advances. 2017;**7**:6957-6965. DOI: 10.1039/C6RA26870F

[19] Lou Z, Li Y, Zhu L, Niu W, Song H, Ye Z, et al. The crystalline/amorphous contact in Cu_2O/Ta_2O_5 heterostructures: Increasing its sunlight-driven overall water splitting efficiency. Journal of Materials Chemistry A. 2017;**5**:2732-2738. DOI: 10.1039/C6TA10728A

[20] Samsudin EM, Hamid SBA. Effect of band gap engineering in anionic-doped TiO_2 photocatalyst. Applied Surface Science. 2017;**391**:326-336. DOI: 10.1016/j.apsusc.2016.07.007

[21] Chen J, Wu G, Wang T, Li X, Li M, Sang Y, et al. Carrier step-by-step transport initiated by precise defect distribution engineering for efficient photocatalytic hydrogen generation. ACS Applied Materials & Interfaces. 2017;**9**:4634-4642. DOI: 10.1021/acsami.6b14700

[22] Low J, Cheng B, Yu J. Surface modification and enhanced photocatalytic CO_2 reduction performance of TiO_2: A review. Applied Surface Science. 2017;**392**:658-686. DOI: 10.1016/j.apsusc.2016.09.093

[23] Jin Z, Liu C, Qi K, Cui X. Photo-reduced Cu/CuO nanoclusters on TiO_2 nanotube arrays as highly efficient and reusable catalyst. Scientific Reports. 2017;**7**:39695. DOI:10.1038/srep39695

[24] Wang T, Lv R, Zhang P, Li C, Gong J. Au nanoparticle sensitized ZnO nanopencil arrays for photoelectrochemical water splitting. Nanoscale. 2015;**7**:77-81. DOI: 10.1039/C4NR03735A

[25] Forrest SR, Thompson ME. Introduction: Organic electronics and optoelectronics. Chemical Reviews. 2007;**107**:923-925. DOI: 10.1021/cr0501590

[26] Zhao YS, Fu H, Peng A, Ma Y, Liao Q, Yao J. Construction and optoelectronic properties of organic one-dimensional nanostructures. Accounts of Chemical Research. 2009;**43**:409-418. DOI: 10.1021/ar900219n

[27] Kelley TW, Baude PF, Gerlach C, Ender DE, Muyres D, Haase MA, et al. Recent progress in organic electronics: Materials, devices, and processes. Chemistry of Materials. 2004;**16**:4413-4422. DOI: 10.1021/cm049614j

[28] Zhao YS, Fu H, Peng A, Ma Y, Xiao D, Yao J. Low-dimensional nanomaterials based on small organic molecules: Preparation and optoelectronic properties. Advanced Materials. 2008;**20**:2859-2876. DOI: 10.1002/adma.200800604

[29] Zang L, Che Y, Moore JS. One-dimensional self-assembly of planar π-conjugated molecules: Adaptable building blocks for organic nanodevices. Accounts of Chemical Research. 2008;**41**:1596-1608. DOI: 10.1021/ar800030w

[30] Li Y, Wang W, Leow WR, Zhu B, Meng F, Zheng L, et al. Optoelectronics of organic nanofibers formed by co-assembly of porphyrin and perylenediimide. Small. 2014;**10**:2776-2781. DOI: 10.1002/smll.201302964

[31] Chen Y, Li A, Huang Z-H, Wang L-N, Kang F. Porphyrin-based nanostructures for photocatalytic applications. Nanomaterials. 2016;**6**:51. DOI:10.3390/nano6030051

[32] Hasobe T. Porphyrin-based supramolecular nanoarchitectures for solar energy conversion. The Journal of Physical Chemistry Letters. 2013;**4**:1771-1780. DOI: 10.1021/jz4005152

[33] Medforth CJ, Wang Z, Martin KE, Song Y, Jacobsen JL, Shelnutt JA. Self-assembled porphyrin nanostructures. Chemical Communications. 2009;**35**:7261-7277. DOI: 10.1039/B914432C

[34] Hasobe T, Imahori H, Kamat PV, Fukuzumi S. Quaternary self-organization of porphyrin and fullerene units by clusterization with gold nanoparticles on SnO_2 electrodes for organic solar cells. Journal of the American Chemical Society. 2003;**125**:14962-14963. DOI: 10.1021/ja0377192

[35] Hasobe T, Saito K, Kamat PV, Troiani V, Qiu H, Solladié N, et al. Organic solar cells. Supramolecular composites of porphyrins and fullerenes organized by polypeptide structures as light harvesters. Journal of Materials Chemistry. 2007;**17**:4160-4170. DOI: 10.1039/B706678C

[36] Noh YY, Kim JJ, Yoshida Y, Yase K. Effect of molecular orientation of epitaxially grown platinum (II) octaethyl porphyrin films on the performance of field-effect transistors. Advanced Materials. 2003;**15**:699-702. DOI: 10.1002/adma.200304005

[37] Che CM, Xiang HF, Chui SSY, Xu ZX, Roy VeAeL, Yan JJ, et al. A high-performance organic field-effect transistor based on platinum (II) porphyrin: Peripheral sSubstituents on porphyrin ligand significantly affect film structure and charge mobility. Chemistry, an Asian Journal. 2008;**3**:1092-1103. DOI: 10.1002/asia.200800011

[38] Endo A, Ogasawara M, Takahashi A, Yokoyama D, Kato Y, Adachi C. Thermally activated delayed fluorescence from Sn4+–porphyrin complexes and their application to organic light emitting diodes—A novel mechanism for electroluminescence. Advanced Materials. 2009;**21**:4802-4806. DOI: 10.1002/adma.200900983

[39] Li B, Li J, Fu Y, Bo Z. Porphyrins with four monodisperse oligofluorene arms as efficient red light-emitting materials. Journal of the American Chemical Society. 2004;**126**:3430-3431. DOI: 10.1021/ja039832y

[40] Marin ML, Santos-Juanes L, Arques A, Amat AM, Miranda MA. Organic photocatalysts for the oxidation of pollutants and model compounds. Chemical Reviews. 2011;**112**:1710-1750. DOI: 10.1021/cr2000543

[41] Kim W, Park J, Jo HJ, Kim H-J, Choi W. Visible light photocatalysts based on homogeneous and heterogenized tin porphyrins. The Journal of Physical Chemistry C. 2008;**112**:491-499. DOI: 10.1021/jp0747151

[42] Craw M, Redmond R, Truscott TG. Laser flash photolysis of haematoporphyrins in some homogeneous and heterogeneous environments. Journal of the Chemical Society, Faraday Transactions 1: Physical Chemistry in Condensed Phases. 1984;**80**:2293-2299. DOI: 10.1039/F19848002293

[43] Imahori H, Ueda M, Kang S, Hayashi H, Hayashi S, Kaji H, et al. Effects of porphyrin substituents on film structure and photoelectrochemical properties of porphyrin/fullerene composite clusters electrophoretically deposited on nanostructured SnO_2 electrodes. Chemistry–A European Journal. 2007;**13**:10182-10193. DOI: 10.1002/chem.200700446

[44] Umeyama T, Mihara J, Tezuka N, Matano Y, Stranius K, Chukharev V, et al. Preparation and photophysical and photoelectrochemical properties of a covalently fixed porphyrin–chemically converted graphene composite. Chemistry–A European Journal. 2012;**18**:4250-4257. DOI: 10.1002/chem.201103843

[45] Kay A, Graetzel M. Artificial photosynthesis. 1. Photosensitization of titania solar cells with chlorophyll derivatives and related natural porphyrins. The Journal of Physical Chemistry. 1993;**97**:6272-6277

[46] Nazeeruddin MK, Humphry-Baker R, Officer DL, Campbell WM, Burrell AK, Grätzel M. Application of metalloporphyrins in nanocrystalline dye-sensitized solar cells for conversion of sunlight into electricity. Langmuir. 2004;**20**:6514-6517. DOI: 10.1021/la0496082

[47] Hori T, Aratani N, Takagi A, Matsumoto T, Kawai T, Yoon MC, et al. Giant porphyrin wheels with large electronic coupling as models of light-harvesting photosynthetic antenna. Chemistry–A European Journal. 2006;**12**:1319-1327. DOI: 10.1002/chem.200501373

[48] Imahori H, Umeyama T, Ito S. Large π-aromatic molecules as potential sensitizers for highly efficient dye-sensitized solar cells. Accounts of Chemical Research. 2009;**42**:1809-1818. DOI: 10.1021/ar900034t

[49] Mathew S, Yella A, Gao P, Humphry-Baker R, Curchod BF, Ashari-Astani N, et al. Dye-sensitized solar cells with 13% efficiency achieved through the molecular engineering of porphyrin sensitizers. Nature Chemistry. 2014;**6**:242-247. DOI: 10.1038/nchem.1861

[50] Martinez-Diaz MV, de la Torre G, Torres T. Lighting porphyrins and phthalocyanines for molecular photovoltaics. Chemical Communications. 2010;**46**:7090-7108. DOI: 10.1039/C0CC02213F

[51] Drobizhev M, Karotki A, Kruk M, Mamardashvili NZ, Rebane A. Drastic enhancement of two-photon absorption in porphyrins associated with symmetrical electron-accepting substitution. Chemical Physics Letters. 2002;**361**:504-512. DOI: 10.1016/S0009-2614(02)00999-5

[52] Suslick KS, Rakow NA, Kosal ME, Chou J-H. The materials chemistry of porphyrins and metalloporphyrins. Journal of Porphyrins and Phthalocyanines. 2000;**4**:407-413. DOI: 10.1002/(SICI)1099-1409(200006/07)4:4<407::AID-JPP256>3.0.CO;2-5

[53] Gao W-Y, Chrzanowski M, Ma S. Metal–metalloporphyrin frameworks: A resurging class of functional materials. Chemical Society Reviews. 2014;**43**:5841-5866. DOI: 10.1039/C4CS00001C

[54] Zheng W, Shan N, Yu L, Wang X. UV–visible, fluorescence and EPR properties of porphyrins and metalloporphyrins. Dyes and Pigments. 2008;**77**:153-157. DOI: 10.1016/j.dyepig.2007.04.007

[55] Shiragami T, Matsumoto J, Inoue H, Yasuda M. Antimony porphyrin complexes as visible-light driven photocatalyst. Journal of Photochemistry and Photobiology C. 2005;**6**:227-248. DOI: 10.1016/j.jphotochemrev.2005.12.001

[56] Gao P, Konrad D, Aghazada S, Nazeeruddin MK. Molecular engineering of functional materials for energy and opto-electronic applications. CHIMIA International Journal for Chemistry. 2015;**69**:253-263

[57] Xue X, Zhang W, Zhang N, Ju C, Peng X, Yang Y, et al. Effect of the length of the alkyl chains in porphyrin meso-substituents on the performance of dye-sensitized solar cells. RSC Advances. 2014;**4**:8894-8900. DOI: 10.1039/C3RA46212A

[58] Ladomenou K, Kitsopoulos T, Sharma G, Coutsolelos A. The importance of various anchoring groups attached on porphyrins as potential dyes for DSSC applications. RSC Advances. 2014;**4**:21379-21404. DOI: 10.1039/C4RA00985A

[59] Liu B, Zhu W, Wang Y, Wu W, Li X, Chen B, et al. Modulation of energy levels by donor groups: An effective approach for optimizing the efficiency of zinc-porphyrin based solar cells. Journal of Materials Chemistry. 2012;**22**:7434-7444. DOI: 10.1039/C2JM16804A

[60] Monteiro CJ, Pereira MM, Azenha ME, Burrows HD, Serpa C, Arnaut LG, et al. A comparative study of water soluble 5, 10, 15, 20-tetrakis (2, 6-dichloro-3-sulfophenyl) porphyrin and its metal complexes as efficient sensitizers for photodegradation of phenols. Photochemical & Photobiological Sciences. 2005;**4**:617-624. DOI: 10.1039/B507597A

[61] Maldotti A, Molinari A, Andreotti L, Fogagnolo M, Amadelli R. Novel reactivity of photoexcited iron porphyrins caged into a polyfluoro sulfonated membrane in catalytic hydrocarbon oxygenation. Chemical Communications. 1998;**24**:507-508. DOI: 10.1039/A706237K.

[62] Tomé JP, Neves MG, Tomé AC, Cavaleiro JA, Soncin M, Magaraggia M, et al. Synthesis and antibacterial activity of new poly-s-lysine– porphyrin conjugates. Journal of Medicinal Chemistry. 2004;**47**:6649-6652. DOI: 10.1021/jm040802v

[63] Caminos DA, Spesia MB, Durantini EN. Photodynamic inactivation of escherichia coli by novel meso-substituted porphyrins by 4-(3-N,N,N-trimethylammoniumpropoxy) phenyl and 4-(trifluoromethyl) phenyl groups. Photochemical & Photobiological Sciences. 2006;**5**:56-65. DOI: 10.1039/B513511G

[64] Carvalho CM, Tome JP, Faustino MA, Neves MG, Tome AC, Cavaleiro JA, et al. Antimicrobial photodynamic activity of porphyrin derivatives: Potential application on medical and water disinfection. Journal of Porphyrins and Phthalocyanines. 2009;**13**:574-577. DOI: 10.1142/S1088424609000528

[65] Murtinho D, Pineiro M, Pereira MM, Gonsalves AMdAR, Arnaut LG, da Graça Miguel M, et al. Novel porphyrins and a chlorin as efficient singlet oxygen photosensitizers for photooxidation of naphthols or phenols to quinones. Journal of the Chemical Society, Perkin Transactions 2. 2000;**12**:2441-2447. DOI: 10.1039/B006583H

[66] Ribeiro SM, Serra AC, Gonsalves AdAR. Covalently immobilized porphyrins on silica modified structures as photooxidation catalysts. Journal of Molecular Catalysis A: Chemical. 2010;**326**:121-127. DOI: 10.1016/j.molcata.2010.05.001

[67] Cai J-H, Huang J-W, Yu H-C, Ji L-N. Fabrication and characterizations of silica composite microspheres immobilized with porphyrins and their photocatalytic properties. Journal of the Taiwan Institute of Chemical Engineers. 2012;**43**:958-964. DOI: 10.1016/j.jtice.2012.05.004

[68] Silva E, Pereira MM, Burrows HD, Azenha M, Sarakha M, Bolte M. Photooxidation of 4-chlorophenol sensitised by iron meso-tetrakis (2,6-dichloro-3-sulfophenyl) porphyrin in aqueous solution. Photochemical & Photobiological Sciences. 2004;**3**:200-204. DOI: 10.1039/B308975D

[69] Rebelo SL, Melo A, Coimbra R, Azenha ME, Pereira MM, Burrows HD, et al. Photodegradation of atrazine and ametryn with visible light using water soluble porphyrins as sensitizers. Environmental Chemistry Letters. 2007;**5**:29-33. DOI: 10.1007/s10311-006-0072-z

[70] Harmon HJ. Photocatalytic demethylation of 2,4,6-trinitrotoluene (TNT) by porphyrins. Chemosphere. 2006;**63**:1094-1097. DOI: 10.1016/j.chemosphere.2005.09.013

[71] Qian D-J, Wenk S-O, Nakamura C, Wakayama T, Zorin N, Miyake J. Photoinduced hydrogen evolution by use of porphyrin, EDTA, viologens and hydrogenase in solutions and Langmuir–Blodgett films. International Journal of Hydrogen Energy. 2002;27:1481–1487.DOI: 10.1016/S0360-3199(02)00104-0

[72] Lazarides T, Delor M, Sazanovich IV, McCormick TM, Georgakaki I, Charalambidis G, et al. Photocatalytic hydrogen production from a noble metal free system based on a water soluble porphyrin derivative and a cobaloxime catalyst. Chemical Communications. 2014;**50**:521-523. DOI: 10.1039/C3CC45025B

[73] Nakazono T, Parent AR, Sakai K. Cobalt porphyrins as homogeneous catalysts for water oxidation. Chemical Communications. 2013;**49**:6325-6327. DOI: 10.1039/C3CC43031F

[74] Natali M, Luisa A, Iengo E, Scandola F. Efficient photocatalytic hydrogen generation from water by a cationic cobalt (II) porphyrin. Chemical Communications. 2014;**50**:1842-1844. DOI: 10.1039/C3CC48882A

[75] Cai J-H, Huang J-W, Zhao P, Ye Y-J, Yu H-C, Ji L-N. Silica microspheres functionalized with porphyrin as a reusable and efficient catalyst for the photooxidation of 1, 5-dihydroxynaphthalene in aerated aqueous solution. Journal of Photochemistry and Photobiology A: Chemistry. 2009;**207**:236-243. DOI: 10.1016/j.jphotochem.2009.07.016

[76] Bedioui F. Zeolite-encapsulated and clay-intercalated metal porphyrin, phthalocyanine and Schiff-base complexes as models for biomimetic oxidation catalysts: An overview. Coordination Chemistry Reviews. 1995;**144**:39-68. DOI: 10.1016/0010-8545(94)08000-H

[77] Skrobot F, Valente A, Neves G, Rosa I, Rocha J, Cavaleiro J. Monoterpenes oxidation in the presence of Y zeolite-entrapped manganese (III) tetra (4-N-benzylpyridyl) porphyrin. Journal of Molecular Catalysis A: Chemical. 2003;**201**:211-222. DOI: 10.1016/S1381-1169(03)00181-X

[78] Johnson Inbaraj J, Vinodu MV, Gandhidasan R, Murugesan R, Padmanabhan M. Photosensitizing properties of ionic porphyrins immobilized on functionalized solid polystyrene support. Journal of Applied Polymer Science. 2003;**89**:3925-3930. DOI: 10.1002/app.12610

[79] Yuan H, Wang B, Lv F, Liu L, Wang S. Conjugated-polymer-based energy-transfer systems for antimicrobial and anticancer applications. Advanced Materials. 2014;**26**:6978-6982. DOI: 10.1002/adma.201400379

[80] Huang G, Guo C-C, Tang S-S. Catalysis of cyclohexane oxidation with air using various chitosan-supported metallotetraphenylporphyrin complexes. Journal of Molecular Catalysis A: Chemical. 2007;**261**:125-130. DOI: 10.1016/j.molcata.2006.08.014

[81] Bozja J, Sherrill J, Michielsen S, Stojiljkovic I. Porphyrin-based, light-activated antimicrobial materials. Journal of Polymer Science Part A: Polymer Chemistry. 2003;**41**:2297-2303. DOI: 10.1002/pola.10773

[82] Pepe E, Abbas O, Rebufa C, Simon M, Lacombe S, Julliard M. Supported photosensitizers for the visible light activation of phenols towards oxygen. Journal of Photochemistry and Photobiology A: Chemistry. 2005;**170**:143-149. DOI: 10.1016/j.jphotochem.2004.06.019

[83] Hequet V, Le Cloirec P, Gonzalez C, Meunier B. Photocatalytic degradation of atrazine by porphyrin and phthalocyanine complexes. Chemosphere. 2000;**41**:379-386. DOI: 10.1016/S0045-6535(99)00474-9

[84] Tkachenko NV, Lemmetyinen H, Sonoda J, Ohkubo K, Sato T, Imahori H, et al. Ultrafast photodynamics of exciplex formation and photoinduced electron transfer in porphyrin–fullerene dyads linked at close proximity. The Journal of Physical Chemistry A. 2003;**107**:8834-8844. DOI: 10.1021/jp035412j

[85] Tsue H, Imahori H, Kaneda T, Tanaka Y, Okada T, Tamaki K, et al. Large acceleration effect of photoinduced electron transfer in porphyrin–quinone dyads with a rigid spacer involving a dihalosubstituted three-membered ring. Journal of the American Chemical Society. 2000;**122**:2279-2288. DOI: 10.1021/ja9900454

[86] Baskaran D, Mays JW, Zhang XP, Bratcher MS. Carbon nanotubes with covalently linked porphyrin antennae: Photoinduced electron transfer. Journal of the American Chemical Society. 2005;**127**:6916-6917. DOI: 101021/ja0508222

[87] Zhu M, Li Z, Xiao B, Lu Y, Du Y, Yang P, et al. Surfactant assistance in improvement of photocatalytic hydrogen production with the porphyrin noncovalently functionalized graphene nanocomposite. ACS Applied Materials & Interfaces. 2013;**5**:1732-1740. DOI: 10.1021/am302912v

[88] Chen D, Wang K, Hong W, Zong R, Yao W, Zhu Y. Visible light photoactivity enhancement via CuTCPP hybridized g-C_3N_4 nanocomposite. Applied Catalysis B: Environmental. 2015;**166**:366-373. DOI: 10.1016/j.apcatb.2014.11.050

[89] Baba A, Matsuzawa T, Sriwichai S, Ohdaira Y, Shinbo K, Kato K, et al. Enhanced photocurrent generation in nanostructured chromophore/carbon nanotube hybrid layer-by-layer multilayers. The Journal of Physical Chemistry C. 2010;**114**:14716-14721. DOI: 10.1021/jp103121m

[90] Das SK, Subbaiyan NK, D'Souza F, Sandanayaka AS, Wakahara T, Ito O. Formation and photoinduced properties of zinc porphyrin-SWCNT and zinc phthalocyanine-SWCNT nanohybrids using diameter sorted nanotubes assembled via metal-ligand coordination and π–π stacking. Journal of Porphyrins and Phthalocyanines. 2011;**15**:1033-1043. DOI: 10.1142/S1088424611003951

[91] Banerjee I, Mondal D, Martin J, Kane RS. Photoactivated antimicrobial activity of carbon nanotube-porphyrin conjugates. Langmuir. 2010;**26**:17369-17374. DOI: 10.1021/la103298e

[92] Xiang Q, Yu J, Jaroniec M. Graphene-based semiconductor photocatalysts. Chemical Society Reviews. 2012;**41**:782-796. DOI: 10.1039/C1CS15172J

[93] Kamat PV. Graphene-based nanoassemblies for energy conversion. The Journal of Physical Chemistry Letters. 2011;**2**:242-251. DOI: 10.1021/jz101639v

[94] Xiang Q, Yu J. Graphene-based photocatalysts for hydrogen generation. The Journal of Physical Chemistry Letters. 2013;**4**:753-759. DOI: 10.1021/jz302048d

[95] Zhang Z, Zhu J, Han Q, Cui H, Bi H, Wang X. Enhanced photo-electrochemical performances of graphene-based composite functionalized by Zn^{2+} tetraphenylporphyrin. Applied Surface Science. 2014;**321**:404-411. DOI: 10.1016/j.apsusc.2014.10.043

[96] Sun J, Meng D, Jiang S, Wu G, Yan S, Geng J, et al. Multiple-bilayered RGO–porphyrin films: From preparation to application in photoelectrochemical cells. Journal of Materials Chemistry. 2012;**22**:18879-17886. DOI: 10.1039/C2JM33900E

[97] Lu Q, Zhang Y, Liu S. Graphene quantum dots enhanced photocatalytic activity of zinc porphyrin toward the degradation of methylene blue under visible-light irradiation. Journal of Materials Chemistry A. 2015;**3**:8552-8558. DOI: 10.1039/C5TA00525F

[98] Xu Y, Liu Z, Zhang X, Wang Y, Tian J, Huang Y, et al. A graphene hybrid material covalently functionalized with porphyrin: Synthesis and optical limiting property. Advanced Materials. 2009;**21**:1275-1279. DOI: 10.1002/adma.200801617

[99] Wang H-X, Zhou K-G, Xie Y-L, Zeng J, Chai N-N, Li J, et al. Photoactive graphene sheets prepared by "click" chemistry. Chemical Communications. 2011;**47**:5747-5749. DOI: 10.1039/C1CC11121C

[100] Ge R, Li X, Kang S-Z, Qin L, Li G. Highly efficient graphene oxide/porphyrin photocatalysts for hydrogen evolution and the interfacial electron transfer. Applied Catalysis B: Environmental. 2016;**187**:67-74. DOI: 10.1016/j.apcatb.2016.01.024

[101] Cao S, Yu J. g-C_3N_4-based photocatalysts for hydrogen generation. The Journal of Physical Chemistry Letters. 2014;**5**:2101-2107. DOI: 10.1021/jz500546b

[102] Wang X, Maeda K, Thomas A, Takanabe K, Xin G, Carlsson JM, et al. A metal-free polymeric photocatalyst for hydrogen production from water under visible light. Nature Materials. 2009;**8**:76-80. DOI: 10.1038/nmat2317

[103] Zheng Y, Liu J, Liang J, Jaroniec M, Qiao SZ. Graphitic carbon nitride materials: Controllable synthesis and applications in fuel cells and photocatalysis. Energy & Environmental Science. 2012;**5**:6717-6731. DOI: 10.1039/C2EE03479D

[104] Niu P, Zhang L, Liu G, Cheng HM. Graphene-like carbon nitride nanosheets for improved photocatalytic activities. Advanced Functional Materials. 2012;**22**:4763-4770. DOI: 10.1002/adfm.201200922

[105] Chen X, Lu W, Xu T, Li N, Qin D, Zhu Z, et al. A bio-inspired strategy to enhance the photocatalytic performance of g-C_3N_4 under solar irradiation by axial coordination with hemin. Applied Catalysis B: Environmental. 2017;**201**:518-526. DOI: 10.1016/j.apcatb.2016.08.020

[106] Zhao G, Pang H, Liu G, Li P, Liu H, Zhang H, et al. Co-porphyrin/carbon nitride hybrids for improved photocatalytic CO_2 reduction under visible light. Applied Catalysis B: Environmental. 2017;**200**:141-149. DOI: 10.1016/j.apcatb.2016.06.074

[107] Vasilopoulou M, Douvas AM, Georgiadou DG, Constantoudis V, Davazoglou D, Kennou S, et al. Large work function shift of organic semiconductors inducing enhanced interfacial electron transfer in organic optoelectronics enabled by porphyrin aggregated nanostructures. Nano Research. 2014;**7**:679-693. DOI: 10.1007/s12274-014-0428-9

[108] Liu H, Xu J, Li Y, Li Y. Aggregate nanostructures of organic molecular materials. Accounts of Chemical Research. 2010;**43**:1496-1508. DOI: 10.1021/ar100084y

[109] Lee SJ, Hupp JT, Nguyen ST. Growth of narrowly dispersed porphyrin nanowires and their hierarchical assembly into macroscopic columns. Journal of the American Chemical Society. 2008;**130**:9632-9633. DOI: 10.1021/ja801733t

[110] Hu J-S, Guo Y-G, Liang H-P, Wan L-J, Jiang L. Three-dimensional self-organization of supramolecular self-assembled porphyrin hollow hexagonal nanoprisms. Journal of the American Chemical Society. 2005;**127**:17090-17095. DOI: 10.1021/ja0553912

[111] Yoon SM, Hwang IC, Kim KS, Choi HC. Synthesis of single-crystal tetra (4-pyridyl) porphyrin rectangular nanotubes in the vapor phase. Angewandte Chemie International Edition. 2009;**48**:2506-2509. DOI: 10.1002/anie.200806301

[112] Guo P, Chen P, Liu M. Porphyrin assemblies via a surfactant-assisted method: From nanospheres to nanofibers with tunable length. Langmuir. 2012;**28**:15482-15490. DOI: 10.1021/la3033594

[113] Atula S. Sonication-assisted supramolecular nanorods of meso-diaryl-substituted porphyrins. Chemical Communications. 2008;**24**:724-726. DOI: 10.1039/B713971C

[114] Wang Z, Medforth CJ, Shelnutt JA. Porphyrin nanotubes by ionic self-assembly. Journal of the American Chemical Society. 2004;**126**:15954-15955. DOI: 10.1021/ja045068j

[115] Wasielewski MR. Self-assembly strategies for integrating light harvesting and charge separation in artificial photosynthetic systems. Accounts of Chemical Research. 2009;**42**:1910-1921. DOI: 10.1021/ar9001735

[116] Sakuma T, Sakai H, Hasobe T. Preparation and structural control of metal coordination-assisted supramolecular architectures of porphyrins. Nanocubes to microrods. Chemical Communications. 2012;**48**:4441-4443. DOI: 10.1039/C2CC30756A

[117] Motoyama S, Makiura R, Sakata O, Kitagawa H. Highly crystalline nanofilm by layering of porphyrin metal-organic framework sheets. Journal of the American Chemical Society. 2011;**133**:5640-5643. DOI: 10.1021/ja110720f

[118] Guo P, Chen P, Ma W, Liu M. Morphology-dependent supramolecular photocatalytic performance of porphyrin nanoassemblies: From molecule to artificial supramolecular nanoantenna. Journal of Materials Chemistry. 2012;**22**:20243-20249. DOI: 10.1039/C2JM33253A

[119] Hasobe T, Sakai H, Mase K, Ohkubo K, Fukuzumi S. Remarkable enhancement of photocatalytic hydrogen evolution efficiency utilizing an internal cavity of supramolecular porphyrin hexagonal nanocylinders under visible-light irradiation. The Journal of Physical Chemistry C. 2013;**117**:4441-4449. DOI: 10.1021/jp400381h

[120] Zhong Y, Wang J, Zhang R, Wei W, Wang H, Lu X, et al. Morphology-controlled self-assembly and synthesis of photocatalytic nanocrystals. Nano Letters. 2014;**14**:7175-7179. DOI: 10.1021/nl503761y

[121] Wang J, Zhong Y, Wang L, Zhang N, Cao R, Bian K, et al. Morphology-controlled synthesis and metalation of porphyrin nanoparticles with enhanced photocatalytic performance. Nano Letters. 2016;**16**:6523-6528. DOI: 10.1021/acs.nanolett.6b03135

[122] Mandal S, Nayak SK, Mallampalli S, Patra A. Surfactant-assisted porphyrin based hierarchical nano/micro assemblies and their efficient photocatalytic behavior. ACS Applied Materials & Interfaces. 2013;**6**:130-136. DOI: 10.1021/am403518d

[123] La DD, Rananaware A, Salimimarand M, Bhosale SV. Well–dispersed assembled porphyrin nanorods on graphene for the enhanced photocatalytic performance. ChemistrySelect. 2016;**1**:4430-4434. DOI: 10.1002/slct.201601001

[124] Tian Y, Martin KE, Shelnutt JY-T, Evans L, Busani T, Miller JE, et al. Morphological families of self-assembled porphyrin structures and their photosensitization of hydrogen generation. Chemical Communications. 2011;**47**:6069-6071. DOI: 10.1039/C1CC10868A

[125] Li C, Park K-M, Kim H-J. Ionic assembled hybrid nanoparticle consisting of tin (IV) porphyrin cations and polyoxomolybdate anions, and photocatalytic hydrogen production by its visible light sensitization. Inorganic Chemistry Communications. 2015;**60**:8-11. DOI: 10.1016/j.inoche.2015.07.016

[126] Chen Y, Zhang C, Zhang X, Ou X, Zhang X. One-step growth of organic single-crystal p–n nano-heterojunctions with enhanced visible-light photocatalytic activity. Chemical Communications. 2013;**49**:9200-9202. DOI: 10.1039/C3CC45169K

[127] Maiti NC, Mazumdar S, Periasamy N. J-and H-aggregates of porphyrin–surfactant complexes: Time-resolved fluorescence and other spectroscopic studies. The Journal of Physical Chemistry B. 1998;**102**:1528-1538. DOI: 10.1021/jp9723372

[128] Marciniak H, Li X-Q, Wurthner F, Lochbrunner S. One-dimensional exciton diffusion in perylene bisimide aggregates. The Journal of Physical Chemistry A. 2010;**115**:648-654. DOI: 10.1021/jp107407p

[129] Jurchescu OD, Baas J, Palstra TT. Effect of impurities on the mobility of single crystal pentacene. Applied Physics Letters. 2004;**84**:3061-3063. DOI: 10.1063/1.1704874

[130] Takeya J, Yamagishi M, Tominari Y, Hirahara R, Nakazawa Y, Nishikawa T, et al. Very high-mobility organic single-crystal transistors with in-crystal conduction channels. Applied Physics Letters. 2007;**90**:102120. DOI: 10.1063/1.2711393

[131] Zhong Y, Wang Z, Zhang R, Bai F, Wu H, Haddad R, et al. Interfacial self-assembly driven formation of hierarchically structured nanocrystals with photocatalytic activity. ACS Nano. 2014;**8**:827-833. DOI: 10.1021/nn405492d

[132] Wang Z, Li Z, Medforth CJ, Shelnutt JA. Self-assembly and self-metallization of porphyrin nanosheets. Journal of the American Chemical Society. 2007;**129**:2440-2441. DOI: 10.1021/ja068250o

[133] Huang C-C, Parasuraman PS, Tsai H-C, Jhu J-J, Imae T. Synthesis and characterization of porphyrin–TiO_2 core–shell nanoparticles as visible light photocatalyst. RSC Advances. 2014;**4**:6540-6544. DOI: 10.1039/C3RA45492D

[134] Chen Y, Li A, Jin M, Wang L-N, Huang Z-H. Inorganic nanotube/organic nanoparticle hybrids for enhanced photoelectrochemical properties. Journal of Materials Science & Technology. 2016. DOI: 10.1016/j.jmst.2016.08.030

[135] Chen Y, Huang Z-H, Yue M, Kang F. Integrating porphyrin nanoparticles into a 2D graphene matrix for free-standing nanohybrid films with enhanced visible-light photocatalytic activity. Nanoscale. 2014;**6**:978-985. DOI: 10.1039/C3NR04908F

[136] Guo P, Chen P, Liu M. One-dimensional porphyrin nanoassemblies assisted via graphene oxide: Sheetlike functional surfactant and enhanced photocatalytic behaviors. ACS Applied Materials & Interfaces. 2013;**5**:5336-5345. DOI: 10.1021/am401260n

[137] Wang D, Pan J, Li H, Liu J, Wang Y, Kang L, et al. A pure organic heterostructure of μ-oxo dimeric iron (iii) porphyrin and graphitic-C_3N_4 for solar H_2 roduction from water. Journal of Materials Chemistry A. 2016;**4**:290-296. DOI: 10.1039/C5TA07278F

[138] Ling MM, Bao Z. Thin film deposition, patterning, and printing in organic thin film transistors. Chemistry of Materials. 2004;**16**:4824-4840. DOI: 10.1021/cm0496117

[139] Zhang X, Jie J, Deng W, Shang Q, Wang J, Wang H, et al. Alignment and patterning of ordered small-molecule organic semiconductor micro-/nanocrystals for device applications. Advanced Materials. 2016;28:2475-2503. DOI: 10.1002/adma.201504206

[140] Kira A, Tanaka M, Umeyama T, Matano Y, Yoshimoto N, Zhang Y, et al. Hydrogen-bonding effects on film structure and photoelectrochemical properties of porphyrin and fullerene composites on nanostructured TiO_2 electrodes. The Journal of Physical Chemistry C. 2007;**111**:13618-13626. DOI: 10.1021/jp0726079

[141] Nagai K, Abe T, Kaneyasu Y, Yasuda Y, Kimishima I, Iyoda T, et al. A full-spectrum visible-light-responsive organophotocatalyst film for removal of trimethylamine. ChemSusChem. 2011;**4**:727-730. DOI: 10.1002/cssc.201100064

[142] Abe T, Tobinai S, Taira N, Chiba J, Itoh T, Nagai K. Molecular hydrogen evolution by organic p/n bilayer film of phthalocyanine/fullerene in the entire visible-light energy region. The Journal of Physical Chemistry C. 2011;**115**:7701-7705. DOI: 10.1021/jp1094992

[143] Imahori H, Norieda H, Nishimura Y, Yamazaki I, Higuchi K, Kato N, et al. Chain length effect on the structure and photoelectrochemical properties of self-assembled monolayers of porphyrins on gold electrodes. The Journal of Physical Chemistry B. 2000;**104**:1253-1260. DOI: 10.1021/jp992768f

[144] Imahori H, Yamada H, Ozawa S, Sakata Y, Ushida K. Synthesis and photoelectrochemical properties of a self-assembled monolayer of a ferrocene–porphyrin–fullerene triad on a gold electrode. Chemical Communications. 1999:1165-1166. DOI: 10.1039/A903237A

[145] Imahori H, Fujimoto A, Kang S, Hotta H, Yoshida K, Umeyama T, et al. Molecular photoelectrochemical devices: Supramolecular incorporation of C_{60} molecules into tailored holes on porphyrin-modified gold nanoclusters. Advanced Materials. 2005;**17**:1727-1730. DOI: 10.1002/adma.200401770

[146] Imahori H, Hasobe T, Yamada H, Nishimura Y, Yamazaki I, Fukuzumi S. Concentration effects of porphyrin monolayers on the structure and photoelectrochemical properties of mixed self-assembled monolayers of porphyrin and alkanethiol on gold electrodes. Langmuir. 2001;**17**:4925-4931. DOI: 10.1021/la010006h

[147] Abe T, Tanno Y, Ebina T, Miyakushi S, Nagai K. Enhanced photoanodic output at an organic p/n bilayer in the water phase by means of the formation of whiskered phthalocyanine. ACS Applied Materials & Interfaces. 2013;**5**:1248-1253. DOI: 10.1021/am302209b

[148] Sandanayaka AS, Murakami T, Hasobe T. Preparation and photophysical and photoelectrochemical properties of supramolecular porphyrin nanorods structurally controlled by encapsulated fullerene derivatives. The Journal of Physical Chemistry C. 2009;**113**:18369-18378. DOI: 10.1021/jp9063577

[149] Klauk H, Halik M, Zschieschang U, Eder F, Schmid G, Dehm C. Pentacene organic transistors and ring oscillators on glass and on flexible polymeric substrates. Applied Physics Letters. 2003;**82**:4175-4177. DOI: 10.1063/1.1579870

[150] Gustafsson G, Cao Y, Treacy G, Klavetter F, Colaneri N, Heeger A. Flexible light emitting diode. Nature. 1992;**357**:477

Chapter 7

Isolation of Porphyrins from Heavy Oil Objects

Makhmut R. Yakubov, Guzalia R. Abilova,
Kirill O. Sinyashin, Dmitry V. Milordov,
Elvira G. Tazeeva, Svetlana G. Yakubova,
Dmitry N. Borisov, Pavel I. Gryaznov,
Nikolay A. Mironov and Yulia Y. Borisova

Additional information is available at the end of the chapter

http://dx.doi.org/10.5772/intechopen.68436

Abstract

The chapter describes the opportunities of extracting porphyrins by polar solvents (acetone, N,N-dimethylformamide (DMF), isopropanol, and acetonitrile) and sulfuric acid from various highly molecular petroleum fractions and residues. It has been found that the predissolution of petroleum objects such as asphaltenes and resins in aromatic solvents allows improving the extraction of porphyrins by means of reducing their association with polycondensed heteroatomic structures. Based on the absorption spectra and mass spectra, primary types of porphyrins in obtained extracts were revealed. The distinctions between porphyrin extractions in resins and asphaltenes were revealed. Sulfuric acid extraction allows producing highly concentrated primary extracts of demetallated porphyrins. The share of porphyrin fractions in obtained extractions was 13.0–24.2 wt%, which depends on the concentration of metal porphyrins in initial asphaltenes and resins.

Keywords: vanadyl porphyrins, asphaltenes, resins, extraction, heavy petroleum residues, vanadium

1. Introduction

Metal porphyrins in oils are primarily represented by vanadyl and nickel porphyrins [1]. The remaining metal-containing compounds of vanadium and nickel are present in the form of chelates with pseudo-porphyrin structures or with porphyrins having atypical substitutes [2]. Due to low concentration of porphyrins having atypical structure, efficient separation and mass-spectrometry with ultrahigh definition are required to determine their structure [3], and the structure of pseudo-porphyrin complexes is still not defined. The paper also reports of oil being present in insignificant amounts in iron and copper porphyrins [4].

The analysis of oil porphyrins in oils and their components by various instrumental methods is impossible due to their low concentration, so usually, fractions enriched with metal porphyrins are used for analysis. All concentration methods for oil porphyrins can be conditionally divided into three groups:

- Complex formation, chemical adsorption, and ion-exchanging chromatography when sufficiently labile chemical bonding is formed between various compounds and oil porphyrins.
- Chemical interaction between various compounds and oil components accompanied by new chemically stable compounds being formed.
- Extraction, separation, and sedimentation by solvents based on various solvability of individual compounds in various solvents.

For light oils having a relatively low concentration of asphaltene-resin substances, the first group of methods is mostly suitable for extraction of oil porphyrins: complex formation with Lewis acids usually represented by waterless metal halides [1]. This method consists in the formation of molecular complexes of metal porphyrins with titanium and iron halides that are insoluble in hydrocarbon systems, with further extraction of these complexes, destruction, and regeneration of metal porphyrins. This method is advantageous because of complete extraction of metal porphyrins when they are contained in trace concentrations.

The second group of methods currently used to extract porphyrins from oil is based on treating oils and oil components with strong acids. When using acids in order to extract oil porphyrins, metal porphyrins are demetaled and transformed into acid phase.

As demetaling agents, acids are used, such as sulfuric [5] and sulfonic acids [6]. A disadvantage of extracting oil porphyrins by using acids is partial destruction of porphyrins (40–80%) and inability to separately measure the concentration and study metal porphyrin complexes of vanadyl and nickel. Acid extraction is also inefficient for oils having low concentration of porphyrins. The advantages of porphyrin extraction with acids include low labor input and an opportunity to directly produce relatively clean concentrates of porphyrin compounds.

The third group includes extraction methods by using solvents that cannot be mixed with oil, with further extract treatment. Extraction methods employing selective solvents are advantageous for soft process conditions, which completely present any chemical transformations. Methanol [7] and N,N-dimethylformamide (DMF) [8] are used as solvents for extraction of metal porphyrins.

The concentration of porphyrins in extracts obtained by any of the above methods allows using them for further analysis, but in some cases, additional treatment of concentrates is needed by means of column chromatography. To identify and measure the concentrations of oil porphyrins, visible and UV-band spectroscopy is used due to three characteristics of absorption bands [9]. The most intensive absorption band also referred to as the Soret band is located at the boundary between the visible and the UV area at 400 nm. Other two bands referred to as α and β bands are located in the area of 570 and 535 nm for vanadyl porphyrins and 575 and 540 nm for nickel porphyrins.

As compared with metal porphyrins, electronic absorption spectra of metal-free porphyrin carry much information on the structure. There are four main spectral types of oil porphyrins differing in the absorption peak intensity at 500, 535, 565, and 620 nm (bands IV, III, II, and I, respectively) [10]. This property is associated with the effects of substitutes at the porphyrin ring periphery. Each of the types is represented with a mixture of various porphyrins differing in the nature and position of substitution (**Figure 1**).

Another informative method to determine the structure of metal porphyrins is mass-spectrometry [11]. Metal porphyrins are present in oil in the form of a continuous series. Most common of them are porphyrins with alkyl substitutes called etioporphyrins (ETIO) with the molecular weight of 375 + 12n (M), and metal porphyrins containing an isocyclic ring, also called deoxophylloerythroetio porphyrins (DPEP) with the molecular weight of 373 + 12n (M-2) [12]. Their ratio in oil is the most important geochemical parameter showing the maturity of oil [13]. Other series (M-2, M-4 …) are present in oil in significantly lower concentrations and are called minor.

To study the concentrates of oil metal porphyrins other methods can be used, such as EPR spectroscopy [14, 15], LDI mass-spectrometry [16], Fourier transform ion cyclotron resonance mass spectrometry [17], and high definition inductively coupled plasma mass spectrometry [18].

Metal porphyrins in oils have a negative effect on catalysts of oil refining processes [19, 20], so currently, oil demetallization methods are used and developed [21–23]. Effective development of methods to remove vanadium and nickel from oil is impossible without having information on the structure and properties of metal porphyrins. Since the complete extraction of oil metal porphyrins from oil objects is complicated due to strong associations with asphaltene molecules, this may cause insufficiently full study of metal porphyrins. This chapter gives a new approach to the extraction of metal porphyrins from asphaltenes and heavy petroleum residues (HPR), which allows increasing the degree of extraction, and it also presents the results obtained in determining the composition and properties of concentrates obtained.

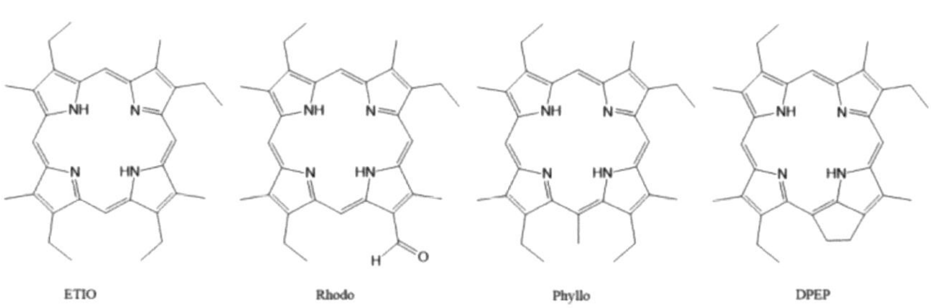

Figure 1. Spectral types of porphyrins.

2. Materials and methods

The study objects were the oil and natural bitumens of various-age deposits of Tatarstan fields and heavy residues of oil refining: vacuum residue (VR) from TAIF-NK OJSC oil refinery; asphalt (A-1) after tar propane deasphalting at the ANK Bashneft OJSC refinery; asphalt (A-2) after tar propane deasphalting at the NK Rosneft refinery; atmospheric residues of heavy oils from the Ashalchinskoe (AR-1) and Zyuzeyevskoe (AR-2) fields.

All oils and natural bitumens were separated from emulsion water and mechanical impurities by centrifugal process. Organic solvents of hch and chda classes were additionally treated and desiccated by employing widely known methods.

Asphaltenes were extracted by using a common methodology by means of diluting with 40× hexane excess with further flushing to remove sedimented resins and oils in a Soxhlet apparatus. Oils and resins were separated by means of column chromatography, using an activated granulated large-pore silica gel as an immobile phase, with the grain size of 0.2–0.5 mm, as well as the hexane/benzene mixture at 85:15 as an eluent.

Vanadium and nickel concentrations in oils and asphaltenes were measured by means of direct flame atomic absorption spectrometry using AAS-1N spectrophotometer, with approved standard samples of metal concentration in oil products used as blank solutions.

To extract asphaltenes by a boiling solvent, a 1 g sample of asphaltenes was placed into a round-bottom flask with back flow condenser, 200 mL of extractant was added, and the mixture was boiled for 1 h. After cool-down, the mixture was filtered. The resulted extract was dried in vacuum. To extract asphaltenes by sedimentation extraction, a 1 g sample of asphaltenes was diluted in 10 mL of benzene. A total volume of 100 mL of extractant was added to the resulting solution, which was then boiled with a back flow condenser for 10 min. After the solution cooldown, asphaltenes were filtered. The resulting extract was dried in vacuum.

The concentration of vanadyl porphyrins in extracts from asphaltenes was calculated for the absorption band of 575 nm according to the following formula:

$$C_{vp} = 0.187h \cdot V/m \cdot l \quad (1)$$

where 0.187 is the conversion factor describing the medium absorption; h is the height of absorption α-band maximum for 575 ± 5 nm; m is the extract sample, g; V is the porphyrin extract volume to be reached, mL; l is the flask thickness, cm.

Matrix-assisted laser desorption/ionization (MALDI) mass spectra of extracts from asphaltenes were obtained by UltraFlex III TOF/TOF mass-spectrometer in linear mode. The data were processed by using FlexAnalysis 3.0 software. The sample was ionized by nitrogen laser radiation (wave length of 337 nm) with the energy of 19 eV. Positively charged ions were recorded. A metallic target was used. Sinapinic acid was used as a matrix. Molecular ions of VPs of various homotypes are presented as peaks with the weight of 373 + 14n amu and 375 + 14n amu (where n is the number of methylene groups in pendent groups). The share of each homotype was calculated by means of internal normalization by using the peak intensity of molecular ions.

Primary porphyrinic extracts were obtained from heavy petroleum residues solutions in benzene with the concentration of 10 wt% by treating with 10× excess of concentrated sulfuric acid. The extract was separated by filtering in a Shott funnel and after being neutralized with sodium hydrate water solution, it was retreated with three portions of tetrachloromethane. The lower layer was separated and dried above anhydrous sodium sulfate. After dried solution filtering and solvent stripping, the primary porphyrinic extract was obtained.

Adsorption-chromatographic separation of primary porphyrinic extracts was carried out in a glass column 1:100 cm by using an activated granulated large-port silica gel as an immobile phase and the 0.5% isopropil alcohol solution in benzene as an eluent. The volume of the eluent sampled was 10 mL. The obtained fractions were combined according to eluent colors. Fraction spectra were obtained in Specord UV-VIS spectrophotometer with the range from 400 to 650 nm.

3. Results and discussion

3.1. Vanadyl porphyrin extraction with polar solvents

The schemes applied to extract metal porphyrin complexes from oil objects have some disadvantages that do not allow using them on a large scale. When using liquid extraction, multiple extraction is needed to achieve high degree of extraction. If asphaltenes are used as extraction objects, the number of extraction steps is multiply increased, therefore, frequently the process is performed in Soxhlet apparatuses. Such extraction conditions are explained by the fact that in oil systems, asphaltenes are associated with oils and resins due to multiple intermolecular interactions, such as electrostatic, dipole, and dispersive ones. The molecular weight of resulted aggregates can be 10,000 amu and more. Metal porphyrins tend to establish strong associations with aggregates of such weight. Metal porphyrins can also be captured into the grid of asphaltenes by the type of molecular sieves. In this connection, it becomes difficult to extract metal porphyrin complexes. It is only possible to obtain concentrates enriched with these compounds. Previously, asphaltene extraction by various solvents was used to obtain metal porphyrin concentrates [24]. It occurred that no more than 60% of the total amount of vanadyl porphyrins is extracted during asphaltene extraction. The extraction process takes place at the solvent boiling temperature in order to increase the extractant solvability. This allows increasing the extraction degree of metal porphyrins, but it also contaminates the extract with highly molecular heteroatomic components.

To increase the extraction degree of metal porphyrins, we proposed using sedimentation extraction. Unlike the direct extraction from asphaltene, this approach suggests preliminary dissolution of asphaltene in a small amount of good solvent. Asphaltene dissolution allows the partial destroying of the intermolecular links inside aggregates that results in some metal porphyrins being liberated. This is followed by metal porphyrin extraction by the solvent excess having low solvability toward asphaltenes, but good solvability toward metal porphyrins. As a result, asphaltenes are sedimented, and metal porphyrins remain in the solution. Polar nonaromatic solvents have low solvability toward asphaltene-tarry components of oil and high solvability metal porphyrins.

To assess the efficiency of metal porphyrin extraction from an asphaltene solution by using sedimentation extraction, the obtained results were compared with the results of standard extraction of metal porphyrins from dry asphaltenes. Benzene was used as an asphaltene solvent; four solvents belonging to different classes of organic compounds were used as polar solvents: acetone, DMF, isopropanol, and acetonitrile.

As an extraction object, asphaltenes of the Ashalchnskoe field oil were used. The extracts obtained in the form of toluene solutions were spectrophotometered within the range of 400–630 nm. Absorption spectra show clear absorption bands at 530 (α-band) and 575 nm (β-band) typical of vanadyl porphyrins (**Figure 2**). A 550 nm band belonging to nickel porphyrins is not recorded in spectra.

When asphaltenes are treated with a boiling solvent, the maximum extract yield is obtained when using isopropanol (**Figure 3**). When using acetone and DMF, extract yields are lower, and in case of acetonitrile, no extract is formed. When using sedimentary extraction from asphaltene solutions, the maximum extract yield occurs with DMF. The sedimentary extraction used for all solvents under study allows reaching higher extract yields and concentrations of vanadyl porphyrins in them as compared to boiling solvent extraction (**Figure 4**).

The most common types of metal porphyrins in oil are etio- and desoxophylloerytroetioporphyrins (DPEP). To assess their ratio in the DMF extract from oil asphaltenes, matrix-assisted laser desorption/ionization (MALDI) is used. (Etio-type porphyrins have the molecular weight of 375 + 14n, and DPEP-type porphyrins – 375 + 14n.) Based on the intensity of peaks, the ratio of Σetio/ΣDPEP equaled 0.73.

MALDI can be used to assess the substitution nature in the porphyrinic ring (**Figure 5**). In both series of porphyrins, alkyl substitutes on the ring periphery contain 6–19 atoms of carbon. Porphyrins of etio-series contain C26–C39 homotypes with the maximum concentration at m/z = 529, which corresponds to C31 homotype that contains alkyl substitutes with

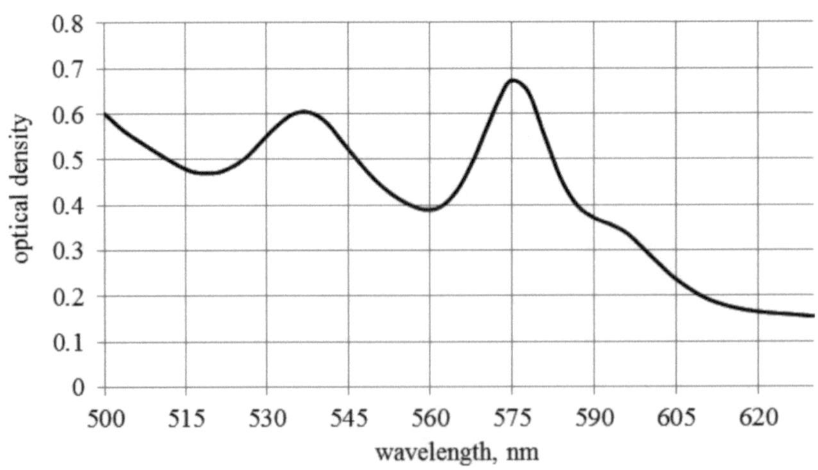

Figure 2. Visible absorption spectrum of asphaltene extract.

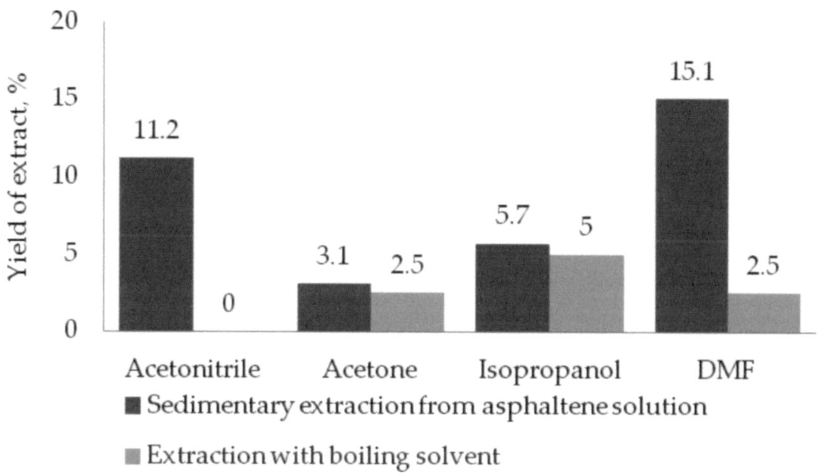

Figure 3. Yield of vanadyl porphyrin extracts from asphaltenes.

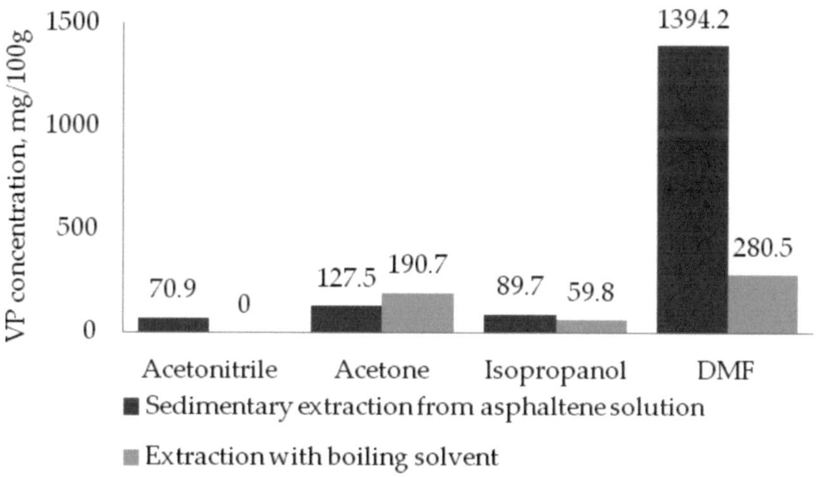

Figure 4. Concentration of vanadyl porphyrin in asphaltenes extracts.

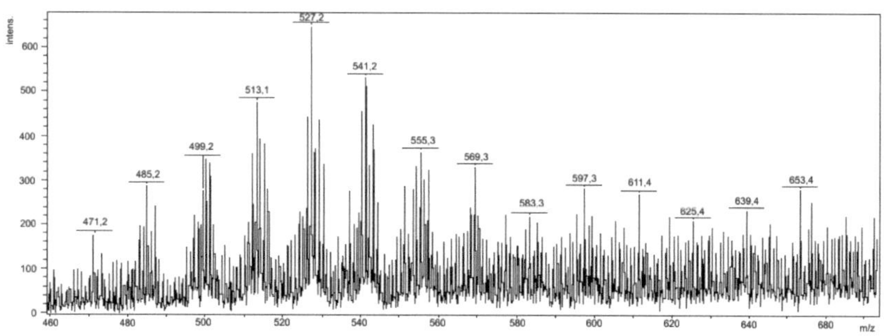

Figure 5. MALDI mass spectrum of vanadyl porphyrin extract from asphaltenes.

high number of carbon 11 atoms. Porphyrins of DPEP-series contain C28–C41 homotypes with the maximum concentration at m/z = 529, which corresponds to C31 homotype that contains alkyl substitutes with high number of carbon 9 atoms.

3.2. Porphyrin extraction with sulfuric acids from asphaltenes and resins

Another methodological approach to the extraction of porphyrin complexes from oils and their components is acid extraction. Extraction methods described in the literature are adaptable and cannot be used for large-scale extraction. The extraction process also takes much time. Furthermore, all the above methodologies were developed for light oils with a low concentration of asphaltene-tarry substances. Meal porphyrin acid extraction from asphaltene-tarry substances is almost not described.

At the first stage, it seems necessary to reveal the most efficient acid extractant with the maximum yield of porphyrin extract. Extraction conditions may have a heavy effect on the results; first of all, this refers to temperature and duration. For a preliminary assessment of acid extraction capabilities, heavy oil asphaltenes from the Zyuzeyevskoe field were used. Concentrated hydrochloric, phosphorous, and sulfuric acids were used as extractants. As in case of polar solvent extraction, a 10% solution in benzene was used to reduce the association of metal porphyrins with asphaltenes, and the process itself was maintained at the room temperature. For phosphorous and hydrochloric acids, asphaltenes do not develop into the acid phase. When treating asphaltene solutions with sulfuric acids, an extractant and a benzene-insoluble residue is formed. A difference from the method currently applied to produce metal-free porphyrin from asphaltenes consists in the fact that demetaling of metal porphyrins occurs simultaneously with their extraction from asphaltenes. The need for preliminary extraction of metal porphyrins is avoided. Due to slurry formation, a centrifugal process and further filtering in a Schott's funnel were used to segregate the extract and the residue. The obtained extract was neutralized with a 20% cooled-down solution of sodium hydrate until reaching a neutral reaction. Tetrachloromethane was extracted from the resulting water solution. The yield of the primary extract after solvent stripping was 9%.

There are no metal porphyrin bands of 530 and 575 nm in the absorption spectrum in the visible area for the primary extract, and there are bands typical of free porphyrin bases (**Figure 6**). This testifies that when sulfuric acid acts on metal porphyrins, they are demetaled and metal-free porphyrins are formed. In this manner, concentrated sulfuric acid is the most optimal extractant to extract and demetaleted porphyrin from oil asphaltenes and resins.

To define the composition of porphyrin extracts in case of sulfuric acid extraction, asphaltenes and resins of heavy oils from Ashalchinskoe (TN-1 asphaltenes and resins) and Zyuzeyevskoe fields (TN-2 asphaltenes and resins) were used. Extraction was carried out according to the scheme described above. In case of sulfuric acid extraction, as for asphaltene extraction, an extract of porphyrins and an insoluble residue are formed. The results obtained for the yield of extracts are summarized in **Table 1**.

The amount of porphyrin extracted from resins or asphaltenes of heavy oils with increased vanadium concentration varies within 7.9–13.0 wt%. If oils are compared individually, the

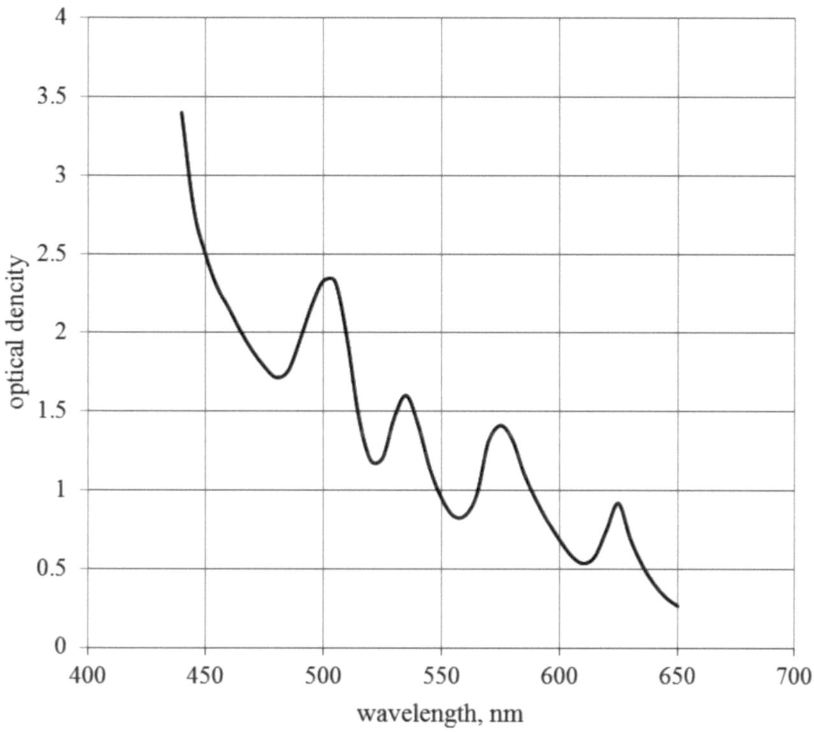

Figure 6. Visible absorption spectrum of sulfuric acid extract of asphaltenes from Zyuzeyevskoe oil field.

porphyrin extract yield from resins as compared to that from asphaltenes is 3–4 wt% higher in both cases. When the vanadium concentration both in resins and asphaltenes increases, the yield of extracts is also increased. If the vanadium concentration in asphaltenes differs by two times, the relative extract yield increase will be about 14% just as in resins where the vanadium concentration difference is even higher (3.33 times), and the extract yield is increased by relative 15% only. In this manner, for asphaltenes and resins where the vanadium concentration will exceed the values as compared to the objects under study, it is unlikely that the porphyrin extract yield will be significantly increased.

To obtain data for the composition and types of porphyrins in extracts, silica gel adsorption chromatographic separation was used. A mixture of 0.5% isopropyl alcohol and 99.5% of

	Yield of extract, %	V content in object of extraction, %
Asphaltenes TN-1	7.9	0.186
Asphaltenes TN-2	9.0	0.384
Resins TN-1	11.3	0.042
Resins TN-2	13.0	0.140

Table 1. Yield of sulfuric acid extracts.

benzene was applied as eluent [25]. A total volume of 10 mL of liquid was sampled during elution. To decrease the number of fractions analyzed against the absorption spectra in the visible band, the obtained solutions were combined visually according to the color change (**Table 2**). A further study of absorption spectra in the visible band confirmed that this approach can be applied, since the differences in spectra allow identifying the types of porphyrins. To all colored fractions, except for the first and the last one, there are four absorption bands registered, having various intensity at 620, 565, 535, and 500 nm (bands I–IV), according to which a specific spectral type can be assigned to porphyrins.

The first (oil-like) and the last (resin-like) colored fractions obtained after separation of asphaltene extracts do not show absorption bands of metal-free porphyrin. In resin extract separation, the resin-like fraction is the first to eluted, followed by the oil-like fraction. These fractions being present in the sulfuric acid extracts are related to occluded oils and resins in asphaltenes.

The data for the yield of fractions after chromatographic separation of the sulfuric acid extract of asphaltenes are given in **Table 3**. The total concentration of oil-like and resin-like fractions in the extract reaches 53.2 wt%. Some part of the extract is not eluted and remains on the silica gel. The concentration of porphyrin fractions in extracts of resins is higher than in those from asphaltenes.

Phyllo-type porphyrins are predominant in asphaltene extracts. Apart from etio- and phyllo-porphyrins, there are also rhodo- and DPEP found in asphaltene extracts. Reduced concentration of DPEP in asphaltene extracts as compared to solvent extraction allows suggesting the isocyclic ring destruction in the porphyrins of this type during acid extraction.

Unlike asphaltenes, the porphyrins of resin extracts contain only these etio- and phyllo-types, which is the primary difference in the composition of porphyrin extracts of resins and asphaltenes. Etio-type of porphyrins is predominant in resin extracts. No rhodo-type porphyrins contained in resins is probably related to co-sedimentation of porphyrins with asphaltenes during their extraction due to polar groups presented in them and, consequently, lower solubility in hexane. As for asphaltenes, the most probable reason for no DPEP in extracts from resins can be the isocyclic ring destruction during sulfuric acid extraction.

Fraction number	Solution color	The order of decreasing intensity of the absorption bands	Spectral type of porphyrins
1	Cherry	III→IV→II→I	Rhodo
2	Dark orange	IV→I→II→III	DPEP
3	Dark red	IV→II→II→I	ETIO
4	Red	IV→II→III→I	Phyllo

Table 2. Spectral types of petroleum demetalleted porphyrins.

	Yield of fraction, %			
	Asphaltenes TN-1	Asphaltenes TN-2	Resins TN-1	Resins TN-2
Rhodo + DPEP	2.3	4.2	0	0
ETIO	2.7	7.1	6.8	11
ETIO + Phyllo	1.9	2.9	5.7	7.2
Phyllo	6.1	10.0	3.5	4.4
Oil-like	41.0	46.4	57.5	54.0
Resin-like	12.2	7.0	17.9	15.5
Residue on a column	33.8	22.4	8.6	7.9

Table 3. Yield of fractions after column chromatography of asphaltenes and resins porphyrinic extracts.

3.3. Porphyrin extraction with sulfuric acids from the solution of heavy petroleum residues

Since the extraction of individual resins and asphaltenes is a hard task, their industrial concentrates are used for porphyrin extraction—residual products of oil refining. First of all, these are the vacuum residue (tar) and asphalts from tar deasphaltizing with the total concentration of asphaltenes and resins being 50–70 wt% depending on the initial oil composition. For the vanadium concentration of 200–500 ppm in some heavy oils of Tatarstan, Samara, and Ulyanovsk region fields, the total concentration of vanadium and nickel in residual products of oil refining will be 1000 ppm and more.

As the objects of study, heavy petroleum residues of existing productions and atmospheric residues obtained in laboratory conditions (>350°C) from heavy oils with increased vanadium concentration were used. Density, component composition, and vanadium and nickel concentration were measured for all heavy petroleum residues (HPR) (**Table 4**).

Determining the concentration of these metals allows the preliminary assessing of the concentration of metal porphyrins in initial objects. The vanadium concentration in A1 and AR-2 is 9.2–9.3 times higher than that of nickel. V/Ni is also 9.8 times higher for A-2, but for

HPR	ρ, g/cm^3	Content, %			Metals, ppm	
		Hydrocarbones	Resins	Asphaltenes	V	Ni
VR	1.0035	41.7	46.8	11.5	280	58
A-1	1.0762	45.6	40.7	13.7	310	63
A-2	1.1113	24.9	49.1	26.0	470	48
AR-1	1.0075	44.4	47.0	8.4	330	36
AR-2	1.0085	40.3	48.2	11.5	970	104

Table 4. Density and composition of HPR.

VR and A-1, the vanadium concentration is about five times higher than the nickel concentration. In this way, a potential share of nickel porphyrins for the selected HPRs can be about 10–20% of the vanadium porphyrin concentration.

During extraction, the most part of HPRs is transformed into an insoluble finely divided black product. As a result of sulfuric acid exposure, the composition is greatly altered primarily because of newly formed sulfonic and sulfoxide groups, which is represented by the absorption growth in the area of 1030 cm^{-1} and 1200–1400 cm^{-1} in IR bands as compared to initial objects.

The yield of primary porphyrin extracts for the selected HPRs varies within 6.9–12.9% equivalent to the weight of the initial oil stock (**Table 5**). The maximum extract yield is found for AR-2 where the vanadium and nickel concentration are also maximal. For AR-1, rather high yield of the primary extract is also found; however, the vanadium and nickel concentrations are relatively low. Thus, the total concentration of vanadium and nickel in HPRs is no determinant for forecasting the yield of porphyrin extracts in case of sulfuric acid extraction.

Based on absorption spectra analysis in the visible band, there are metal-free porphyrins found in all obtained extracts with simultaneous dissipation of characteristic absorption bands of vanadyl and nickel porphyrins, which testify demetallization of metal porphyrins during extraction. To characterize porphyrins in obtained extracts, adsorption chromatographic separation was used with further analysis of electronic spectra.

The results of adsorption chromatographic separation of primary porphyrin extracts (**Table 5**) show that about 70–75% include various oil-like and resin-like fractions containing no porphyrins, with some part of them not eluted by the recommended solvent mixture and remaining adsorbed on the silica gel.

A comparative analysis of IR spectra allowed identifying that oil-like and resin-like fractions have the same structural and group composition as resins and oils obtained when analyzing the composition of initial HPRs. The results of chromatographic separation of primary extracts show that the share of porphyrin fractions is 13.0–24.2%. The absorption spectrum analysis

	Yield, %				
	VR	**A-1**	**A-2**	**AR-1**	**AR-2**
Primary extract	7.6	6.8	8.5	11.7	13.1
Rhodo + DPEP	3.7	2.7	3.1	2.3	4.2
ETIO	5.9	6.9	5.6	2.7	7.1
ETIO + Phyllo	3.2	7.1	6.5	1.9	2.9
Phyllo	2.3	2.1	1.7	6.1	10.0
Oil-like	50.0	43.6	54.1	49.0	46.4
Resin-like	25.2	27.3	19.3	27.2	18.0
Residue on a column	9.7	10.3	11.2	11.8	11.4

Table 5. Yield of fractions after column chromatography of primary porphyrinic extracts.

in the visible band for all porphyrin fraction shows that there are all four types of porphyrins (**Figure 3**). Phyllo- and etio-porphyrins are predominant. In AR-1 and AR-2 extracts, the share of phyllo-porphyrins is significantly higher. One of the reasons for reduced concentration of DPEP in sulfuric acid extracts can be the isocyclic ring destruction when exposed to sulfuric acid. The total yield of porphyrin fractions for both the primary extract composition and in-equivalent to the initial HPRs prove no unambiguous correlation with the total vanadium and nickel concentration in the initial feed.

Metal-free porphyrins have four absorption bands in electronic spectra whose intensity depends on the type of porphyrins, so it is complicated to precisely measure the concentration of metal-free porphyrin in the obtained fractions. As an indirect method to assess the concentration of porphyrins in the extract, a comparative analysis of vanadium and nickel concentration reduction in HPRs can be used by analyzing the vanadium and nickel concentration in the extract by means of atomic absorption spectroscopy. As a result, it has been found that the sulfuric acid extract composition is extracted from 62.4 to 81.1% of vanadium contained in initial HPRs, with full extraction of nickel. Probably, the vanadium extraction from asphaltenes is incomplete, since the share of vanadium in the extract is inversely proportional to the concentration of asphaltenes in HPRs. Correspondingly, some vanadyl porphyrins remain in the insoluble residue. Since the molecular weight of vanadyl and nickel porphyrins is 10 times higher as compared to the atomic mass of vanadium and nickel, the potential concentration of metal porphyrins in HRP can be assessed, which is approximately evaluated as the total concentration increased by ten times.

A similar level of porphyrin concentration in concentrates cannot be achieved when extracted by such polar solvents as DMF or acetone with further single chromatography. It is especially important that the maximum concentration of porphyrins in obtained concentrates is reached when using heavy petroleum atmospheric residue (AR-2) as a feed, with increased vanadium and nickel concentration and simultaneously minimal ratio of asphaltenes and resins. In perspective, it is possible that such oils can be regarded as a stock to produce cheap natural porphyrins for using primarily as dyes and catalysts.

4. Conclusion

It has been found that extraction of metal porphyrins by polar solvents from a solution of asphaltenes or heavy oil residues in aromatic solvents allows increasing the level of extraction of metal porphyrins into the extract due to the destruction of aggregates of metal porphyrins and asphaltenes as compared to the direct extraction of asphaltenes or residues. In asphaltene extraction by polar solvents, the DMF appeared to be the best extractant. In the DMF extract from asphaltenes of heavy petroleum with increased vanadium concentration, vanadyl asphaltenes of etio- and DPEP-types were primarily found. The maximum distribution of the molecular weight for both types is found in vanadyl porphyrin homotype C31 containing nine carbon atoms in alkyl substitutes.

Using sulfuric acid for porphyrin extraction allows producing highly concentrated extracts of demetalated porphyrins without the need for preliminary concentration. The analysis of porphyrin fractions obtained after chromatographic separation of the primary extract of asphaltenes shows predominantly etio- and phyllo-types of porphyrins and smaller concentrations of rhodo- and DPEP-types. Only etio- and phyllo-types of porphyrins were found in resin extracts. In asphaltene extraction, it was found that less than 75% of vanadium is extracted to the extract. In this way, some part of vanadyl porphyrins rigidly bound by a σ-linkage with polycondensed structures of asphaltenes is not extracted by using extraction methods which may be a restriction for petroleum demetallization processes. The concentration of porphyrins when switching from clean asphaltenes and resins to heavy residues is almost indiscernible, which allows proving the selectivity of the sulfuric acid extraction method. This is why the sedimentary extraction method can be used in analyzing oil objects in order to obtain additional information for the composition of metal porphyrins.

Acknowledgements

This study was supported by the Russian Science Foundation (Project No. 15-13-00139).

Author details

Makhmut R. Yakubov*, Guzalia R. Abilova, Kirill O. Sinyashin, Dmitry V. Milordov, Elvira G. Tazeeva, Svetlana G. Yakubova, Dmitry N. Borisov, Pavel I. Gryaznov, Nikolay A. Mironov and Yulia Y. Borisova

*Address all correspondence to: yakubovmr@mail.ru

A.E. Arbuzov Institute of Organic and Physical Chemistry, Kazan Scientific Center, Russian Academy of Sciences, Kazan, Russian Federation

References

[1] Dechaine G, Gray M. Chemistry and association of vanadium compounds in heavy oils and bitumen, and implications for their selective removal. Energy and Fuels. 2010;**24**:2795-2808. DOI: 10.1021/ef100173j

[2] Lewan M, Maynard J. Factors controlling enrichment of vanadium and nickel in the bitumen of organic sedimentary rocks. Geochimica et Cosmochimica Acta. 1982;**46**:2547-2559. DOI: 10.1016/0016-7037(82)90377-5

[3] Zhao X, Shi Q, Gray M, Xu C. New vanadium compounds in Venezuela heavy crude oil detected by positive-ion electrospray ionization Fourier transform ion cyclotron resonance mass spectrometry. Nature. 2014;**4**:1-6. DOI: 10.1038/srep05373

[4] Speight J. The Chemistry and Technology of Petroleum. New York: Marcel Dekker; 1980. 957 p

[5] Chu X, Shen M, Li B. Effect and mechanism research of removing nickel and vanadium porphyrins from model oil by chemical agent. Petroleum Processing and Petrochemicals. 2010;**41**:19-22

[6] Ali M, Perzanowski H, Bukhari A, Al-Haji A. Nickel and vanadyl porphyrins in Saudi Arabian crude oils. Energy and Fuels. 1993;**7**:179-184. DOI: 10.1021/ef00038a003

[7] Yin C, Stryker J, Gray M. Separation of petroporphyrins from asphaltenes by chemical modification and selective affinity chromatography. Energy and Fuels. 2009;**23**:2600-2605. DOI: 10.1021/ef801059y

[8] Liu T, Lu J, Zhao X, Zhou Y, Wei Q, Xu C, Shi Q. Distribution of vanadium compounds in petroleum vacuum residuum and their transformations in hydrodemetallization. Energy and Fuels. 2015;**29**:2089-2096. DOI: 10.1021/ef502352q

[9] Freeman D, Saint Martin D, Boreham C. Identification of metalloporphyrins by third-derivative UV/Vis diode array spectroscopy. Energy and Fuels. 1993;**7**:194-199

[10] Smith K. General features of the structure and chemistry of porphyrin compounds. In: Smith K, editor. Porphyrins and Metalloporphyrins. New York: Elsevier Scientific Publishing Company; 1975. pp. 1-28

[11] Zhao X, Liu Y, Xu C, Yan Y, Zhang Y, Zhang Q, Shi Q. Separation and characterization of vanadyl porphyrins in Venezuela Orinoco heavy crude oil. Energy and Fuels. 2013;**27**:2874-2882. DOI: 10.1021/ef400161p

[12] Baker E, Palmer S. Geochemistry of porphyrins. In: Dolphi D, editor. The Porphyrins. New York: Academic Press; 1978. pp. 486-552

[13] Barwise A. Metal complexes in fossil fuels. In: Filby R, Branthaver J, editors. ACS Symposium Series. Vol. 344. American Chemical Society: Washington, DC: 1987. pp. 100-109

[14] Gilinskaya L. EPR spectra of V(IV) complexes and the structure of oil porphyrins. Journal of Structural Chemistry. 2008;**49**:245-254. DOI: 10.1007/s10947-008-0120-6

[15] Trukhan S, Yudanov V, Gabrienko A, Subramani V, Kazarian S, Martyanov O. In situ electron spin resonance study of molecular dynamics of asphaltenes at elevated temperature and pressure. Energy and Fuels. 2014;**28**:6315-6321. DOI: 10.1021/ef5015549

[16] Xu H, Yu D, Que G. Characterization of petroporphyrins in Gudao residue by ultraviolet visible spectrophotometry and laser desorption ionization-time of flight mass spectrometry. Fuel. 2005;**84**:647-652. DOI: 10.1016/j.fuel.2004.06.034

[17] Liu H, Mu J, Wang Z, Ji S, Shi Q, Guo A, Che K, Lu J. Characterization of vanadyl and nickel porphyrins enriched from heavy residues by positive-ion electrospray ionization FT-ICR mass spectrometry. Energy and Fuels. 2015;**29**:4803-4813. DOI: 10.1021/acs.energyfuels.5b00763

[18] Qian K, Mennito A, Edwards K, Ferrughelli D. Observation of vanadyl porphyrins and sulfur-containing vanadyl porphyrins in a petroleum asphaltene by atmospheric pressure photonionization Fourier transform ion cyclotron resonance mass spectrometry. Rapid Communications in Mass Spectrometry. 2008;**22**:2153-2160. DOI: 10.1002/rcm.3600

[19] Vogelaar B, Eijsbouts S, Bergwerff J, Heiszwolf J. Hydroprocessing catalyst deactivation in commercial practice. Catalysis Today. 2010;**154**:256-263. DOI: 10.1016/j.cattod.2010.03.039

[20] Furimsky E, Massoth F. Deactivation of hydroprocessing catalysts. Catalysis Today. 1999;**52**:381-495

[21] Abdrabo A, Husein M. Method for converting demetallization products into dispersed metal oxide nanoparticles in heavy oil. Energy and Fuels. 2012;**26**:810-815. DOI: 10.1021/ef201819j

[22] Shang H, Liu Y, Shi J, Shi Q, Zhang W. Microwave-assisted nickel and vanadium removal from crude oil. Fuel Processing Technology. 2016;**142**:250-257. DOI: 10.1016/j.fuproc.2015.09.033

[23] Shiraishi Y, Hirai T, Komasawa I. Novel demetalation process for vanadyl and nickel porphyrins from petroleum residue by photochemical reaction and liquid - liquid extraction. Industrial & Engineering Chemistry Research 2000;**39**:1345-1355. DOI: 10.1021/ie990809o

[24] Milordov D, Usmanova G, Yakubova S, Yakubov M, Romanov G. Comparative analysis of extractive methods of porphyrin separation from heavy oil asphaltenes. Chemistry and Technology of Fuels and Oils. 2013;**3**:232-238. DOI: 10.1007/s10553-013-0435-7

[25] Yakubov M, Milordov D, Yakubova S, Borisov D, Gryaznov G, Usmanova G. Sulfuric acid assisted extraction and fractionation of porphyrins from heavy petroleum residuals with a high content of vanadium and nickel. Petroleum Science and Technology. 2015;**33**:992-998. DOI: 10.1080/10916466.2015.1030078

Control of Fluorescence and Photosensitized Singlet Oxygen-Generating Activities of Porphyrins by DNA: Fundamentals for "Theranostics"

Kazutaka Hirakawa

Additional information is available at the end of the chapter

http://dx.doi.org/10.5772/67882

Abstract

The purpose of this chapter is the brief review of the fundamental study of porphyrin "theranostics" by DNA. Porphyrins have been studied as photosensitizer for photodynamic cancer therapy. The activity control of fluorescence emission and photosensitized singlet oxygen generation by porphyrins using the interaction with DNA is the initial step in achieving theranostics. To control these photochemical activities, several types of electron donor–connecting porphyrins were designed and synthesized. The theoretical calculations speculated that the photoexcited state of these porphyrins can be deactivated via intramolecular electron transfer, forming a charge–transfer state. The electrostatic interaction between the cationic porphyrin and DNA predicts a rise in the energy of the charge–transfer state, leading to the inhibition of electron transfer quenching. Pyrene– and anthracene–connecting porphyrins showed almost no fluorescence in an aqueous solution. Furthermore, these porphyrins could not photosensitize singlet oxygen generation. These porphyrins bind to a DNA groove through an electrostatic interaction, resulting in the increase of fluorescence intensity. The photosensitized singlet oxygen–generation activity of DNA–binding porphyrins could also be confirmed. On the other hand, several other porphyrins could not demonstrate the activity control properties. To realize effective activity control, a driving force of more than 0.3 eV is required for the porphyrins.

Keywords: cationic porphyrin, DNA, singlet oxygen, electron transfer, fluorescence

1. Introduction

"Theranostics" [1–3] is a relatively new technical term that includes the meanings of therapeutics and diagnostics [4–9]. The purpose of this review is an introduction of examples of theranostics using porphyrins. Porphyrins can emit relatively strong fluorescence in the wavelength range of visible light and generate singlet oxygen ($^1O_2$), an important reactive oxygen species [10]. Singlet oxygen is generated through energy transfer from the triplet excited (T_1) state of the photosensitizer to the ground state of oxygen molecules ($^3O_2$) [11–13]. Fluorescence imaging is the fundamental mechanism of photodynamic diagnosis (PDD) [14], and $^1O_2$ is the important reactive species for photodynamic therapy (PDT) [15]. PDT is a less-invasive and promising treatment for cancer and other nonmalignant conditions [4–9, 15]. In general, a mechanism of PDT is the oxidation of biomacromolecules, including DNA and proteins, by $^1O_2$, which is generated through energy transfer from the excited photosensitizer to oxygen molecules. Porphyrins have been extensively studied as photosensitizers of PDT. Porfimer sodium [16] and talaporfin sodium [17] are especially important clinical drugs used in PDT (**Figure 1**). The control of the photoexcited state of porphyrins by targeting molecules or surrounding environments is the fundamental mechanism of theranostics. In this chapter, the fundamental studies about DNA-targeting porphyrin theranostics are introduced. DNA is a potentially important target molecule of PDT. Indeed, many DNA-targeting drugs have been studied and reported [18–20].

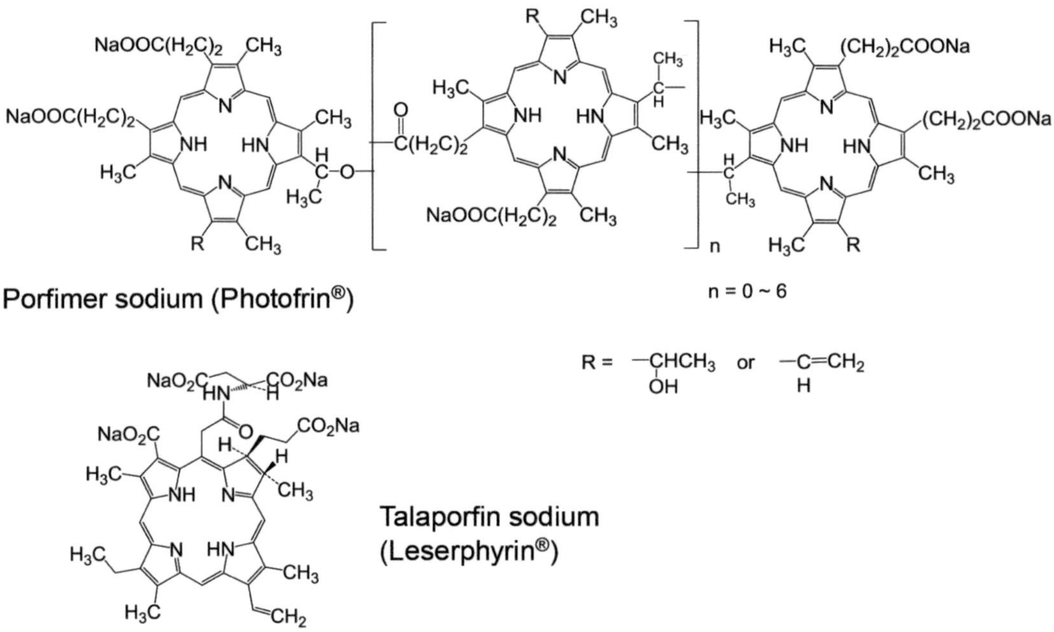

Figure 1. Structures of examples of PDT photosensitizers, porfimer sodium and talaporfin sodium.

http://dx.doi.org/10.5772/67882

1.1. Photodynamic therapy

Photodynamic therapy is a promising and less-invasive treatment for cancer [4-9, 15]. Porphyrins are used as photosensitizers of PDT (**Figure 2**). The abovementioned porphyrins, porfimer sodium [16] and talaporfin sodium [17], are especially important photosensitizers. Under visible light irradiation, especially long wavelength visible light (wavelength > 650 nm), an administered porphyrin photosensitizer generates $^1O_2$ through energy transfer to an oxygen molecule (the type II mechanism) [21]. Since human tissue has a relatively high transparency for visible light, especially red light, visible light rarely demonstrates side effects. Critical targets of $^1O_2$ include mitochondria and enzyme proteins; DNA is also an important target for PDT [22–26]. In general, the $^1\Sigma_g^+$ state of $^1O_2$ ($^1O_2(^1\Sigma_g^+)$) is mainly formed through the energy transfer from the T_1 state of photosensitizers. This state of $^1O_2$ has relatively high energy, about 1.6 eV, corresponding to the ground state; however, the lifetime is very short (several picoseconds). The $^1O_2(^1\Sigma_g^+)$ is rapidly converted to the $^1\Delta_g$ state ($^1O_2(^1\Delta_g)$), which has a relatively long lifetime (several microseconds). Therefore, $^1O_2(^1\Delta_g)$ is a more important reactive oxygen species of PDT. In this chapter, $^1O_2$ indicates $^1O_2(^1\Delta_g)$ without explanation. This biomacromolecule damage induces apoptosis and/or necrosis. Apoptosis, a programed death of cancer cells, is considered the main mechanism of PDT [15, 27]. Necrosis also contributes to the mechanism of cell death in the case of severe damage of biomacromolecules by a high dose of photosensitizers and intense photoirradiation [15]. In the case of DNA-targeting PDT, $^1O_2$ selectively oxidizes guanines. The main oxidized product of guanine is 8-oxo-7,8-dihydrodeoxyguanine [28–30].

1.2. Aminolevulinic acid

One of the most important practical applications of theranostics is the method using the administration of 5-aminolevulinic acid (5-ALA, see **Figure 3**) [31–33]. Although the strategy of 5-ALA theranostics is different from the activity control of the photosensitizer by target molecules mentioned in this chapter, this method is important for cancer theranostics. 5-ALA is the source of protoporphyin IX (PPIX) in human cells. In the normal cell, PPIX is converted into iron porphyrin, which cannot emit fluorescence. However, in cancer cells,

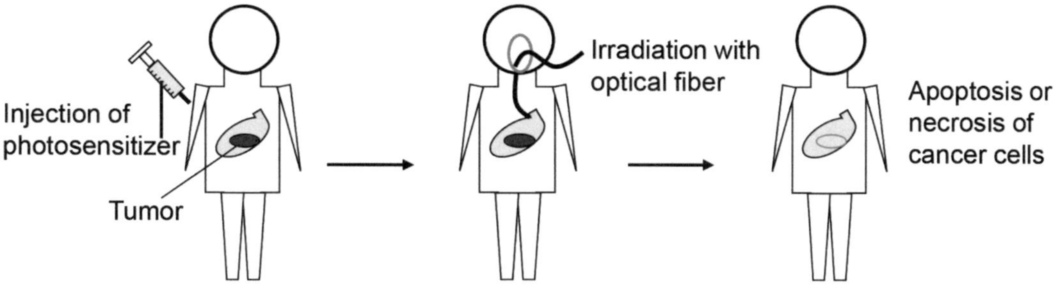

Figure 2. A general procedure of PDT.

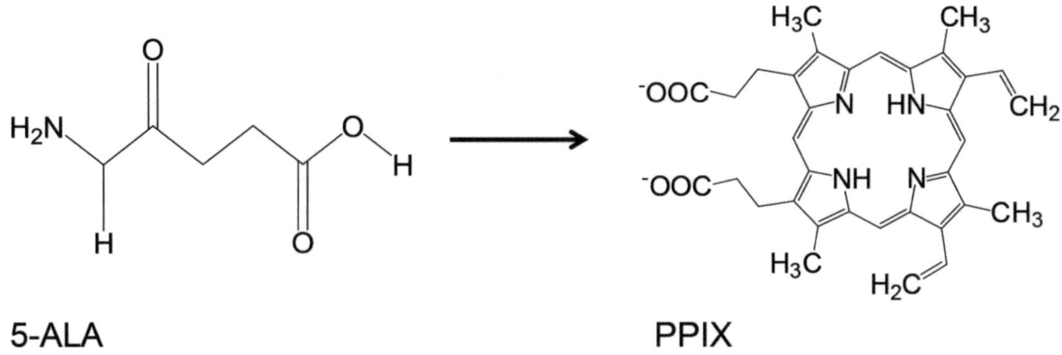

Figure 3. PPIX formation from 5-ALA.

PPIX is selectively concentrated. Several mechanisms for this cancer-selective concentration of PPIX have been speculated [34, 35]. Because PPIX demonstrates relatively strong red fluorescence around 650 nm and under blue light irradiation around 450 nm, this phenomenon can be applied to cancer diagnosis. Indeed, the diagnosis of 5-ALA is clinically applied to the treatment of cancer, for example, malignant brain tumors [36, 37] and bladder cancer [38]. Furthermore, PPIX can photosensitize $^1O_2$ generation. Although the efficiency of $^1O_2$ generation by free PPIX is relatively low, the $^1O_2$-generating activity of PPIX can be increased depending on the environment [39]. These properties of 5-ALA and PPIX can be used in cancer theranostics.

1.3. Strategy of porphyrin theranostics with target biomolecules

Figure 4 shows the energy diagram of the relaxation process of photoexcited porphyrins and theranostics. The singlet excited (S_1) state of the photosensitizer (Sens*(S_1)) is formed by photoirradiation. In the OFF state, without the target biomacromolecules, the S_1 state is rapidly quenched, and the excitation energy is dispersed as heat. For example, intramolecular

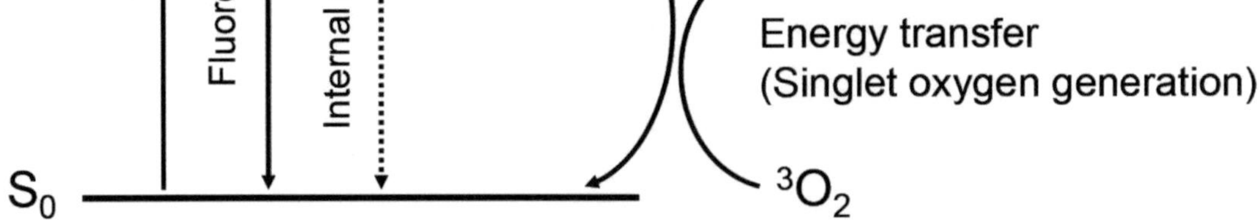

Figure 4. An energy diagram of the relaxation process of photoexcited porphyrin.

electron transfer is a convenient pathway for the quenching to control photochemical activity. In the presence of target molecules, the interaction between the photosensitizer (Sens) and the target molecule inhibits the intramolecular electron transfer. The S_1 state with target molecules can emit fluorescence (ON state). In the case of porphyrin, the quantum yield of fluorescence (Φ_f) is almost 10% for a relatively intense case. In addition, the intersystem crossing proceeds with a relatively large quantum yield (Φ_T); more than 50% is a sufficient value for the Φ_T. These processes are expressed by the following equations:

$$\text{Sens} + h\nu \rightarrow \text{Sens}^*(\text{S}_1) \tag{1}$$

$$\text{Sens}^*(\text{S}_1) \rightarrow \text{Sens} + \text{heat (Activity: OFF)} \tag{2}$$

$$\text{Sens}^*(\text{S}_1) \rightarrow \text{Sens} + h\nu\ \text{(Activity: ON)} \tag{3}$$

$$\text{Sens}^*(\text{S}_1) \rightarrow \text{Sens}^*(\text{T}_1)\ \text{(Activity: ON)} \tag{4}$$

$$\text{Sens}^*(\text{T}_1) + {}^3\text{O}_2 \rightarrow {}^1\text{O}_2\ \text{(Activity: ON)} \tag{5}$$

where Sens*(T_1) is the T_1 state of the photosensitizer. **Figure 5** shows the scheme of the activity control of photosensitizer by DNA. In the case of DNA, several forms of the binding interaction can be speculated [40–43]. For example, an electrostatic interaction can switch the activity of photosensitizers.

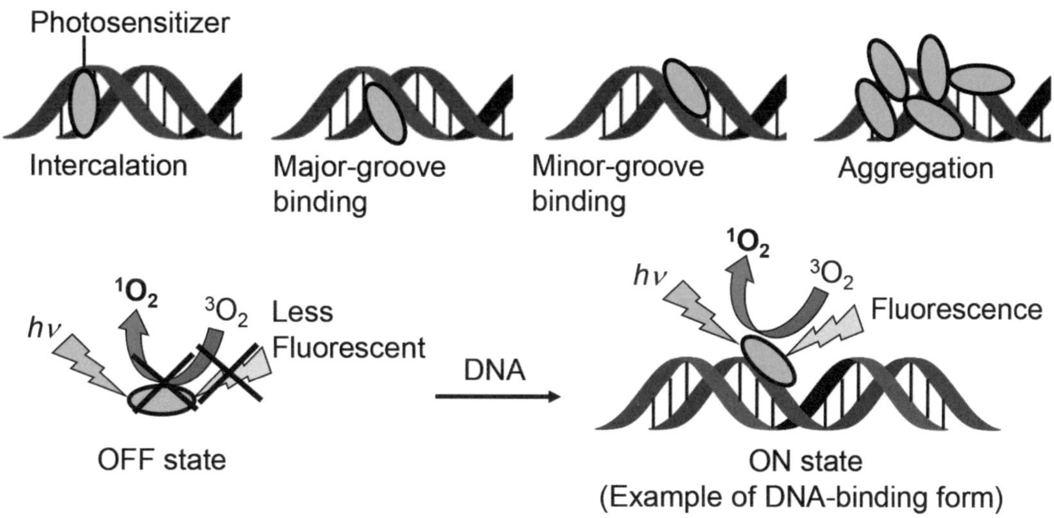

Figure 5. Scheme of the binding interaction between photosensitizers and DNA and the activity switching of photosensitizers through the interaction with DNA.

2. Control of fluorescence and $^1O_2$-generating activity of alkaloids by DNA

Photosensitized DNA damage is an important process in medical applications of photochemical reactions [44, 45]. In this section, the activity control of naturally occurring photosensitizers

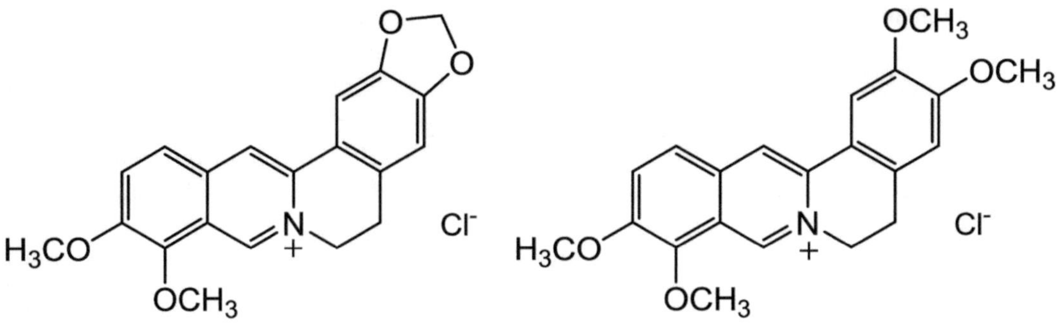

Figure 6. Structures of berberine (left) and palmatine (right).

is introduced. Berberine and palmatine are alkaloids (**Figure 6**). These molecules barely emit fluorescence. The S_1 state of these alkaloids deactivates within 40~50 ps through intramolecular electron transfer in aqueous solution [46–48]. Since these alkaloids are cationic compounds, in the presence of DNA, an anionic polymer, berberine and palmatine spontaneously bind to the DNA strand through electrostatic interaction. Indeed, it was reported that berberine preferentially binds to adenine-thymine–rich minor grooves [49]. The minor groove bindings of berberine and palmatine could be speculated from molecular mechanics calculation [48]. The interaction between these alkaloids and DNA was investigated using oligonucleotides of the adenine-thymine sequence (AATT: $d(AAAATTTTAAAATTTT)_2$) and the guanine-containing sequence (AGTC: $d(AAGCTTTGCAAAGCTT)_2$) [48]. The apparent binding constant can be easily estimated from the absorption spectral change of these alkaloids, and the reported values are relatively high [48]. The fluorescence intensity of berberine and palmatine was markedly increased in the presence of DNA. The Φ_f and the fluorescence lifetimes (τ_f) of berberine and palmatine were markedly increased through interaction with DNA (**Table 1**).

Furthermore, the $^1O_2$-generation activity of berberine and palmatine was markedly enhanced by DNA. In aqueous solution, berberine and palmatine hardly photosensitize $^1O_2$ generation.

Alkaloid	DNA	Φ_f	τ_f/ns (ratio)			Φ_Δ
Berberine	Without	<0.001	0.05			nd
	AATT	0.093	0.30 (0.30)	3.7 (0.42)	11.9 (0.28)	0.066
	AGTC	0.043	0.12 (0.60)	1.6 (0.32)	8.0 (0.08)	0.036
Palmatine	Without	<0.001	0.04			nd
	AATT	0.054	0.16 (0.39)	2.3 (0.45)	6.9 (0.16)	0.044
	AGTC	0.031	0.14 (0.54)	1.4 (0.37)	5.9 (0.09)	0.030

The fluorescence properties were examined in a 10-mM sodium phosphate buffer (pH = 7.6). The Φ_Δ values were determined in deuterium oxide. These values were reported in the literature [48].

Table 1. Fluorescence and photosensitized $^1O_2$-generating activities of berberine and palmatine in the absence or presence of DNA.

However, in the presence of DNA, the near-infrared emission at around 1270 nm, assigned to the radiative deactivation of $^1O_2$ into its ground state, was clearly observed under photoirradiation of these alkaloids. The estimated quantum yield of $^1O_2$ generation (Φ_Δ) using the reference compound, methylene blue (Φ_Δ = 0.52) [50], depended on the sequence and decreased for the guanine-containing sequence (**Table 1**). These characteristics are the fundamental mechanisms of theranostics. The theranostics mechanism of berberine and palmatine can be explained as follows:

1. The photoexcited states of these compounds are rapidly quenched through intramolecular electron transfer. These alkaloids consist of the *iso*-quinoline moiety and dialkoxybenzene moiety (**Figure 7**). The *iso*-quinoline moiety can fluoresce and photosensitize $^1O_2$ generation, and the dialkoxybenzene moiety can act as an electron-donating site.
2. The electrostatic interaction with DNA increases the Gibbs free energy (ΔG) of the intramolecular electron transfer. In addition, the hydrophobic environment of the DNA strand [51, 52] is unfavorable for the intramolecular electron transfer. Consequently, the lifetime of the S_1 state becomes markedly long compared with that without DNA.
3. Fluorescence intensity and the intersystem crossing yield are increased, resulting in the enhancement of energy transfer to the oxygen molecule to generate $^1O_2$.

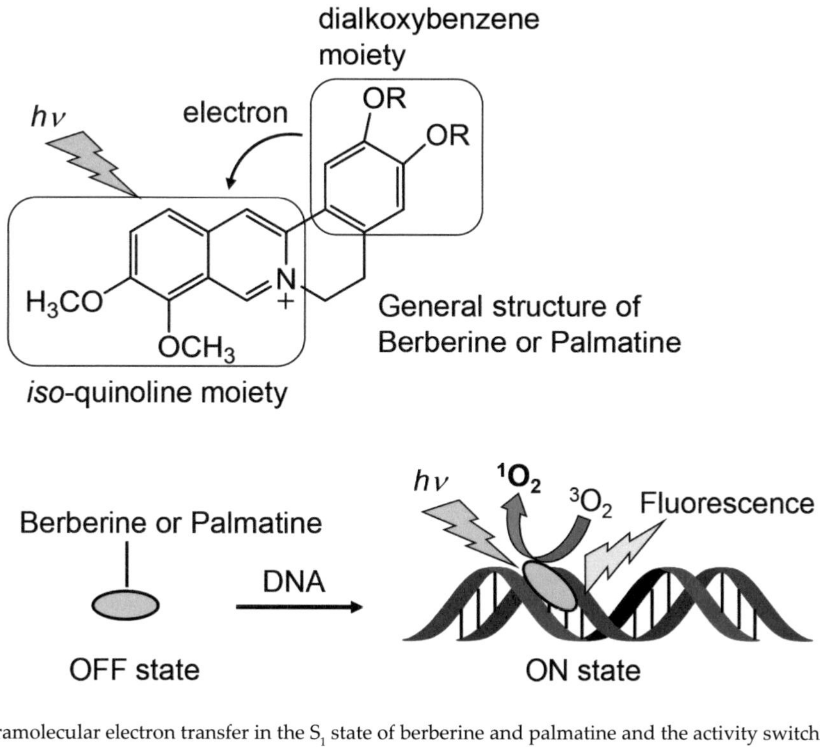

Figure 7. Intramolecular electron transfer in the S_1 state of berberine and palmatine and the activity switching by DNA.

3. DNA-targeting porphyrin theranostics

The abovementioned mechanisms of berberine and palmatine can be applied to porphyrin theranostics. For this purpose, cationic porphyrins are useful because they can be incorporated into the cell nucleus and can photosensitize cellular DNA damage [53]. Furthermore, cationic porphyrins can bind to a DNA strand through electrostatic interaction, similar to berberine and palmatine. For example, anionic water-soluble porphyrin PPIX hardly induces cellular and isolated DNA damage, whereas tetrakis(*N*-methyl-4-pyridinio) porphyrin (TMPyP, see **Figure 8**) effectively photosensitizes the guanine-specific oxidation of cellular and isolated DNA through $^1O_2$ generation. Thus, electron donor-connecting cationic porphyrins were designed and synthesized to realize porphyrin theranostics.

3.1. Binding interaction with DNA and cellular and isolated DNA-damaging activity of water-soluble porphyrins

The effect of a DNA microenvironment on the photosensitized reaction of water-soluble porphyrin derivatives, TMPyP and its zinc complex (ZnTMPyP, see **Figure 8**), was reported [42]. The main driving force of DNA binding is electrostatic interaction. The binding form between these porphyrins and DNA depends on the concentration ratio of porphyrins and DNA bases. In the presence of a sufficient concentration of DNA, TMPyP mainly intercalates to the DNA strand, whereas ZnTMPyP binds to the DNA groove. An electrostatic interaction with DNA raises the redox potential of the binding porphyrins, resulting in suppression of the photoinduced electron transfer from an electron donor to the DNA-binding porphyrins, whereas the electron transfer from the porphyrins to the electron acceptor was enhanced.

Cellular DNA damage by photoirradiated water-soluble porphyrins, TMPyP and PPIX was examined [53]. TMPyP and PPIX induced apoptosis in the human leukemia HL-60 cell

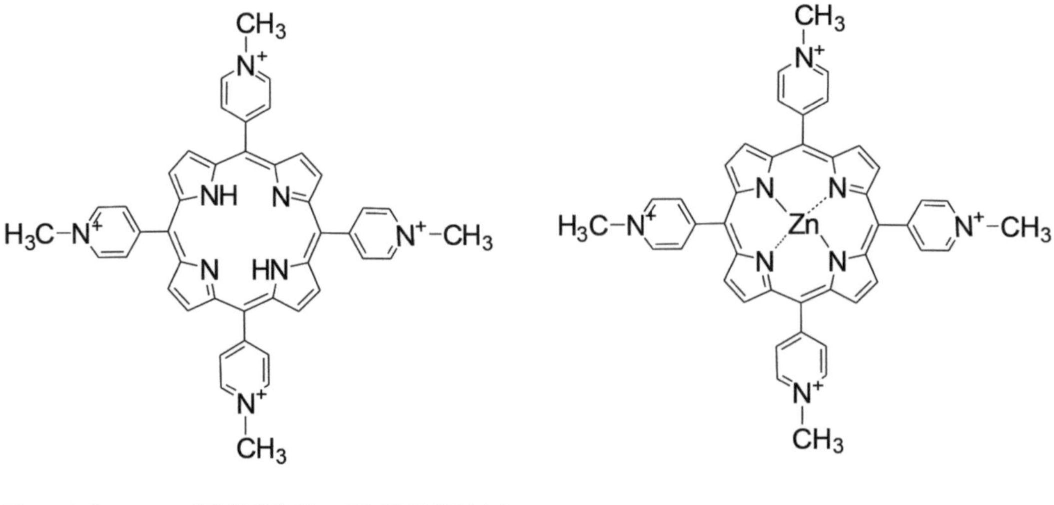

Figure 8. Structures of TMPyP (left) and ZnTMPyP (right).

http://dx.doi.org/10.5772/67882

under photoirradiation [53]. TMPyP is incorporated in the cell nucleus and photosensitizes cellular DNA oxidation, whereas PPIX hardly demonstrates cellular DNA-damaging ability. In the case of an isolated DNA fragment, photoexcited TMPyP effectively oxidized most guanine residues, whereas little or no DNA damage was observed in the PPIX case [53]. Consequently, a TMPyP cationic porphyrin should be useful as a DNA-targeting photosensitizer.

3.2. Design and synthesis of electron donor-connecting porphyrin

Molecular orbital (MO) calculation suggests that pyrene-connecting TMPyP (PyTPP, see **Figure 9**) can be used for porphyrin theranostics in a DNA microenvironment [54]. **Figure 9** shows the optimized structures of PyTPP and AnTPP and their highest-occupied MOs (HOMO). The binding action of PyTPP into the DNA major groove was suggested, and the apparent association constants, estimated from the relationship between the absorbance change and the DNA concentration, are relatively large (1.0×10^6 M^{-1} and 8.3×10^5 M^{-1} for AATT and AGTC, respectively). The fluorescence spectrum and its lifetime measurements showed that this porphyrin demonstrates almost no fluorescence in

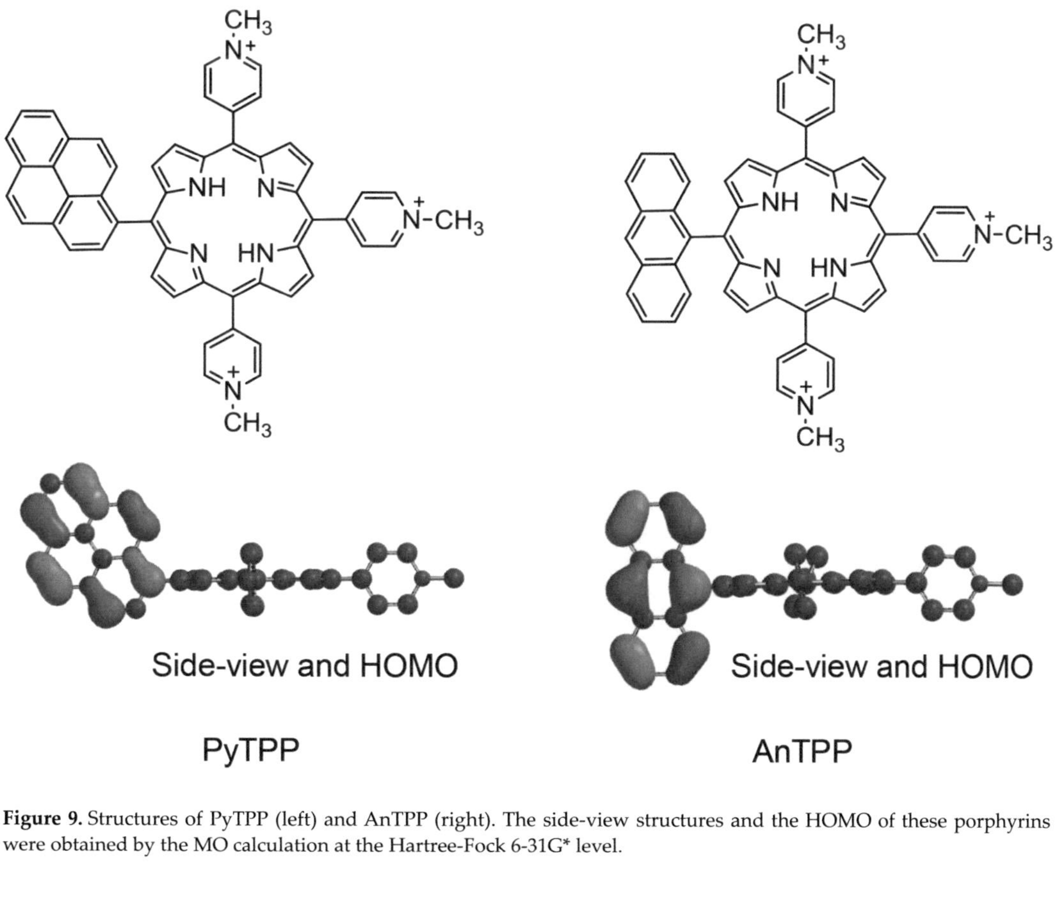

Figure 9. Structures of PyTPP (left) and AnTPP (right). The side-view structures and the HOMO of these porphyrins were obtained by the MO calculation at the Hartree-Fock 6-31G* level.

aqueous solution (Φ_f < 0.001, see **Table 2**) because of the rapid intramolecular electron transfer. The electron-accepting ability of the porphyrin moiety is decreased by the electrostatic interaction with DNA. In the presence of DNA, the fluorescence intensity was markedly increased (Φ_f is 0.12 and 0.10 in the presence of 50-μM base pairs AATT and AGTC, respectively). In addition, the typical near-infrared emission spectrum of $^1O_2$ was clearly observed during the photoexcitation of PyTPP with DNA, whereas the emission was not observed without DNA. The estimated Φ_Δ by PyTPP-DNA was 0.051 and 0.038 in the presence of 50-μM base pairs AATT and AGTC, respectively. In conclusion, the S_1 state of PyTPP is effectively quenched by the pyrenyl moiety. The interaction with DNA suppresses this electron transfer, leading to the enhancement of fluorescence emission. The intersystem crossing is also enhanced and makes $^1O_2$ generation possible.

Porphyrin	DNA	Φ_f	τ_f/ns (ratio)		Φ_Δ
PyTPP [54]	Without	<0.001	0.04		nd
	AATT	0.12	12.0		0.051
	AGTC	0.10	10.6 (0.62)	2.8 (0.38)	0.038
AnTPP [55]	Without	<0.001	0.04		nd
	AATT	0.098	10.4 (0.88)	3.6 (0.12)	0.22
	AGTC	0.077	10.6 (0.79)	2.8 (0.21)	0.17

The fluorescence properties and the Φ_Δ values were examined in a 10-mM sodium phosphate buffer (pH = 7.6). These values were reported in the literature [54, 55].

Table 2. Fluorescence and photosensitized $^1O_2$-generating activities of PyTPP and AnTPP in the absence or presence of DNA.

3.3. Improvement of the activity control using anthracene

In the abovementioned case of PyTPP, Φ_f can be recovered to a value comparable to that of TMPyP. However, Φ_Δ is significantly smaller than that of TMPyP. A relatively small Φ_Δ value might be due to the self-oxidation of PyTPP through the photosensitized $^1O_2$ generation. Since an electron donor is easily oxidized by $^1O_2$, the connection of the electron donor tends to decrease the apparent yield of $^1O_2$ generation. $^1O_2$ may oxidize the pyrene moiety through the Diels-Alder reaction. To avoid this self-oxidation, anthracene-connecting TMPyP (AnTPP, see **Figure 9**) was designed and synthesized [55]. The optimized structure of AnTPP according to MO calculation suggested that oxidation of the anthracene moiety directly connecting at the mesoposition of the porphyrin is difficult because of steric hindrance, resulting in recovery of the $^1O_2$ yield. In addition, the MO calculation indicated the steric rotational hindrance of the anthracene moiety around the mesoposition of the porphyrin, which keeps the two π-electronic systems nearly orthogonal to each other. This calculation also showed that the activity control of fluorescence and $^1O_2$ generation of this porphyrin through an interaction with DNA is possible.

http://dx.doi.org/10.5772/67882

In aqueous solution, AnTPP barely demonstrates fluorescence emission ($\Phi_f < 0.001$) and $^1O_2$ generation (**Table 2**). The observed fluorescence lifetime (<40 ps) indicates the rapid intramolecular electron transfer in the S_1 state of the porphyrin moiety of AnTPP. AnTPP also binds to the DNA strand, mainly the minor groove, and the reported association constant is relatively large (~10^6 M^{-1}). DNA-binding AnTPP demonstrates a relatively strong fluorescence and long fluorescence lifetime comparable to those of the reference porphyrin without an electron donor. Furthermore, the $^1O_2$-generating activity of AnTPP is recovered by DNA. The estimated values of Φ_Δ relative to that of methylene blue are 0.22 and 0.17 for the AATT- and AGTC-binding forms of AnTPP, respectively (**Table 2**). The observed values of Φ_Δ are significantly larger than those of PyTPP. These results suggest that the $^1O_2$-generating activity of AnTPP has improved due to the inhibition of self-oxidation by the generated $^1O_2$.

3.4. Phenanthrene-connecting cationic porphyrin

Phenanthrene was also used as the electron donor of the cationic porphyrin [56]. However, the activity control of the phenanthrene-connecting porphyrin (PhenTPP, see **Figure 10**) was not successful. The MO calculation showed the HOMO location on the phenanthryl moiety of PhenTPP and predicted the similarity of this porphyrin property to the abovementioned PyTPP and AnTPP. However, the observed values of Φ_f and τ_f without DNA are 0.028 and 5.8 ns (89%) and 2.7 ns (11%), respectively, indicating insufficient quenching of the S_1 state by phenanthrene. Furthermore, the estimated value of Φ_Δ by PhenTPP without DNA is large (0.38). Consequently, the activity control of this type of porphyrin by phenanthrene is not appropriate. This result can be explained by the relatively small driving force of the intramolecular electron transfer ($-\Delta G$ = 0.18 eV). The driving force dependence of this electron transfer is discussed in the next section in detail.

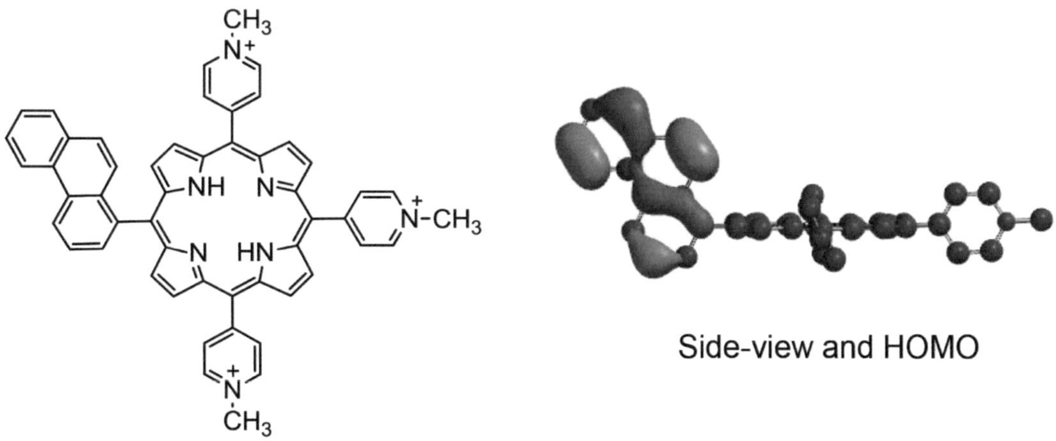

Figure 10. A structure of PhenTPP. The side-view structure and the HOMO of PhenTPP (right) were obtained by the MO calculation at the Hartree-Fock 6-31G* level.

4. Factors governing the activity control of the photochemical property of the electron donor-connecting porphyrin

As mentioned above, the controls of fluorescence intensity and $^1O_2$-generating activities of the cationic porphyrin connecting to the pyrenyl and anthryl groups by DNA could be successfully established. On the other hand, in the case of phenanthrylporphyrin, the S_1 state of this porphyrin could not be deactivated through intramolecular electron transfer because the electron-donating property of the phenanthryl moiety was insufficient [56]. To investigate the factors governing the activity control of the electron donor-connecting porphyrins, two types of electron donor-connecting porphyrins, *meso*-(1-naphthyl)-tris(*N*-methyl-*p*-pyridinio)porphyrin (1-NapTPP) and *meso*-(2-naphthyl)-tris(*N*-methyl-*p*-pyridinio)porphyrin (2-NapTPP) (**Figure 11**), were designed and synthesized [57].

These naphthylporphyrins, 1-NapTPP and 2-NapTPP, spontaneously bind to double-stranded DNA [57]. The electrostatic force between cationic porphyrins and the anionic DNA strand, as well as the hydrophobic interaction, can be speculated as the driving force of the binding interaction. In the presence of relatively small concentrations of DNA, these naphthylporphyrins aggregate around the DNA strand because their water solubility is relatively low. In the presence of a sufficient concentration of DNA, these naphthylporphyrins can form

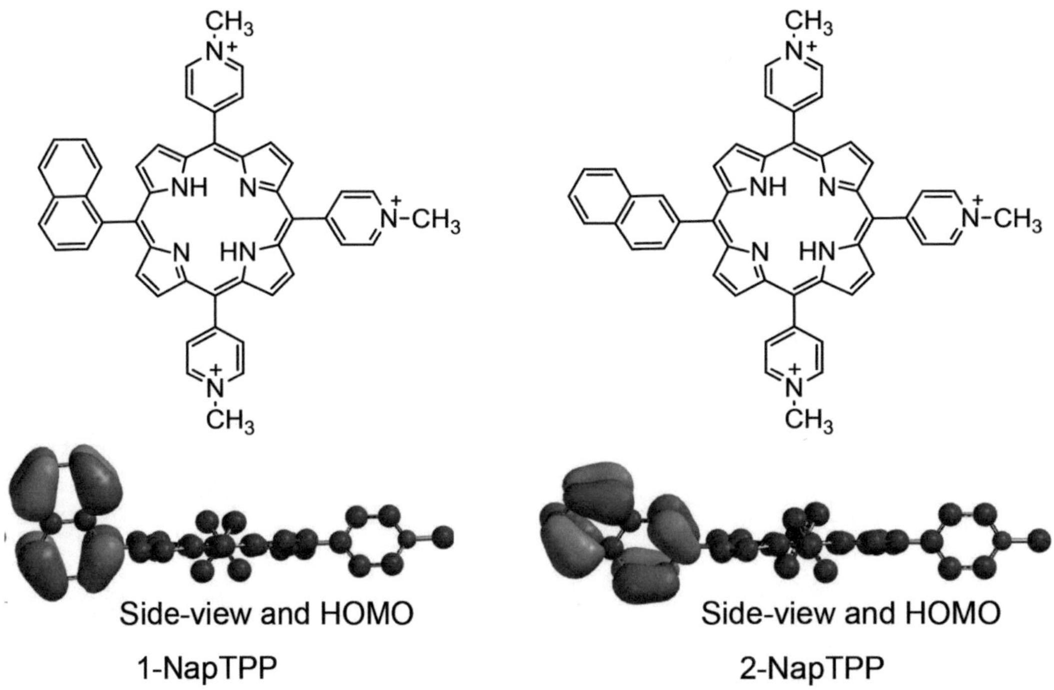

Figure 11. Structures of 1-NapTPP (left) and 2-NapTPP (right). The side-view structures and the HOMO of these porphyrins were obtained by the DFT calculation at the B3LYP/6-31G* level.

a stable complex with the DNA strand. The estimated binding constants were relatively large (more than 10^6 M^{-1}). The binding constants for those of the adenine-thymine sequence only were larger than those of the guanine-cytosine-containing sequences.

Similar to the other electron donor-connecting cationic porphyrin cases, the calculations by the density functional treatment (DFT) demonstrated that the photoexcited states of these naphthylporphyrins are deactivated through intramolecular electron transfer from their naphthalene moieties to the S_1 states of the porphyrin moieties [57]. However, the S_1 state of these porphyrins was hardly quenched by their naphthalene moieties. The Φ_Δ values of these naphthylporphyrins are also relatively large without DNA (**Table 3**). The orthogonal position of these naphthalene moieties and the porphyrin rings and the relatively small values of $-\Delta G$ of the intramolecular electron transfer (0.11 and 0.07 eV for 1- and 2-NapTPP, respectively) are not appropriate for electron-transfer quenching. The relationship between the estimated intramolecular electron transfer rate constants (k_{ET}), which are reported in the literature [57], and the driving force ($-\Delta G$ values) is plotted using the reported values and shown in **Figure 12**. The plots were analyzed by Marcus theory [58, 59] using the following equation:

$$k_{ET} = \sqrt{\frac{4\pi^3}{h^2 \lambda K_B T}} V^2 \exp\frac{-(\Delta G^{\circ} + \lambda)^2}{4\lambda K_B T} \tag{6}$$

where h is Planck's constant, λ is the reorganization energy, K_B is the Boltzmann constant, V is the electronic coupling matrix element, and T is the absolute temperature. Observed several components of the τ_f for 1- and 2-NapTPP suggest the different conformations. Therefore, the different V values were considered to explain slow electron transfer and relatively fast electron transfer. The analyzed values of V were significantly smaller than those of other directly connecting electron donor-acceptor molecular systems [60–62], suggesting that the interaction between the electron donor and the porphyrin ring is small, possibly due to the orthogonal structure. This plot suggests that a $-\Delta G$ of more than 0.3 eV is required for effective quenching through electron transfer in these types of porphyrin systems.

Porphyrin	DNA	Φ_f	τ_f/ns (ratio)			Φ_Δ
1-NapTPP	Without	0.030	6.1 (0.76)	3.7 (0.22)	0.2 (0.02)	0.26
	AATT	0.062	12.3 (0.95)	2.2 (0.03)	0.1 (0.02)	0.20
	AGTC	0.048	11.3 (0.89)	4.1 (0.09)	0.1 (0.02)	0.19
2-NapTPP	Without	0.030	3.5 (0.94)	1.3 (0.06)		0.43
	AATT	0.092	11.7 (0.89)	5.8 (0.10)	0.9 (0.01)	0.46
	AGTC	0.072	10.5 (0.76)	4.9 (0.23)	0.8 (0.01)	0.37

The fluorescence properties and the Φ_Δ values were examined in a 10-mM sodium phosphate buffer (pH 7.6). These values were reported in the literature [57].

Table 3. Fluorescence and photosensitized $^1O_2$-generating activities of 1-NapTPP and 2-NapTPP in the absence or presence of DNA.

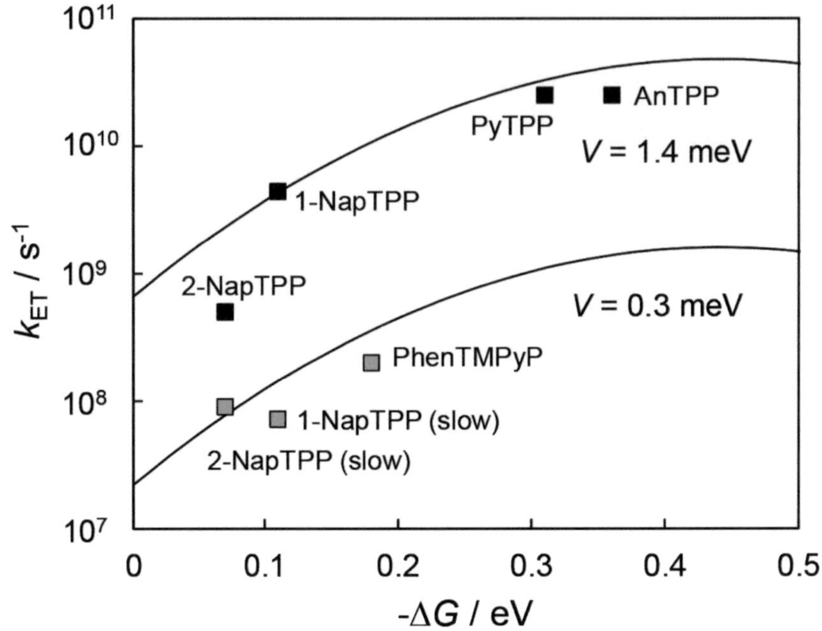

Figure 12. Relationship between the electron transfer rate and the driving force. The plots of 1-NapTPP (slow) and 2-NapTPP (slow) were calculated by using the components of their long fluorescence lifetime. These curves were calculated by the Marcus equation using two appropriate values of *V*. This relationship is reported in the literature [57].

5. Conclusions

Naturally occurring photosensitizers, berberine and palmatine, demonstrate important photochemical properties. In aqueous solution, the S_1 state of these compounds was rapidly quenched through an intramolecular electron transfer. These compounds bind to a DNA strand through electrostatic interaction, resulting in inhibition of electron transfer-mediated quenching. This interaction makes the fluorescence emission and $^1O_2$ generation by these compounds possible. A similar mechanism can be applied to the cationic porphyrin. TMPyP cationic porphyrins can be incorporated into the cell nucleus and can photosensitize guanine-specific oxidation by $^1O_2$ generation, leading to apoptosis. Therefore, the electron donor-connecting TMPyP porphyrins can be considered as model photosensitizers for theranostics. For example, PyTPP and AnTPP were designed and synthesized. The activity control of fluorescence and $^1O_2$ generation by these cationic porphyrins could be successfully established. However, the activity control of phenanthrene- and naphthalene-connecting porphyrins is insufficient because of their slow intramolecular electron transfer rate. These results suggest that a driving force of more than 0.3 eV is required for sufficiently fast electron transfer in similar porphyrin types. These studies demonstrate the possibility of porphyrin theranostics through control of the S_1 state of the porphyrin ring by the electron-donating moiety and interaction with DNA, one of the most important target biomacromolecules for cancer therapy.

Acknowledgements

The author wishes to thank Professor Shigetoshi Okazaki and Professor Toru Hirano (Hamamatsu University School of Medicine), Professor Yoshinobu Nishimura and Professor Tatsuo Arai (University of Tsukuba), Emeritus Professor Yoshio Nosaka (Nagaoka University of Technology), Mr. Takashi Yamada, Ms. Mari Harada, Mr. Yusuke Ito, and Mr. Makoto Taguchi (Shizuoka University) for their collaborations. These works were partially supported by JSPS KAKENHI from the Ministry of Education, Culture, Sports, Science and Technology of the Japanese Government.

Author details

Kazutaka Hirakawa

Address all correspondence to: hirakawa.kazutaka@shizuoka.ac.jp

Department of Engineering, Graduate School of Integrated Science and Technology, Shizuoka University, Hamamatsu, Shizuoka, Japan

Department of Optoelectronics and Nanostructure Science, Graduate School of Science and Technology, Shizuoka University, Hamamatsu, Shizuoka, Japan

References

[1] Rai P, Mallidi S, Zheng X, Rahmanzadeh R, Mir Y, Elrington S, Khurshid A, Hasan T. Development and applications of photo-triggered theranostic agents. Advanced Drug Delivery Reviews. 2010;**62**:1094–1124. doi:10.1016/j.addr.2010.09.002

[2] Ai X, Mu J, Xing B. Recent advances of light-mediated theranostics. Theranostics. 2016;**6**:2439–2457. doi:10.7150/thno.16088

[3] Albert K, Hsu HY. Carbon-based materials for photo-triggered theranostics applications. Molecules. 2016;**21**:1585. doi:10.3390/molecules21111585

[4] Chilakamarthi U, Giribabu L. Photodynamic therapy: past, present and future. The Chemical Records. 2017;**17**:1–29. doi:10.1002/tcr.201600121

[5] Dolmans DEJGJ, Fukumura D, Jain RK. Photodynamic therapy for cancer. Nature Reviews Cancer. 2003;**3**:380–387. doi:10.1038/nrc1071

[6] Castano AP, Mroz P, Hamblin MR. Photodynamic therapy and anti-tumour immunity. Nature Reviews Cancer. 2006;**6**:535–545. doi:10.1038/nrc1894

[7] Wilson BC, Patterson MS. The physics, biophysics and technology of photodynamic therapy. Physics in Medicine and Biology. 2008;**53**:R61–R109. doi:10.1088/0031-9155/53/9/R01

[8] Collins HA, Khurana M, Moriyama EH, Mariampillai A, Dahlstedt E, Balaz M, Kuimova MK, Drobizhev M, Yang VXD, Phillips D, Rebane A, Wilson BC, Anderson HL. Blood-vessel closure using photosensitizers engineered for two-photon excitation. Nature Photonics. 2008;**2**:420–424. doi:10.1038/nphoton.2008.100

[9] Calixto GM, Bernegossi J, de Freitas LM, Fontana CR, Chorilli M. Nanotechnology-based drug delivery systems for photodynamic therapy of cancer: a review. Molecules. 2016;**21**:342. doi:10.3390/molecules21030342

[10] Lang K, Mosinger J, Wagneroviá DM. Photophysical properties of porphyrinoid sensitizers non-covalently bound to host molecules; models for photodynamic therapy. Coordination Chemistry Reviews. 2004;**248**;321–350. doi:10.1016/j.ccr.2004.02.004

[11] DeRosa MC, Crutchley RJ. Photosensitized singlet oxygen and its applications. Coordination Chemistry Reviews. 2002;**233–234**:351–371. doi:10.1016/S0010-8545(02)00034-6

[12] Schweitzer C, Schmidt R. Physical mechanisms of generation and deactivation of singlet oxygen. Chemical Reviews. 2003;**103**:1685–1758. doi:10.1021/cr010371d

[13] Ogilby PR. Singlet oxygen: there is indeed something new under the sun. Chemical Society Reviews. 2010;**39**:3181–3209. doi:10.1039/B926014P

[14] Almerie MQ, Gossedge G, Wright KE, Jayne DG. Photodynamic diagnosis for detection of peritoneal carcinomatosis. Journal of Surgical Research. 2015;**195**:175–187. doi:10.1016/j.jss.2015.01.009

[15] Li B, Lin L, Lin H, Wilson BC. Photosensitized singlet oxygen generation and detection: recent advances and future perspectives in cancer photodynamic therapy. Journal of Biophotonics. 2016;**9**:1314–1325. doi:10.1002/jbio.201600055

[16] Moghissi K, Dixon K, Stringer M, Thorpe JA. Photofrin PDT for early stage oesophageal cancer: long term results in 40 patients and literature review. Photodiagnosis and Photodynamic Therapy. 2009;**6**:159–166. doi:10.1016/j.pdpdt.2009.07.026

[17] Wang S, Bromley E, Xu L, Chen JC, Keltner L. Talaporfin sodium. Expert Opinion on Pharmacotherapy. 2010;**11**:133–140. doi:10.1517/14656560903463893

[18] Tørring T, Toftegaard R, Arnbjerg J, Ogilby PR, Gothelf KV. Reversible pH-regulated control of photosensitized singlet oxygen production using a DNA i-motif. Angewandte Chemie International Edition. 2010;**49**:7923–7925. doi:10.1002/anie.201003612

[19] Op de Beeck M, Madder A. Sequence specific DNA cross-linking triggered by visible light. Journal of the American Chemical Society. 2012;**134**:10737–10740. doi:10.1021/ja301901p

[20] Tørring T, Helmig S, Ogilby PR, Gothelf KV. Singlet oxygen in DNA nanotechnology. Accounts of Chemical Research. 2014;**47**:1799–806. doi:10.1021/ar500034y

[21] Foote CS. Definition of type I and type II photosensitized oxidation. Photochemistry and Photobiology. 1991;**54**:659. doi:10.1111/j.1751-1097.1991.tb02071.x

[22] Casas A, Di Venosa G, Hasan T, Al Batlle. Mechanisms of resistance to photodynamic therapy. Current Medicinal Chemistry. 2011;**18**:2486–2515. doi:10.2174/092986711795843272

[23] Dumont E, Monari A. Understanding DNA under oxidative stress and sensitization: the role of molecular modeling. Frontiers in Chemistry. 2015;**3**:43. doi:10.3389/fchem.2015.00043

[24] Nam G, Rangasamy S, Ju H, Samson AA, Song JM. Cell death mechanistic study of photodynamic therapy against breast cancer cells utilizing liposomal delivery of 5,10,15,20-tetrakis(benzo[b]thiophene) porphyrin. Journal of Photochemistry and Photobiology B: Biology. 2017;**166**:116–125. doi:10.1016/j.jphotobiol.2016.11.006

[25] Tabrizi L, Chiniforoshan H. New Ru(II) pincer complexes: synthesis, characterization and biological evaluation for photodynamic therapy. Dalton Transactions. 2016;**45**:18333–18345. doi:10.1039/C6DT03502G

[26] Boodram S, Bullock JL, Rambaran VH, Holder AA. The use of inorganic compounds in photodynamic therapy: improvements in methods and photosensitizer design. Recent Patents on Nanotechnology. 2017;**11**:3–14. doi:10.2174/1872210510666160425121512

[27] Soriano J, Mora-Espí I, Alea-Reyes ME, Pérez-García L, Barrios L, Ibáñez E, Nogués C. Cell death mechanisms in tumoral and non-tumoral human cell lines triggered by photodynamic treatments: apoptosis, necrosis and parthanatos. Scientific Reports. 2017;**7**:41340. doi:10.1038/srep41340

[28] Burrows CJ, Muller JG. Oxidative nucleobase modifications leading to strand scission. Chemical Reviews. 1998;**98**:1109–1151. doi:10.1021/cr960421s

[29] Kawanishi S, Hiraku Y, Oikawa S. Mechanism of guanine-specific DNA damage by oxidative stress and its role in carcinogenesis and aging. Mutation Research. 2001;**488**:65–76. doi:10.1016/S1383-5742(00)00059-4

[30] Hiraku Y, Ito K, Hirakawa K, Kawanishi S. Photosensitized DNA damage and its protection via a novel mechanism. Photochemistry and Photobiology. 2007;**83**:205–212. doi:10.1562/2006-03-09-IR-840

[31] Namikawa T, Yatabe T, Inoue K, Shuin T, Hanazaki K. Clinical applications of 5-aminolevulinic acid-mediated fluorescence for gastric cancer. World Journal of Gastroenterology. 2015;**21**:8769–8775. doi:10.3748/wjg.v21.i29.8769

[32] Ishikawa T, Takahashi K, Ikeda N, Kajimoto Y, Hagiya Y, Ogura S, Miyatake S, Kuroiwa T. Transporter-mediated drug interaction strategy for 5-aminolevulinic acid (ALA)-based photodynamic diagnosis of malignant brain tumor: molecular design of ABCG2 inhibitors. Pharmaceutics. 2011;**3**:615–635. doi:10.3390/pharmaceutics3030615

[33] Harmatys KM, Musso AJ, Clear KJ, Smith BD. Small molecule additive enhances cell uptake of 5-aminolevulinic acid and conversion to protoporphyrin IX. Photochemistry and Photobiological Sciences. 2016;**15**:1408–1416. doi:10.1039/C6PP00151C

[34] Yang X, Palasuberniam P, Kraus D, Chen B. Aminolevulinic acid-based tumor detection and therapy: molecular mechanisms and strategies for enhancement. International Journal of Molecular Sciences. 2015;**16**:25865–25880. doi:10.3390/ijms161025865

[35] Ishizuka M, Abe F, Sano Y, Takahashi K, Inoue K, Nakajima M, Kohda T, Komatsu N, Ogura S, Tanaka T. Novel development of 5-aminolevurinic acid (ALA) in cancer diagnoses and therapy. International Immunopharmacology. 2011;**11**:358–365. doi:10.1016/j.intimp.2010.11.029

[36] Guyotat J, Pallud J, Armoiry X, Pavlov V, Metellus P. 5-Aminolevulinic acid-protoporphyrin IX fluorescence-guided surgery of high-grade gliomas: a systematic review. Advances and Technical Standards in Neurosurgery. 2016;**43**:61–90. doi:10.1007/978-3-319-21359-0_3

[37] Huang Z, Shi S, Qiu H, Li D, Zou J, Hu S. Fluorescence-guided resection of brain tumor: review of the significance of intraoperative quantification of protoporphyrin IX fluorescence. Neurophotonics. 2017;**4**: 011011. doi:10.1117/1.NPh.4.1.011011

[38] Mowatt G, N'Dow J, Vale L, Nabi G, Boachie C, Cook JA, Fraser C, Griffiths TR. Photodynamic diagnosis of bladder cancer compared with white light cystoscopy: systematic review and meta-analysis. International Journal of Technology Assessment in Health Care. 2011;**7**:3–10. doi:10.1017/S0266462310001364

[39] Ding H, Sumer BD, Kessinger CW, Dong Y, Huang G, Boothman DA, Gao J. Nanoscopic micelle delivery improves the photophysical properties and efficacy of photodynamic therapy of protoporphyrin IX. Journal of Controlled Release. 2011;**151**:271–277. doi:10.1016/j.jconrel.2011.01.004

[40] Khan GS, Shah A, Zia-ur-Rehman, Barker D. Chemistry of DNA minor groove binding agents. Journal of Photochemistry and Photobiology B: Biology. 2012;**115**:105–118. doi:10.1016/j.jphotobiol.2012.07.003

[41] Hamilton PL, Arya DP. Natural product DNA major groove binders. Natural Product Repots. 2012;**29**:134–143. doi:10.1039/c1np00054c

[42] Hirakawa K, Nakajima S. Effect of DNA microenvironment on photosensitized reaction of water soluble cationic porphyrins. Recent Advances in DNA & Gene Sequences. 2014;**8**:35–43. doi:10.2174/2352092208666141013231434

[43] Thulasiram B, Devi CS, Kumar YP, Aerva RR, Satyanarayana S, Nagababu P. Correlation between molecular modelling and spectroscopic techniques in investigation with DNA binding interaction of ruthenium(II) complexes. Journal of Fluorescence. 2017;**27**:587–594. doi:10.1007/s10895-016-1986-x

[44] Hirakawa K. DNA damage through photo-induced electron transfer and photosensitized generation of reactive oxygen species. In: Kimura H, Suzuki A, editors. New research on DNA damage. Nova Science Publishers Inc.; 2008, pp. 197–219. ISBN 978-1-60456-581-2

[45] Hirakawa K, Ota K, Hirayama J, Oikawa S, Kawanishi S. Nile blue can photosensitize DNA damage through electron transfer. Chemical Research in Toxicology. 2015;**27**:649–655. doi:10.1021/tx400475c

[46] Hirakawa K, Kawanishi S, Hirano T. The mechanism of guanine specific photooxidation in the presence of berberine and palmatine: activation of photosensitized singlet oxygen generation through DNA-binding interaction. Chemical Research in Toxicology. 2005;**18**:1545–1552. doi:10.1021/tx0501740

[47] Hirakawa K, Hirano T. The microenvironment of DNA switches the activity of singlet oxygen generation photosensitized by berberine and palmatine. Photochemistry and Photobiology. 2008;**84**:202–208. doi:10.1111/j.1751-1097.2007.00220.x

[48] Hirakawa K, Hirano T, Nishimura Y, Arai T, Nosaka Y. Dynamics of singlet oxygen generation by DNA-binding photosensitizers. The Journal of the Physical Chemistry B. 2012;**116**:3037–3044. doi:10.1021/jp300142e

[49] Mazzini S, Bellucci MC, Mondelli R. Mode of binding of the cytotoxic alkaloid berberine with the double helix oligonucleotide d(AAGAATTCTT)$_2$. Bioorganic and Medicinal Chemisry. 2003;**11**:505–514. doi:10.1016/S0968-0896(02)00466-2

[50] Usui Y, Kamogawa K. A standard system to determine the quantum yield of singlet oxygen formation in aqueous solution. Photochemistry and Photobiology. 1974;**19**:245–247. doi:10.1111/j.1751-1097.1974.tb06506.x

[51] Barawkar DA, Ganesh KN. Fluorescent d(CGCGAATTCGCG): characterization of major groove polarity and study of minor groove interactions through a major groove semantophore conjugate. Nucleic Acids Research. 1995;**23**:159–164. PMCID: PMC306644

[52] Jin R, Breslauer KJ. Characterization of the minor groove environment in a drug-DNA complex: bisbenzimide bound to the poly[d(AT)].poly[d(AT)]duplex. Proceedings of the National Academy of Sciences of the United States of America. 1988;**85**:8939–8942. PMCID: PMC282622

[53] Tada-Oikawa S, Oikawa S, Hirayama J, Hirakawa K, Kawanishi S. DNA damage and apoptosis induced by photosensitization of 5,10,15,20-tetrakis (N-methyl-4-pyridyl)-21H,23H-porphyrin via singlet oxygen generation. Photochemistry and Photobiology. 2009;**85**:1391–1399. doi:10.1111/j.1751-1097.2009.00600.x

[54] Hirakawa K, Harada M, Okazaki S, Nosaka Y. Controlled generation of singlet oxygen by a water-soluble meso-pyrenylporphyrin photosensitizer through interaction with DNA. Chemical Communications. 2012;**48**:4770–4772. doi:10.1039/c2cc30880k

[55] Hirakawa K, Nishimura Y, Arai T, Okazaki S. Singlet oxygen generating activity of an electron donor-connecting porphyrin photosensitizer can be controlled by DNA. The Journal of the Physical Chemistry B. 2013;**117**:13490–13496. doi:10.1021/jp4072444

[56] Hirakawa K, Ito Y, Yamada T, Okazaki S. Relaxation process of the photoexcited state and singlet oxygen generating activity of water-soluble meso-phenanthrylporphyrin in a DNA microenvironment. Rapid Communication in Photoscience. 2014;**3**:81–84. doi:10.5857/RCP.2014.3.4.81

[57] Hirakawa K, Taguchi M, Okazaki S. Relaxation process of photoexcited meso-naphthylporphyrins while interacting with DNA and singlet oxygen generation. The Journal of Physical Chemistry B. 2015;**119**:13071–13078. doi:10.1021/acs.jpcb.5b08025

[58] Marcus RA. On the theory of oxidation-reduction reactions involving electron transfer. I. The Journal of Chemical Physics. 1956;**24**:966–978. doi:10.1063/1.1742723

[59] Marcus RA, Sutin N. Electron transfers in chemistry and biology. Biochimica et Biophysica Acta. 1985;**811**:265–322. doi:10.1016/0304-4173(85)90014-X

[60] Wasielewski MR. Photoinduced electron transfer in supramolecular systems for artificial photosynthesis. Chemical Reviews. 1992;**92**:435–461. doi:10.1021/cr00011a005

[61] Kobori Y, Yamauchi S, Akiyama K, Tero-Kubota S, Imahori H, Fukuzumi S, Norris Jr. JR. Primary charge-recombination in an artificial photosynthetic reaction. Proceedings of the National Academy of Sciences of the United States of America. 2005;**102**:10017–10022. doi:10.1073/pnas.0504598102

[62] Kubo M, Mori Y, Otani M, Murakami M, Ishibashi M, Yasuda M, Hosomizu K, Miyasaka H, Imahori H, Nakashima S. Ultrafast photoinduced electron transfer in directly linked porphyrin-ferrocene dyads. The Journal of Physical Chemistry A. 2007;**111**:5136–5143. doi:10.1021/jp071546b

Chapter 9

Porphyrins and Phthalocyanines: Photosensitizers and Photocatalysts

Rodica-Mariana Ion

Additional information is available at the end of the chapter

http://dx.doi.org/10.5772/intechopen.68654

Abstract

The porphyrins and phthalocyanines are among the most representative macrocycles in synthetic chemistry, characterized by specific molecular structures, responsible and associated with their particular properties. They have high versatility and an adequate variability introduced either by macrocycle metallation or by different substituents at its periphery called meso-substituents. The porphyrins could find, together with the synthetic phthalocyanines, various scientific and technological applications in chemical and photochemical areas. They can serve as catalysts and photocatalysts in different reactions, or as photosensitizers in photodynamic therapy of cancer, due to their high efficiency to generate free radicals and excited state species. The aim of this chapter is to achieve a better understanding of the complex and various properties of the porphyrins and phthalocyanines in different solvents or in heterogeneous media, trying to draw the relationships between these topics with their possible applications in catalytic, photocatalytic area (photodegradation of Kraft-lignin, photooxidation of different unsaturated hydrocarbons, photodegradation of different organic aquatic pollutants, oxirans polymerization) and some preliminary results about the heterogeneous porphyrin sensitizers in the photodynamic therapy area. Their photobleaching mechanisms in different conditions are also discussed.

Keywords: porphyrins, phthalocyanines, catalysis, photocatalysis, photosensitization, photodynamic therapy

1. Introduction

Porphyrins and phthalocyanines are synthetic dyes characterized by the specific molecular structure which is associated with their particular spectral and photoelectric properties. They can serve as photosensitizers in photodynamic therapy of cancer, as catalysts and photocatalysts in different reactions, as good systems to generate free radical and excited state species [1]. Porphyrins (P) constitute a class of the molecules which contain four pyrrole rings linked by the methane carbon

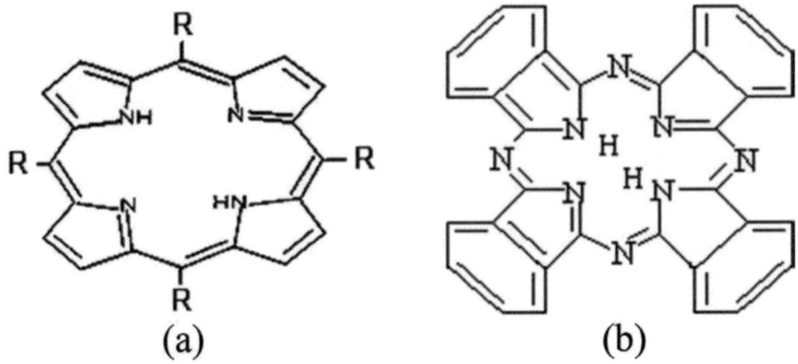

Figure 1. The general chemical structure of the (a) porphyrins and (b) phthalocyanines.

bridges, whereas the phthalocyanine molecules (Pc) are composed of four indole units—pyrrole rings linked by nitrogen atoms conjugated with benzene rings [2], (**Figures 1a** and **b**).

A large group of porphyrins could exist as free bases and metallo complexes, which can be obtained by introducing some metals in the center of the pyrrole rings. Different structures could be obtained by attaching the peripheral groups to the outer rings of the methane bridges or isoindol units, respectively. With a more symmetrical structure than the free bases, the metalloporphyrins present a reduced structure bands in the absorption spectra, called beta and alpha, the first being at longer wavelengths than the last one. Their ratio is known as a stability indicator. When $\beta \approx \alpha$, the metal forms a stable square-planar complex with porphyrin (Ni, Pd); when $\alpha \approx \beta$, the generated complexes show a strong instability. For the porphyrins containing diamagnetic metals, the triplet lifetimes are high (Mg, Cd, Zn), while those porphyrins with paramagnetic metals (Cu, Ni), will have short triplet lifetimes [3]. By comparison with porphyrins, the the phthalocyanines are more stable and their lifetimes during different reactions are longer [4].

The phthalocyanines (Pcs) hold a special position due to their interesting electronic and physicochemical characteristics, and could be organize into different condensed systems [5–7]. Pcs are planar aromatic macrocycles constituted by four isoindole units linked together through nitrogen atoms. The internal and external positions of the fused benzene ring are also commonly known as α- and β-positions, respectively. Their 42 π-electrons are distributed over 32 carbon and 8 nitrogen atoms, but the electronic delocalization mainly takes place on the inner ring, which is constituted by 16 atoms and 18 π-electrons, the outer benzene rings maintaining their electronic structure [8].

2. Porphyrins and phthalocyanines in catalytic processes

A series of metallo-porphyrins and metallo-phthalocyanines shows an increased catalytic activity in different chemical/photochemical processes:

- Photodegradation of Kraft-lignin [9, 10];
- Photooxidation of different unsaturated hydrocarbons [11];
- Photodegradation of different organic aquatic pollutants [12].

In all these processes, the porphyrins are used either in organic solutions or supported on different inorganic supports in order to increase the catalytic activity and their stability (photostability) (due to their strong interaction between the support and the complex) [13, 14].

The necessary conditions for their catalytic capacity are as follows: (i) the metallic ion from the macrocycle has to be redox active, i.e., two or higher oxidation states [15]; (ii) free coordination positions could be available to coordinate different molecules to the metallic ion [16]; (iii) the coordination sphere of the metallic ion must be labile and must have sufficiently high exchange rates for the catalyst. Many authors found that the organic radical (the substituents) and the π character, especially the dative π effect of coordination, play an important role in the catalytic activity of such compounds [17].

2.1. Degradation of Kraft-lignin

Lignins and lignosulfonates are formed as by-products in pulping processes [18], but they have not found wide-scale industrial applications. One reason could be the price and technological scheme, which are pretty complex, and the necessary catalysts for such processes are not very stable and with a modest turnover number. Under such circumstances, the efforts to find new and efficient catalysts are increasing. The photochemical degradation of lignin can be achieved in solution with porphyrins supported on metallic oxides (Al_2O_3, SiO_2, $Ti0_2$, ZnO, and WO_3) as photocatalysts, leading to some useful products, like vanillin, syringyl, and cinnamyl derivatives [9, 19]. An example is cobalt-5,10,15,20-p-tetraphenylporphyrin (Co(II) TPP) supported on the above-mentioned metallic oxides. Co(II)TPP supported on TiO_2 was the best photocatalyst used for the photodecomposition of Kraft-lignin. The metallic oxides suspended in aqueous solutions are recognized as the most widely used photocatalysts for many interesting reactions, since the photoinduced hole and electron pairs formed on the surface of the semiconductor particle can act as oxidizing and reducing agents, respectively [20]. Also, they could improve the stability and catalytic activity as metallo-porphyrins and metallo-phthalocyanines. The active species participating at such photocatalytic process are mentioned in **Table 1**.

Metallic oxide	Oxidized forms of Co(II)TPP
SiO_2	(Co(III)TPP)+
TiO_2	(CO(III)TPP+, (Co(II)TPP)
ZnO	(CO(III)TPP+, (Co(II)TPP)
WO_3	(Co(III)TPP)++

Table 1. The oxidized forms of Co(II)TPP.

In all these cases, by adsorption, Co(II)TPP suffers a strong interaction between metallic oxide and the porphyrin ring, like an isoenergetic electron transfer from porphyrin to the metallic oxide conduction band. By supporting on these metallic oxides, the new oxidized forms of Co-TPP could appear [21]. Vanillin formation from Kraft-lignin is a favorable reaction, because vanillin and its derivatives are widely used in perfumery and in pharmaceutical applications and also because Kraft-lignin is a by-product in the industrial preparation of pulp and paper.

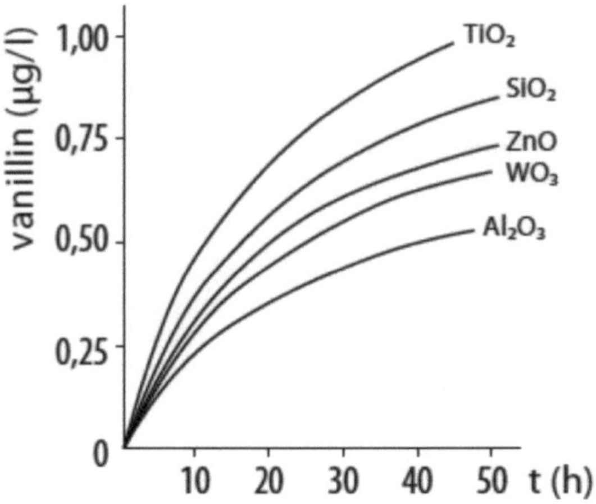

Figure 2. The kinetic diagram for vanillin generation during the photocatalytic degradation of Kraft-lignin with Co(II) TPP supported on metallic oxides as catalysts.

The photocatalytic degradation of lignin has been investigated by using TiO_2, known as one of the best photocatalysts, generating good concentrations of vanillin (**Figure 2**). These photocatalysts are acting either by energy transfer or by initial radical abstraction, after a reaction mechanism shown below, adapted after [22].

$$\mathrm{Co-TPP/TiO_2 \rightarrow Co-TPP/TiO_2(i) + 1/2\ O_2} \tag{1}$$

$$\mathrm{Co-TPP/TiO_2(i) \rightarrow Co-TPP/TiO_2{}^{+}(i) + e^- + OH\bullet} \tag{2}$$

$$\mathrm{Co-TPP/TiO_2{}^{+}(i) + h\nu \rightarrow {}^1Co-TPP^* + Ti^{2+}} \tag{3}$$

$$\mathrm{Ti^{2+} + nOH\bullet \rightarrow Ti(OH)_2{}^{(n-2)-}} \tag{4}$$

$$\mathrm{{}^1Co-TPP^* \rightarrow {}^3Co-TPP^*} \tag{5}$$

$$\mathrm{{}^1Co-TPP^* + {}^3O_2 \rightarrow Co-TPP + {}^1O_2} \tag{6}$$

$$\mathrm{{}^1O_2 + RH \rightarrow ROOH} \tag{7}$$

$$\mathrm{Ti(OH)_2^{(n-2)} + Co-TPP \rightarrow Co-TPP^* + nOH\bullet + Ti^{2+}} \tag{8}$$

$$\mathrm{Co-TPP/TiO_2 + h\nu \rightarrow e^- + Co-TPP^* + Ti+} \tag{9}$$

$$\mathrm{O_2(ads) + e^- \rightarrow O_2^{-\bullet}(ads)} \tag{10}$$

$$\mathrm{O_2^{-\bullet} + RH \rightarrow RCO\bullet + HO-} \tag{11}$$

$$\mathrm{2RCO\bullet + HO- \rightarrow RCHO + O_2} \tag{12}$$

2.2. Catalytic/photocatalytic oxidation

The iron porphyrins (**Figure 3**) are efficient catalysts both in epoxidations and in hydroxylations reactions, by using either of some oxygen donors or molecular oxygen in the presence of one reductant agent.

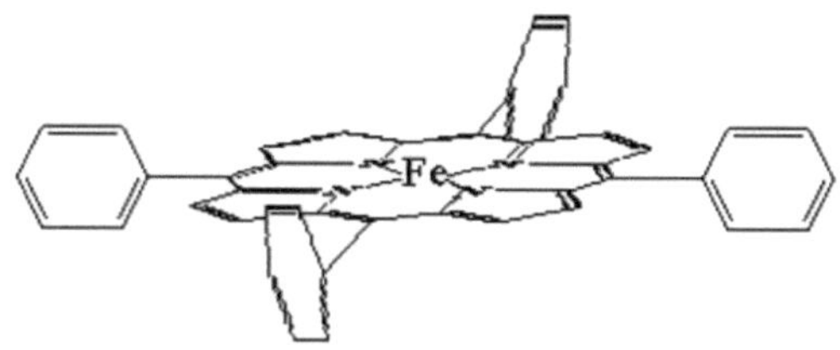

Figure 3. The structure of iron-5,10,15,20-tetra-phenyl-porphyrin (Fe(III)TPP).

$$R - H + O_2 + 2e^- + 2H^+ \rightarrow R - OH + H_2O \tag{13}$$

The cycloalkanes are oxidated by molecular oxygen under normal conditions with good yields, without some reductant agents and by using iron(III) meso-tetra (2,6-dichloro-phenyl)-porphyrin ($Cl_2Fe(III)TPP$), with light irradiation λ = 350–450 nm, leading to cyclooctenes, by the reaction mechanism shown in **Figure 4**. The axial ligand of the central metal could be OH or different halogen ions.

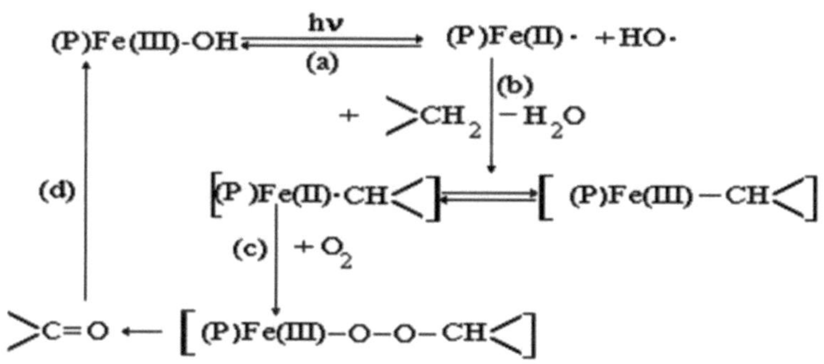

Figure 4. The oxidation cycloalkane oxidation mechanism with (OH Fe(III)P).

Fe(III)(TSPP) supported on TiO_2 is a new catalyst which, if suspended in a hydrocarbon and irradiated with λ = 365 nm, yielded selective oxidation products (for example, the cyclohexan and cyclohexene oxidation) (**Table 2**).

Cyclohexane oxidation with TiO_2 leads in principle to cyclohexene and CO_2. When TiO_2 is complexed with silan, a pronounced decreasing of catalytic activity occurs. In the presence of the system TiO_2-Sil-Fe(III)(TSPP), the cyclohexanol amount significantly increases. Iron cation could easily coordinate in the reduced form, forming an oxygen molecule and a superoxide complex [23], which is the essential form in the hydrocarbon monoooxigenations (**Figure 5**).

The final oxidation products appear only due to the final decomposition of the complex hydrocarbon radical—peroxide [24, 25].

2.3. Catalytic/photocatalytic epoxidation

The porphyrin μ-oxo-dimers are recognized as the best catalysts in the olefins chemical epoxidation reactions; however, there are reported some results for few unsaturated organic substrates (styrene and dodecene) which could support both chemical and photochemical mechanism.

Catalyst	Substrate	Products	Φ_{ox}
TiO_2	Cyclohexane	-one	0.09
		-ol	–
		CO_2	0.024
	Cyclohexene	-one	0.57
		-ol	0.01
		-Oxide	0.01
		CO_2	0.004
TiO_2-sil-Fe(III)(TSPP)	Cyclohexane	-one	0.03
		-ol	0.01
		CO_2	–
	Cyclohexene	-one	0.79
		-ol	0.48
		-oxide	0.01
		CO_2	0.0027

Table 2. The catalytic photooxidation of cyclohexane and cyclohexene on modified TiO_2.

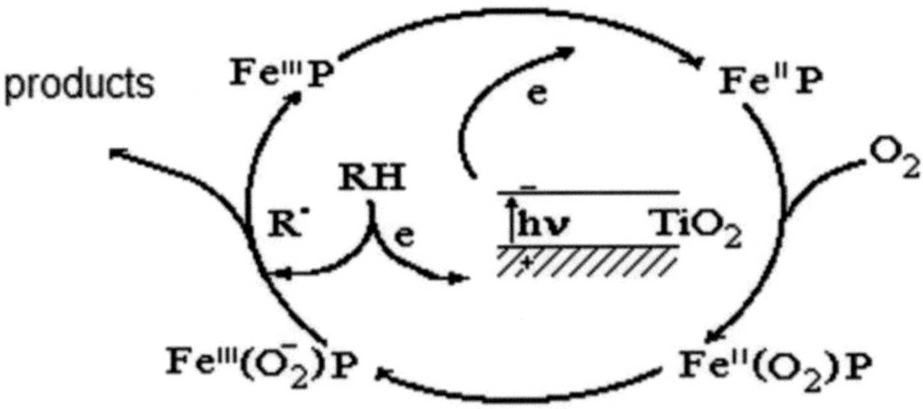

Figure 5. The scheme of the catalytic hydroxylation.

The responsible mechanisms involve the metal inside of the macrocycle and oxygen bond between the two porphyrin macrocycles. (**Figure 6**). Styrene supports an epoxidation reaction with higher yield than dodecene due to its own aromatic structure. For dodecene, the epoxide content of 3–8% has been obtained for Mn compounds, higher than 1.5–6.5% obtained with Fe compounds (better catalytic efficiency by the photochemical pathway than by the chemical one) [26]. The reaction mechanism is shown in **Figures 7** and **8**.

The high efficacy of Mn-μ-oxo-dimers could be explained by means of the high valence state of Mn (IV), which could contribute to the electron transfer and to different oxidation states of this

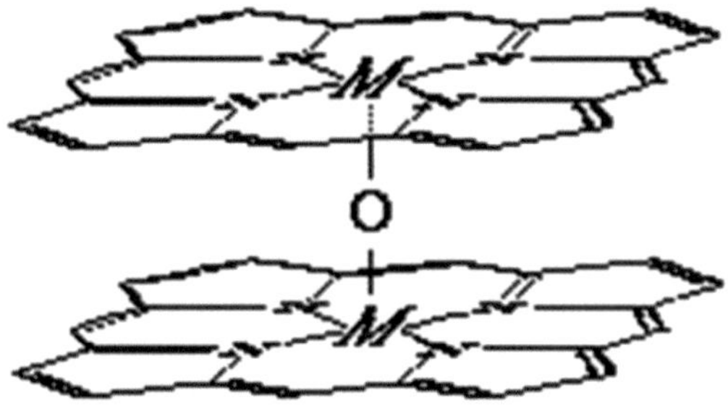

Figure 6. The structure of the porphyrinic μ-oxo-dimer.

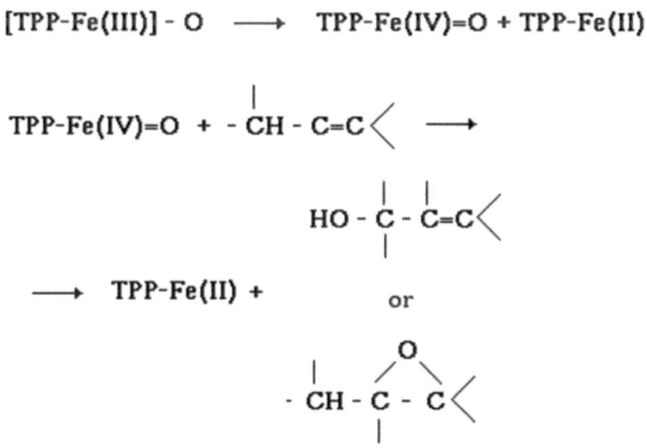

Figure 7. The general reaction mechanism of olefins photo-epoxidation.

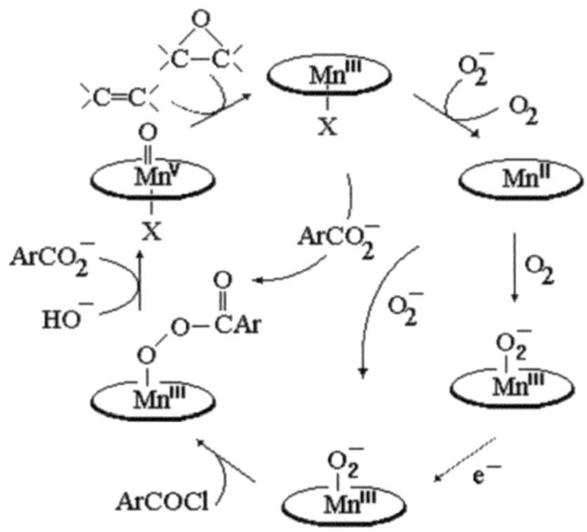

Figure 8. The catalytic mechanism of the epoxidation reaction.

metal coordinated to porphyrins (**Table 3**). The dissociation rate of the μ-oxo-bridge is bigger for 5,10,15,20-p-tetra-naphthyl-porphyrin (TNP) derivatives than for TPP derivatives, this fact contributing to a higher concentration of active species [27].

At the dodecene epoxidation, a higher catalytic efficiency of manganese oxo-dimers could be evidenced. From all the unsaturated hydrocarbons, the most efficient epoxidated is styrene compound, due to the π-π strong interactions between the phenyl ring of styrene and the large macrocycle of porphyrins [28]. The epoxide content is higher for TNP derivatives than to TPP derivatives. The greater dissociation rate of oxo-bridge is caused by some internal tensions inside of the macrocycle responsible for the subsequent catalytic activity. In photochemical epoxidations, a higher photocatalytic activity for Mn complexes could be observed due to the Mn(IV) state [29].

2.4. Photooxidation reaction with porphyrins as sensitizers

The photooxidation reactions ("ene" reaction) are characteristic reactions of singlet oxygen with different substrates (alchenes, dienes, aromatic compounds, and heterocycles) [30]. The generation of $^1O_2$ molecules occurs through the following mechanism:

$$P(S_o) + h\nu \text{-------}> P(S_1) \xrightarrow{\text{ISC}} P(T_1) \tag{14}$$

$$P(T_1) + {}^3O_2 \text{------}> P(S_o) + {}^1O_2 \tag{15}$$

Compounds with a triplet state energy lower than the excitation energy for $^1\Delta_g\, O_2$ (7900 cm^{-1}) are generally unable to transfer their excitation energy (the excitation energy is necessary to be

Catalyst/hydrocarbon	Epoxide concentration (%)
Dodecene	
$(TPP\text{-}Fe)_2O$	43.1
$(TNP\text{-}Fe)_2O$	3.27
$(TPP\text{-}Mn)_2O$	15.46
$(TNP\text{-}Mn)_2O$	17.45
Styrene	
$(TPP\text{-}Fe)_2O$	27.34
$(TNP\text{-}Fe)_2O$	35.56
$(TPP\text{-}Mn)_2O$	25.45
$(TNP\text{-}Mn)_2O$	21.17
Dodecene (irradiated)	
$(TPP\text{-}Fe)_2O$	82.03
$(TNP\text{-}Fe)_2O$	62.02
$(TPP\text{-}Mn)_2O$	55.74
$(TNP\text{-}Mn)_2O$	11.43

Table 3. The concentrations of epoxide obtained by chemical and photochemical conditions in the presence of different porphyrinic μ-oxo-dimeric catalysts.

approximately 13,200 cm^{-1}) to the ground state of molecular oxygen ($^3\Sigma_g^+$ O_2) and to generate $^1\Delta_g$ O_2. The energy of the sensitizer S_1 state (with a lifetime higher than 500 ns) could be transferred to the molecular oxygen, which in fact is a triplet state, but only if the energy gap $S_1 \rightarrow T_1$ is higher than 7900 cm^{-1}. When this energy gap is lower than 7900 cm^{-1}, the compound could be a valuable candidate for type I sensitizers, but only if the singlet and/or triplet states is higher than 500 ns, necessary for electron transfer reactions. By means of this reaction, some new hydroperoxides or peroxides could be obtained, the most efficient sensitizers being porphyrins in homogeneous or heterogeneous solutions. The porphyrins could be used as sensitizers, but only those with the triplet lifetimes >10^{-6} s can generate singlet oxygen. If a mixture of iso-amilenes (85% 2-methyl-2-butene and 15% 2-methyl-1-butenă) (**Figure 9** and **Table 4**) is irradiated with a mercury lamp or with sunlight, an alcoholic mixture as a precursor for isoprene is obtained [31]. The yields for such photooxidation reaction under different conditions are shown in **Tables 5** and **6**.

The same plant and processing scheme has been used for isoamilene dimers, with the following compositions: 23%, 3,4,5,5-tetramethyl-2-hexene; 6%, 2,3,4,4-tetramethyl-1-hexene; 15%, 3,5,5-trimethyl-2-heptene; 37%, 3,4,4,5-tetramethyl-2-hexene; 17%, unidentified). Their structures are shown in **Figure 10**.

By using some metallo-porphyrins supported on different metallic oxides, at different solar irradiances, it has been obtained at not very high HP concentrations, the best being observed for divalent metallo-porphyrins, but with the disadvantages of their price and limited stability (**Table 7**).

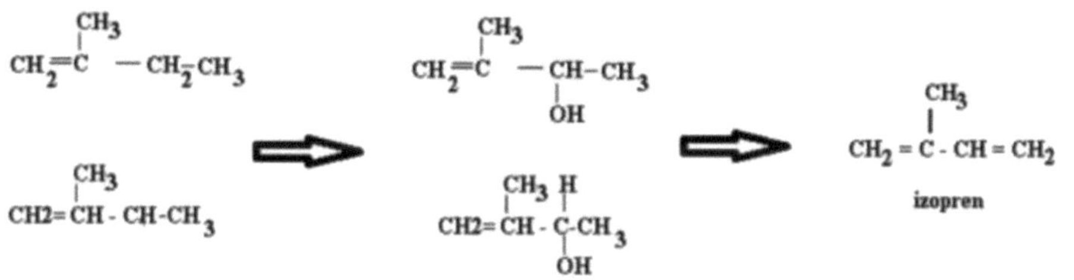

Figure 9. The scheme of isoamylene photooxidation to isoprene via hydroperoxides.

Component	Concentration (%)
Acetone	0.08
2-Methyl-2,3-epoxybutan	2.37
t-Amylic alcohol	5.57
sec-Amylenic alcohol	5.68
2-Methyl-1-buten-3-ona	3.80
2-Methyl-2-buten-1-al	1.28
HP t-amylenic	41.47
HP-s-amylenic	29.68
HP unidentified	5.93
Unidentified products	4.14

Table 4. Isoamylene HP composition.

Molecular weight	Compound	Composition (%)
70	2-Methyl-1-butene	15.76
70	2-Methyl-2-butene	83.30
58	Acetone	0.01
86	2-Methyl-2,3-epoxybutan	0.04
88	t-Amylic alcohol	0.01
86	t-Amylenic alcohol	0.06
86	sec-Amylenic alcohol	0.01

Table 5. Isoamylene composition.

Photosensitizer	Lamp 125 W	Sun
	HP/hour	
TPP	0.9	4.5
TNP	1.007	5.5
Al_2O_3	0.46	2.5
SiO_2	0.2	1.2
TPP/SiO_2	1.18	6
TPP/Al_2O_3	1.21	6

Table 6. The concentration variation of isoamylenes hydroperoxides (HP) in the presence of different sensitizers.

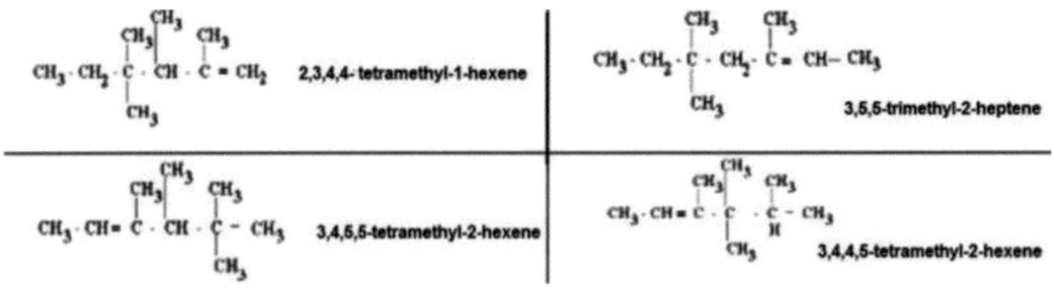

Figure 10. The structure of isoamylene components.

The photosensitizing efficiency order for the tested metallo-porphyrins is

$$Ni = Cu < Pb < Zn < Cd < Mg = Pd \quad (16)$$

Certainly, the strong difference between the activity of homogeneous sensitizer and that belonging to a heterogeneous one [32–34] should be mentioned (**Figure 11**). Similarly, the sensitizing activity obtained with a lamp irradiation and sunlight should be pointed out (**Figure 12**).

Sensitizer	Irradiation time (h)/irradiance (W/cm^2)					
	0/860	0.5/860	1/880	2/880	4/790	6.5/750
NiTPP/SiO_2	0.06	0.275	0.5005	0.3905	0.5	0.72
SnTPP/SiO_2	0.83	0.94	0.5	0.72	0.83	1.05
PbTPP/SiO_2	0.17	0.61	0.72	0.83	0.94	1.16
MgTPP/SiO_2	0.39	0.72	0.72	1.04	1.27	1.49
ZnTPP/SiO_2	0.17	0.39	0.39	0.39	0.83	1.16
CuTPP/SiO_2	0.275	0.39	0.5	0.5	0.83	1.05
PdTPP/SiO_2	0.1705	0.39	0.94	0.83	1.27	1.27

Table 7. Concentration variation of C_{10} HP.

2.5. Photocatalytic oxidation of o-nitro-phenol on heterogeneous organic semiconductors

The most common pollutants include a wide range of aliphatic and aromatic halogenated compounds, different types of herbicides, mercaptans, and other groups unaccountable industrial organic compounds. Sulfur-based compounds such as mercaptans, alkaline sulphides, sulphites, and alkaline thyosulphates constitute byproducts in industrial processes, such as reform Processing, petroleum processing, etc. [35].

The photochemical processes for destroying aquatic pollutants have been used in the last decade as a viable alternative for wastewater purification. Organic pollutants' direct photolysis is induced by light irradiation with λ = 290–400 nm. New technologies such as enhanced oxidation processes (EOP) or advanced oxidation processes (AOP) were able to convert pollutants into useful chemicals. They are described as oxidative processes with full oxidation

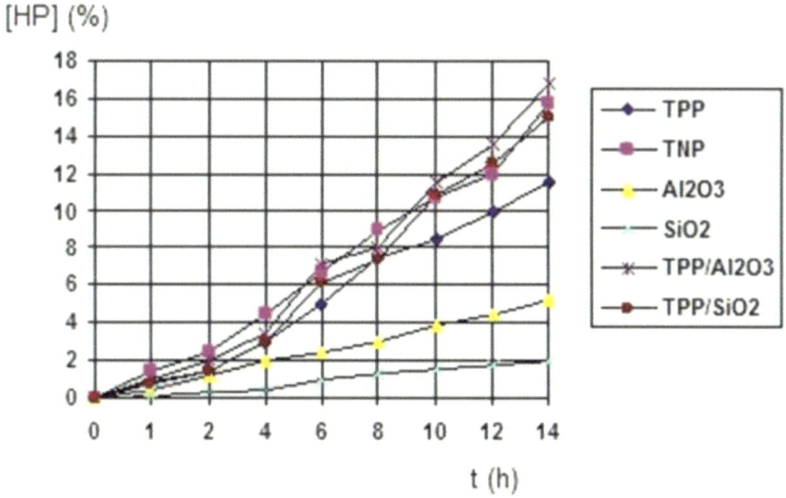

Figure 11. The kinetic for C10 fractions oxidation (lamp irradiation).

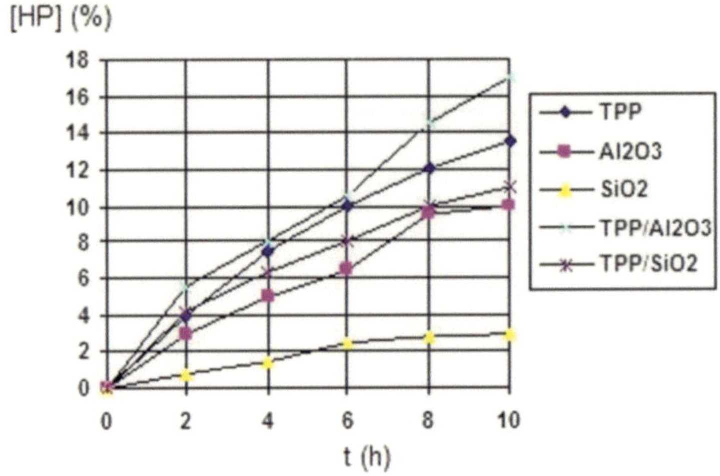

Figure 12. The kinetic for C10 fractions oxidation (sun light irradiation).

reactions (mineralization) of pollutants, to give CO_2 and a small amount of HCl, H_2SO_4, and HNO_3. EOP technologies are based on the generation of highly reactive free radicals, such as hydroxyl (OH), which act as initiators [36].

Although the quantum yields of photocatalytic reactions are small, the organic pollutants could be destroyed by heterogeneous photocatalysis by a pseudo zero order.

It is necessary to differentiate between the terminologies:

- Photocatalytic reaction: when the catalyst is photochemically generated, and the reactant is transformed via the thermal conversion to the final product.
- Photoassisted reaction: when the catalyst is also formed via the photochemical conversion and the reactant is interacting by thermal mechanism with the catalyst.
- Catalyzed photoreactions involving a catalytic effect, which is different from that of the photocatalytic reactions, in the sense that the catalyst is not generated via the photochemical reaction, but the active species in the presence of a catalyst lead to the product.
- Photosensitized reactions: when the system is not isolated, $\phi < 1$, the energy must enter each cycle and is transferred to the substrate.

The metal complexes of phthalocyanines are known to play important biological roles as electron mediators for different catalytic processes [37]. Heterogeneous photocatalysis with organic semiconductors is a fast-growing field of applied research, especially for the case of the oxidation processes of organic pollutants [11]. The main advantage is the complete destruction of pollutants to harmless compounds, e.g., carbon dioxide and inorganic acids.

The oxidative photochemical reaction (via hydroxyl radicals, generated through heterogeneous photocatalysts) is one of the most appropriate reaction types for the photochemical degradation process of many aquatic refractory pollutants; even some of them (e.g., nitro-phenols) may degrade very slowly. Some photocatalysts such as zinc (II) 2,9,16,23-tetrasulphophthalocyanine (ZnTSPc) supported on SiO_2 and zinc (II) 2,9,16,23-tetracarboxyphthalocyanine (ZnTCPc) supported on hydrotalcite (HT) have been used for ONP photodegradation. Reactions are 4–10 times faster than those obtained using the same photocatalysts but in water solution (**Figure 13**).

It is well known that upon irradiation with visible light, the non-metallic phthalocyanine complexes and those containing metal ions with filled electron shells (Mg and Zn) or with empty d orbitals (Al) show long lifetimes of the first excited triplet state [2, 38, 39]. All these compounds are good sensitizers for electron or energy transfer reactions [40–43]. Oxygen in its ground triplet state interacts with the excited triplet electronic state of the complexes yielding to singlet oxygen [44–46].

Nitrophenols have been recognized as priority pollutants. The allowed concentrations from 1 to 20 ppb [47] in lakes and fresh water reservoirs justify the search for new purification methods having low cost and practical value.

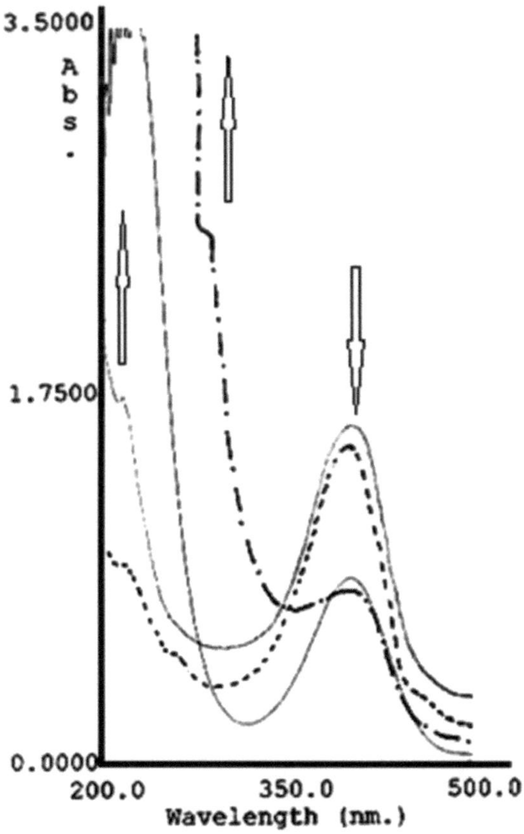

Figure 13. The changing of the ONP durng the photodegradation reaction.

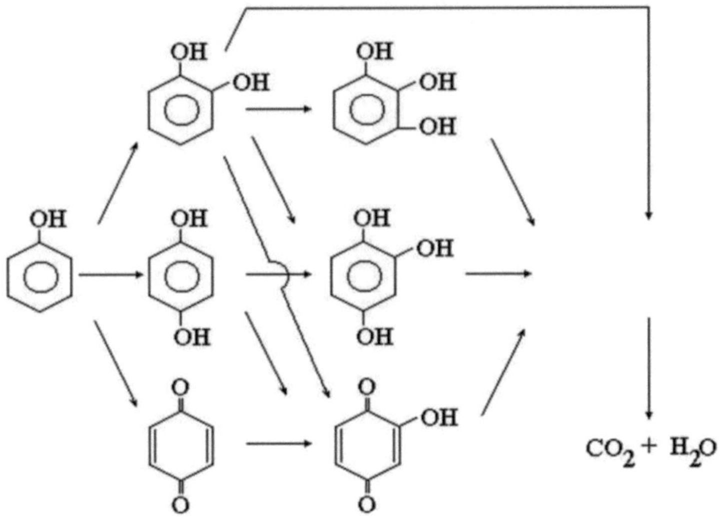

Figure 14. The reaction mechanism of ONP degradation.

The direct photodegradation of nitrophenols has not been found to be an effective method for destroying them using UV light in a homogeneous solution (**Figure 14**) [48].

Many efforts have been done in the area of porphyrins and phthalocyanines supporting them on different substrates (metallic oxides and cellulose) knowing that longer photochemical stability is essentially important for their longer activity as luminescent materials [49].

Microcrystalline cellulose (MC) can form hydrogen bonds, both within its own structure and with other molecules that may remain attached to the polymer chains by localized interactions (**Figure 15**).

In order to obtain more stable and more efficient photocatalysts for o-nitrophenol degradation, some porphyrins, such as zinc-5,10,15,20-sulphonato-phenyl-porphyrin (ZnTSPP) and zinc-4,8,18,22-tetra-sulphonato-phthalocyanine (ZnTSPc) (**Figures 16** and **17**), could be entrapped into the polymer chains of some cellulose derivatives, such as microcrystalline cellulose, their adsorption efficacy being evaluated by the state diffuse reflectance technique (GSDR).

The anionic porphyrins and phthalocyanines are rather closely packed in the submicroscopical pores of MC. It has been shown that cellulose is composed of amorphous and crystalline

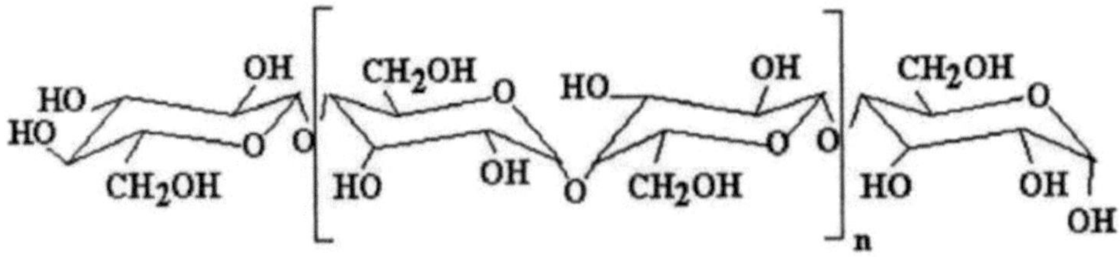

Figure 15. The microcrystalline cellulose structure.

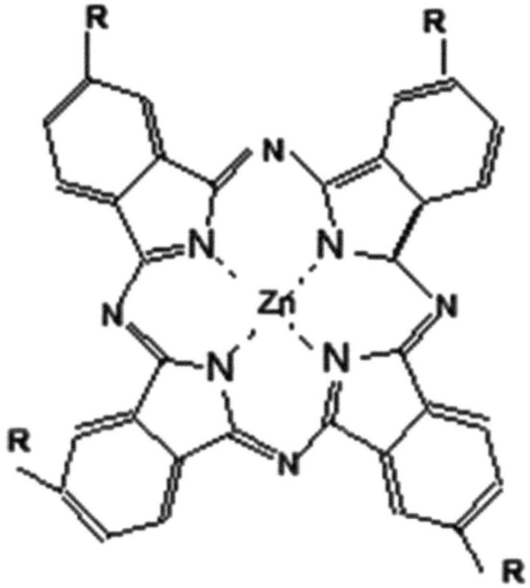

Figure 16. The structure of ZnTSPc.

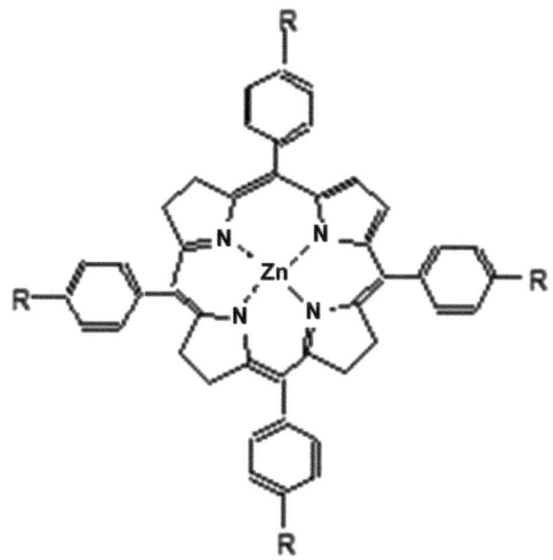

Figure 17. The structure of ZnTSPP.

domains. When MC is added to the solution, cellulose-to-cellulose hydrogen bonds are replaced by cellulose-to-solvent bonds due to the strong interactions with glycosidic chain segments, thus providing the swelling of the polymer. Tetrasulfonated porphyrin or phthalocyanine complexes could be adsorbed into microcrystalline cellulose by entrapping between the glycoside chains in the crystalline area of MC for low concentrations of complexes or by entrapping in amorphous domains for high concentrations of these complexes.

Due to the diffusion of the pollutant and O_2 toward the ZnTSPc molecules entrapped in the modified MC, due to an increase in the aggregation degree of the complexes on the solid support. The specific structure of MC prevents a larger degree of aggregation of ZnTSPP and ZnTSPc molecules because of their location in the intracrystalline voids and galleries of the supports. Also, MC being a strongly hydrophylic support coadsorbs water. This effect improves the diffusion of reactants and products. Another possible reason for this behavior is that the residual positive charge in MC interacts with the oppositely charged voluminous organic ions.

Negatively charged porphyrins and phthalocyanine complexes could form molecular associates as dimers and oligomers even in diluted solutions [50]. The well-defined structures of MC prevent the aggregation of the phthalocyanine molecules because of their location in the intracrystalline voids and galleries of the supports. Also, MC being strongly hydrophilic supports the coadsorption of water in the galleries and cavities, which improves the diffusion of reactants and products. This is the major reason for the lower catalytic activity of the phthalocyanine supported on MC. Another reason for this is residual positive charge of MC, which is able to interact with the oppositely charged voluminous organic ions [51]. The concentration-time profile for the degradation of ONP using a series of four catalysts is shown in **Figures 18** and **19**.

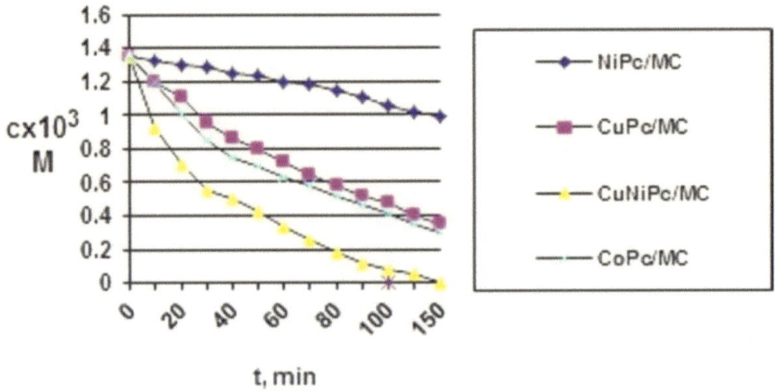

Figure 18. The kinetic graphical representation of ONP photocatalytic degradation.

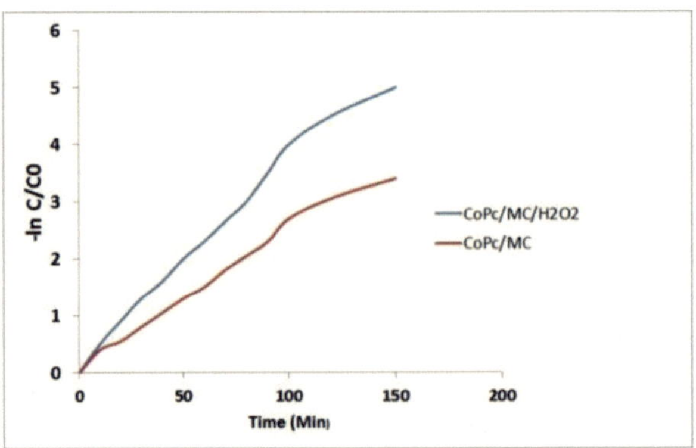

Figure 19. Graphical determination of apparent rate constant for ONP photocatalytical degradation.

It was possible to conclude that the efficacy of the studied MPc complexes as photocatalysts toward the degradation of ONP is in the order CuNiPc/MC > CoPc/MC > CuPc/MC > NiPc/MC.

The influence of hydrogen peroxide and light on the kinetic and reaction mechanism of ONP photodegradation was investigated. Reactions are 4–10 times faster in the presence of hydrogen peroxide when compared to those obtained using the same photocatalysts but in water solution (in the absence of hydrogen peroxide). The activity of heterogeneous catalysts depends on the nature of the supports and decrease from Co to the complex CuNi (**Table 8**).

The combined effect of light, H_2O_2, and photocatalysts is higher than the summed individual effects, due to the higher capacity of the UV/H_2O_2 system for OH radical generation.

Photocatalyst	$k_{app} \times 10^3$ (min^{-1})	$k_{app} \times 10^3$ (s^{-1})
CoPc/MC/H_2O_2	40	0.666
CoPc/MC	25	0.416
CuPc/MC/H_2O_2	18.33	0.305
CuPc/MC	13	0.216
NiPc/MC/H_2O_2	11	0.205
NiPc/MC	9	0.189
CuNiPC/MC/H_2O_2	7	0.165
CuNiPc/MC	5	0.123

Table 8. The rate constants for ONP photocatalytic degradation.

3. Porphyrins in polymerization processes

3.1. Living polymerization of oxirans

Living polymerization is important for obtaining polymers with uniform molecular weight. Aluminum porphyrins are excellent initiators for ring-opening polymerization of acrylic monomers, lactones, cyclic carbonates, and conjugated vinylic monomers, such as methacrylic esters, and have the advantage of uniform molecular weight and a thin molecular weight distribution [52]. For example, diethyl aluminum chloride (Et_2AlCl) induces the epoxides polymerization, and the $\alpha,\beta,\gamma,\delta$-tetraphenylporphyrin (TPP)H_2 and Et_2AlCl could be considered a catalytic and very strong system, generating aluminum tetraphenylporphyrin (TPP) AlCl, with a structure or plan of a square pyramid (**Figure 20**).

By the coordination of one monomer molecule, aluminum becomes hexacoordinated, and the final system becomes a square bipyramide (**Figure 21**).

Many other catalysts for the epoxidation reaction could be used in the oxiran epoxidation, for example, aluminum tetra-carboxyphenyl porphyrin (TPP)AlO_2CR, aluminum alcoxyde-tetraphenyl porphyrin, (TPP)AlOR, or the corresponding phenoxide, (TPP)AlOPh, all of them being prepared by the reaction between (TPP)AlEt and a protic compound [53].

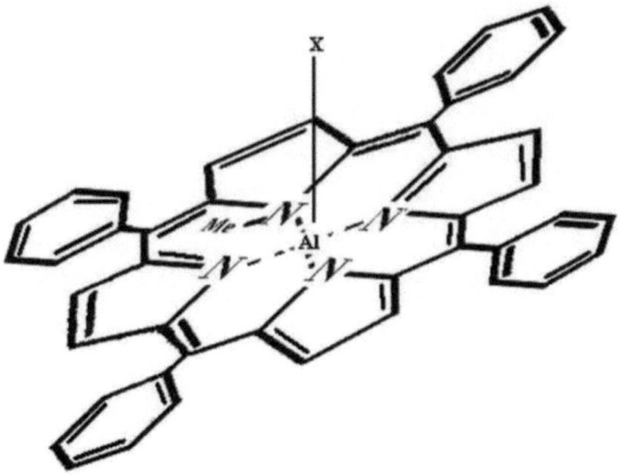

Figure 20. The structure of aluminum 5,10,15,20-tetraphenyl porphyrin.

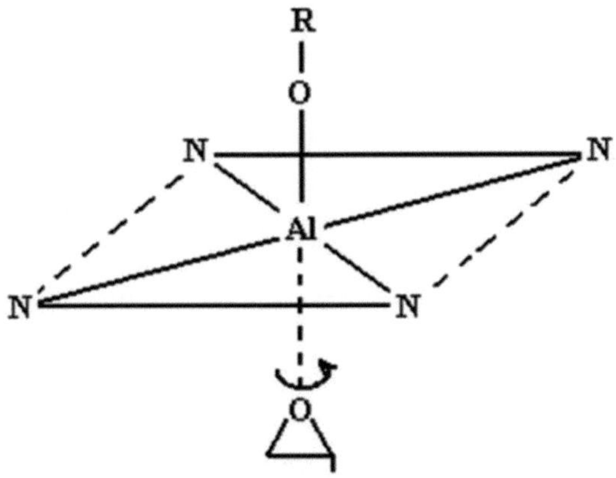

Figure 21. The spatial structure of an aluminum porphyrin.

Monomer	$[M]/[Cat]_0$	Time (h)	Conversion (%)
Ethylene oxide	200	6	100
Propenoxide	200	6	100
1,2-Butenoxide	200	6	100
Epichlorohydrine	400	3	80
Cyclohexenoxide	200	6	85
Cyclopentoxide	200	6	70
Styreneoxide	200	8	13
Izo-butenoxide	200	8	15

Table 9. The epoxides polymerization with (TPP)AlCl [54].

$$(TPP)Al - Et + HX \rightarrow (TPP)Al - X + EtH, \text{ where } X = OR, OPh, O_2CR, \text{etc.} \quad (17)$$

(TPP)AlCl immediately reacts with the epoxide in order to initiate the polymerization.

$$(TPP)AlCl + (TPP)Al\text{–}O\text{–}C\text{–}C\text{–}Cl \rightarrow (TPP)Al\text{–}O\text{–}C\text{–}C\text{–}Cl + (TPP)Al\text{–}(O\text{–}C\text{–}C)_n\text{–}Cl \quad (18)$$

The results obtained from alkylene oxide with the catalytic system (TPP) AlCl are shown in **Table 9**.

4. Applications of porphyrins in photodynamic therapy of cancer

Photodynamic therapy (PDT) consists of three essential components: **photosensitizer** (PS), light and oxygen, is an interesting concept that destroys diseased tissue via light-driven reaction. After administration of a photosensitizer, which is selectively retained by tumor cells, the subsequent irradiation with visible light in the presence of oxygen specifically inactivates neoplastic cells. Antitumor effects of PDT derive from three mechanisms: direct cytotoxic effects on tumor cells, damage to the tumor vasculature, and induction of a robust inflammatory reaction that can lead to development of systemic immunity [55, 56].

For a good correlation between preclinical phase and the possible outcomes from clinical trials, some 3D systems have been created, which have been tested during the preparation of a liposomal suspension of different drugs: folinic acid, oxaliplatin, and 5-fluorouracil loaded with meso-tetra (4-sulphonato phenyl) porphyrin [57].

Cancer treatment is one of the most important topics that are associated with photocatalysis [58]. In PDT, the porphyrins are recognized as ideal sensitizers, because these have good absorption in the NIR region, low toxicity, are better incorporated into different cells, show selective solubility and, in monomeric forms, have good penetration rate through cell membranes [58]. To be efficient, a photosensitizer should be selectively incorporated into a tissue. Following the Jablonski diagram, the first primary photochemical process is represented by the sensitizer excitation from the ground state to the first excited state in a very short time (10^{-15} s), in which the spins of the excited and unexcited electrons are still paired. The excited electron can lose its energy by falling back down to the electronic ground state, giving up most of the energy it has absorbed in a burst of fluorescence. This is a highly probable process, so that the lifetime of the singlet excited state is usually not more than a few nanoseconds (~10^{-9} s). However, it is possible for either one of the electrons to undergo the forbidden process of flipping their spins so that both electrons have parallel spins, giving a triplet excited state of the sensitizer, $^3S^*$.

The generated singlet oxygen and free radicals are very reactive species and can damage proteins, lipids, nucleic acids, and other cellular targets, as important membrane components, responsible for cell damage and death [59, 60].

Three consecutive processes occur during the PDT treatment: initial consumption of oxygen through the photodynamic process, pathophysiologic alterations in regional blood supply (hypoxia), and total vascular occlusion (ischemia).

The following could be used as laser sources: He-Ne laser (λ = 6328 Å, power 180 J/cm^2), Ar ion laser (λ = 514.5 nm and fluence rate less 100 J/cm^2), Ar ion laser (λ = 488 nm, 100 J/cm^2).

Some porphyrins were studied, and 5,10,15,20-tetra-phenyl-p-sulfonato-porphyrin (TSPP) has been filed in concentration of 10^{-4} M on nanostructured compounds: covalently bound to titanium oxide—TSPP-Sil-TiO_2. As an alternative to homogeneous solutions, using a suspension of TiO_2 nanopowder may create premises; this compound is used in studies of cell functionality. The effect of a new drug such as TSPP-Sil-TiO_2 was also studied (**Figure 22**). By coupling with TiO_2, TSPP is losing the proper absorption bands, due to the chemical bonds SO_2-NH, present in the new synthesized system. The particle size and the morphology of TSPP-Sil-TiO_2 nanoparticles in acrylate matrix visible in AFM showed sizes around 30–50 nm. The images of TSPP-Sil-TiO_2 nanoparticles in polymer matrix have been compared with TiO_2 nanoparticles alone. Injected into a mouse tumor (about 0.5 cm), after 2 or 3 days, a further marked antineoplastic effect was obtained [61, 62].

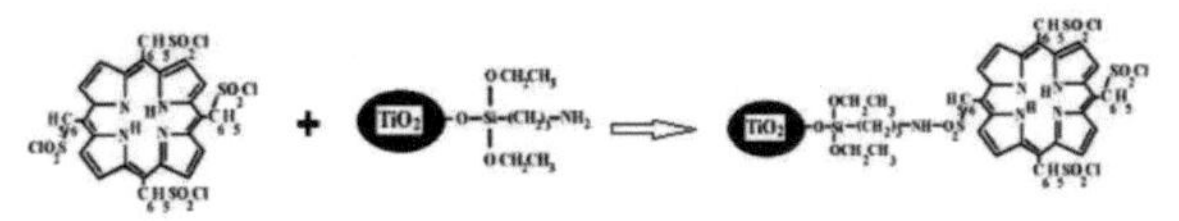

Figure 22. TSPP-sil-TiO_2 structure.

5. Photodegradation reactions of porphyrins

Porphyrins and metallo-porphyrins, due to their electronic spectra in the visible region, and their excited states with relatively long life, are considered to be good photosensitizer agents almost ideal, generating singlet oxygen in the presence of light radiation and atmospheric molecular oxygen [63, 64]. However, as singlet oxygen generator, porphyrins and their complexes metal can be oxidative degraded, generating the tetrapyrrole linear (phlorine) species with distorted photophysical and photochemical properties [65].

Structure and product names for photolysis of porphyrins differ from one porphyrin to another: from biliverdin, the bilivioline, or benzoilbilitriene (**Figure 23**).

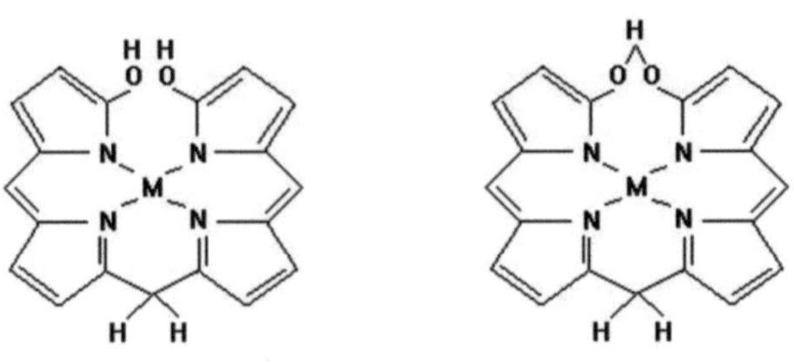

Figure 23. Structure of photolysis products of porphyrins.

The mechanisms underlying the photolysis products could be:

- mononuclear, where photodegradation products were obtained via singlet oxygen atom which is attached;
- bimolecular, where the product of photodegradation is attached to two oxygen atoms;
- mechanism called hydrolytical, where the product of photodegradation is attached to the oxygen atom derived from molecular oxygen and the other coming from a water molecule [66] (see **Figure 24**).

The photodegradation reaction mechanism can be:

- mechanism type I (or mechanism singlet), in which the reactive singlet oxygen species is formed by energy transfer processes:

$$P \xrightarrow{h\nu} {}^{1}P^{*} \xrightarrow{ISC} {}^{3}P^{*} \tag{19}$$

$${}^{3}P^{*}+{}^{3}O_2 \longrightarrow {}^{1}P_o+{}^{1}O_2 \tag{20}$$

$${}^{1}O_2 + P \longrightarrow P^{+\cdot} \tag{21}$$

$$P + P^{+\cdot} \longrightarrow P^{+\cdot} + P' \longrightarrow \text{photodegradation products} \tag{22}$$

- mechanism type II (or free radical mechanism) in which the species are the reactive superoxide anion formed by electron transfer processes:

$$P \xrightarrow{h\nu} {}^{1}P^{*} \longrightarrow {}^{3}P^{*} \tag{23}$$

$${}^{3}P^{*}+{}^{1}P^{*}\text{-------}AH \longrightarrow P^{-\cdot} + P^{+\cdot} \tag{24}$$

$$P^{-\cdot} + {}^{3}O_2 \rightarrow P_o + O_2^{-\cdot} \tag{25}$$

$$P + {}^{3}P^{*} \rightarrow {}^{3}P^{*} + P \rightarrow \text{photodegradation products} \tag{26}$$

The main factors influencing the photolysis reaction of porphyrins are meso-substituent, central metal, axial ligand, solvent, and binary mixture of solvents, and so on.

Among the complex range of disruptive factors, the meso-substituents attached to the porphyrin ring produce one of the weakest effects on photophysical and photochemical properties of porphyrins (**Table 10**). Meso-substituents are generally of aromatic character and inductive load and have an effect on electron system π porphyrin macrocycle. At the same time, a phenomenon of mixing the meso-substituents of the porphyrin macrocycle with the meso-carbon (hereinafter referred to as phenomenon hyperconjugation) leads to reducing the electron-electron repulsion configuration and interaction between them. In order to evaluate

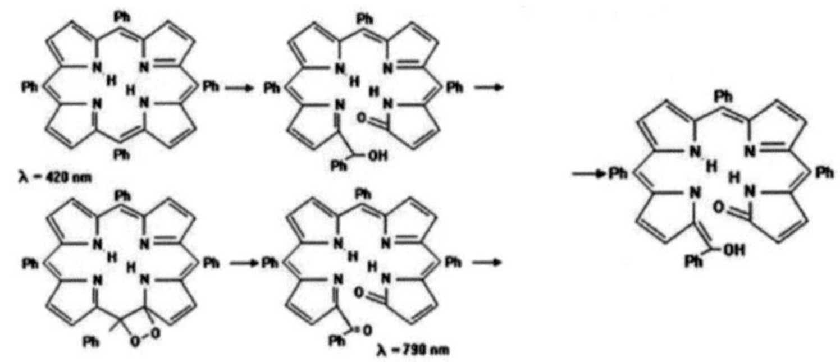

Figure 24. Photodegradation mechanism of porphyrins.

No	R	Porphyrin name	Abbreviation
1.	C_6H_5-H	5,10,15,20-Tetra(p-sulfonato)phenyl porphyrin	TPP
2.	C_6H_5-NH_2	5,10,15,20-Tetra(4-amino-phenyl)porphyrin	TAPP
3.	C_6H_5-NO_2	5,10,15,20-Tetra(4-nitro-phenyl)porphyrin	TNPP
4.	C_6H_5-SO_3H	5,10,15,20-Tetra(4-sulfonato-phenyl)porphyrin	TSPP
5.	C_6H_5-OH	5,10,15,20-Tetra(4-hydroxy-phenyl)porphyrin	THPP
6.	C_6H_5-CH_3	5,10,15,20-Tetra(4-methyl-phenyl)porphyrin	TMPP
7.	C_6H_5-OCH_3	5,10,15,20-Tetra(4-methoxy-phenyl)porphyrin	TMOPP
8.	$C_{10}H_7$-H	5,10,15,20-Tetra(1-naphthyl)porphyrin	TNP
9.	$C_{10}H_7$-SO_3H	5,10,15,20-Tetra(4-sulfonato-naphthyl)porphyrin	TSNP
10.	$C_{10}H_7$-OH	5,10,15,20-Tetra(4-hydroxy-naphthyl)porphyrin	THNP

Table 10. The structure of the meso-substituents and the name of the porphyrins.

the effect of meso-substituent on the spectral properties of porphyrins is important the assessment of donor/acceptor of the substituents (assessed by constant values Hammett (constant substitute) which is a measure of the ability of donor/acceptor substitutes) [67].

For the meso-substituted porphyrins, a rigorous correlation between the Hammett substituent constant and the rate of photodegradation has been noted (**Figure 25**).

For a photostable meso-substituted porphyrin is needed a meso-substituent with small capacity of electron acceptor ($\sigma < 0$), which involves:

- inductive load small effect (from meso-substituent on the porphyrin macrocycle);
- conjugation effect of meso-substituent macrocycle to be large;
- a small hyperconjugation macrocycle-meso-carbon.

- the photodegradation reaction mechanism of meso-substituted porphyrins involving active oxygen species (singlet oxygen, superoxide anion species dioxetane via hydroperoxide).

$$TPP \rightarrow {}^1TPP* \tag{27}$$

$$^1TPP * - - ISC \rightarrow {}^3TPP* \tag{28}$$

$$^3TPP * + {}^3O_2 \rightarrow {}^1O_2 + TPP \tag{29}$$

$$^3TPP * + {}^3O_2 \rightarrow (TPP^{+\cdot}....O_2^{-\cdot}) \rightarrow TPP^{+\cdot} + O_2^{-\cdot} \tag{30}$$

$$TPP^{+\cdot} + O_2^{-\cdot} \rightarrow (TPPO_2) \text{ dioxetan} \tag{31}$$

$$TPP^{+\cdot} + O_2 \rightarrow (TPP - O_2)^{+\cdot} \tag{32}$$

$$TPP + (TPP - O_2)^{+\cdot} \rightarrow TPPO_2 + TPP^{+\cdot} \tag{33}$$

$$TPP^{+\cdot} + O_2^{-\cdot} \rightarrow (TPP....{}^1O_2) \rightarrow TPPO_2 \tag{34}$$

$$TPP^{+\cdot} + TPP^{+\cdot} \rightarrow TPP + TPP^{+\cdot} \tag{35}$$

$$TPP^{+\cdot} + O_2^{-\cdot} \rightarrow (TPP - O_2)^{+\cdot} \rightarrow (TPPO_2) \tag{36}$$

$$(TPPO_2) \rightarrow \text{photodegradation products} \tag{37}$$

They have been identified by UV-vis (**Figure 26**), IR (**Figure 27**), and mass spectrometry (**Figure 28**).

The optimal parameters that must be taken into account in the photolysis reaction of metalloporphyrins are a high electronegativity of the central metal, a high oxidation potential, the

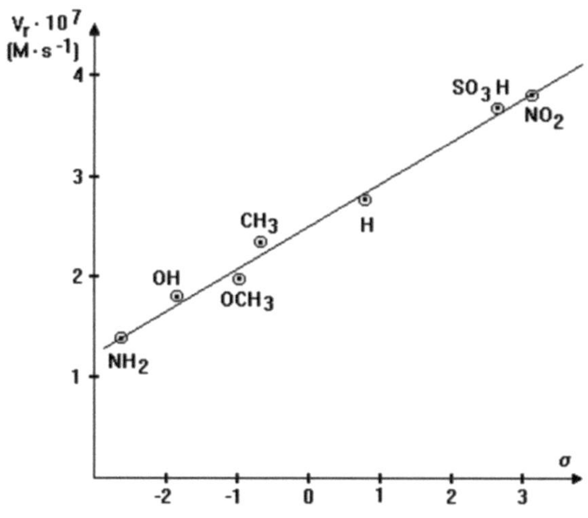

Figure 25. Correlation between the photodegradation rate and Hammett constant.

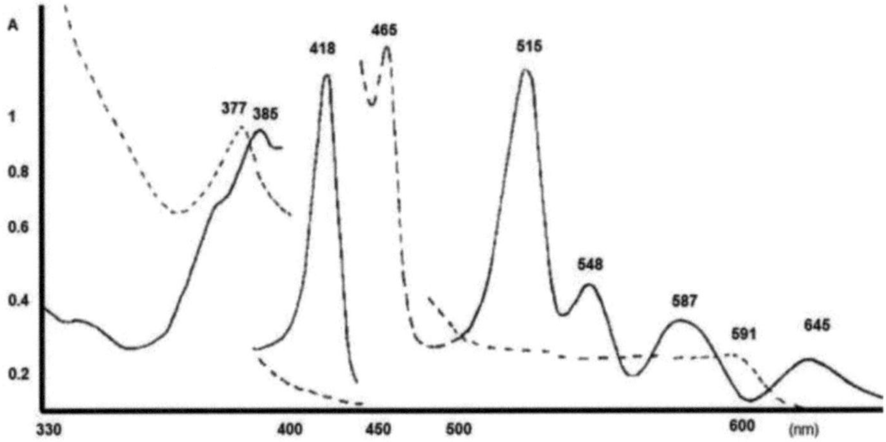

Figure 26. UV-vis spectra of TPP (___) and its photodegradation product (….).

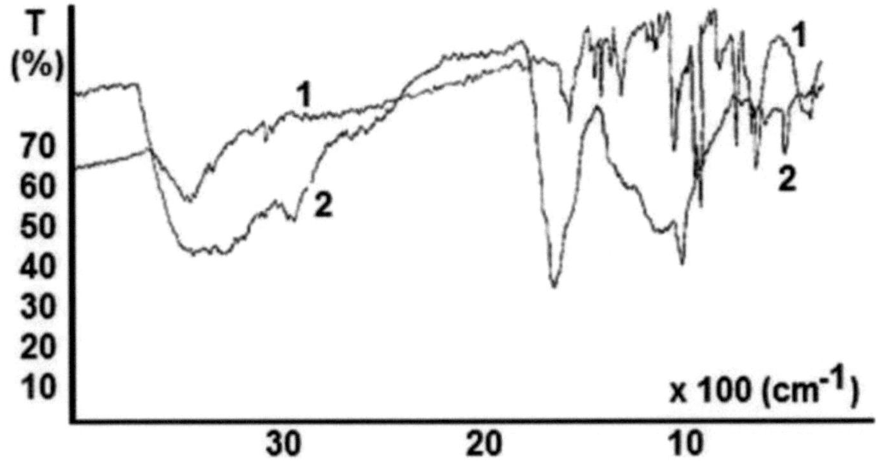

Figure 27. IR spectra of TPP (1) and its photodegradation product (2).

occupancy of d orbitals low, β and α band intensities ratio to be as low as possible (assuming that a smaller energy band β) [67, 68].

In the context of finding parameters that reflect the order of photodegradation rate variation, the following are taken into consideration: the lifetime of the first excited singlet states, lower metallo-porphyrins that easily photodegraded the metallo-complexes exhibiting strong fluorescence, and the photodegradation mechanism that could involve the dioxetane species that decomposition leads to ketones and peroxides (**Figure 29**).

In the case of metallo-porphyrins with trivalent metal and axial ligand Cl, regardless of the meso-substituent attached to the macrocycle, the photodegradation mechanism involves the following:

http://dx.doi.org/10.5772/intechopen.68654

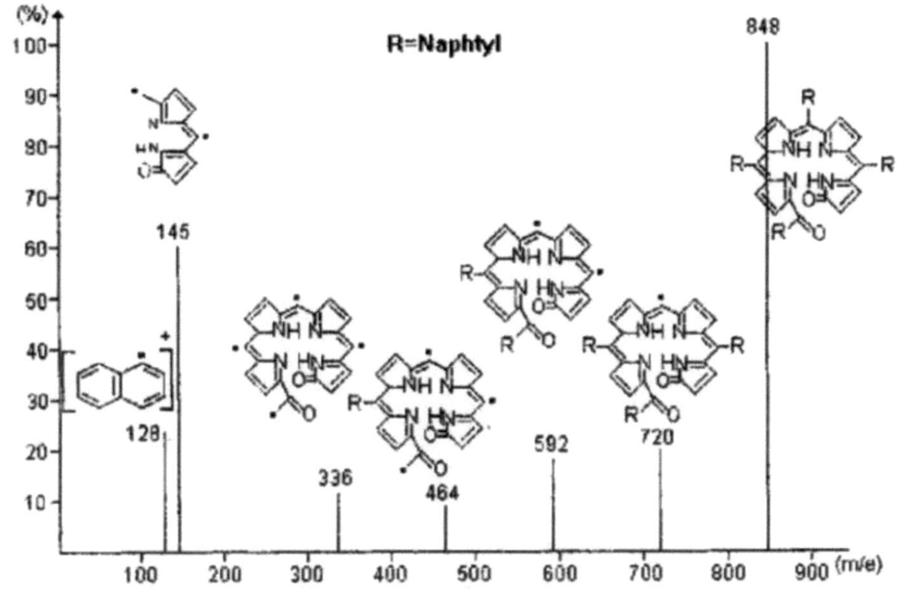

Figure 28. Mass fragments of TNP and its photodegradation product.

A homolytic break of the Cl-metal bond followed by irreversible degradation of the porphyrin macrocycle [69]. If instead of the axial Cl ligand, there are different other ligands (methyl, ethyl, octyl, and dodecyl), the trivalent metal ion is transformed from pentacoordinated into tetracoordinated by forming μ-oxo-dimers (identified by ^{27}Al-NMR spectrometry) (**Figure 30**) and by formation of peroxides and ketones, as shown in the following diagram:

$$\text{TPP AlX} \rightarrow \text{X}\bullet + \text{Al(III)TPP} \tag{38}$$

$$\text{X}\bullet + \text{X}\bullet \rightarrow \text{X} - \text{OOH} \rightarrow \text{X} - \text{C} = \text{O} \tag{39}$$

$$\text{AlTPP}\bullet + {}^3\text{O}_2 \rightarrow \text{O}[(\text{TPP})\text{Al}]_2 + {}^1\text{O}_2 \tag{40}$$

$$\text{O} - [\text{AlTPP}]_2 \Leftrightarrow {}^1[\text{Al}(\text{TPP})_2\text{O}]* \tag{41}$$

$${}^1[\text{Al}(\text{TPP})_2\text{O}] \rightarrow \text{AlOTPP} + \text{AlTPP} \tag{42}$$

$$\text{AlTPP} + \text{AlOTPP} \rightarrow (\text{AlTPP})_2\text{O} \tag{43}$$

$$\text{AlTPP}\bullet \rightarrow {}^1\text{AlTPP}* \tag{44}$$

$${}^1\text{AlTPP} * + {}^3\text{O}_2 \rightarrow (\text{AlTPP}\ldots.\ \text{O}_2^{-\bullet}) \tag{45}$$

$$\text{AlTPP} + {}^3\text{O}_2 \rightarrow \text{O}_2^{-\bullet} + \text{AlTPP} \tag{46}$$

$$\text{X}\bullet + {}^3\text{O}_2 \rightarrow \text{X} - \text{OOH} \tag{47}$$

$$X\bullet + \text{solvent} \rightarrow XH \quad (48)$$

$$HX + O_2^{-\bullet} \rightarrow XOO\bullet + H- \quad (49)$$

$$XOO\bullet + XH \rightarrow XOOX + H- \quad (50)$$

The photostability of aluminum porphyrins increases up to axial ligands with eight carbon atoms and declines thereafter [70]. Obviously, one of the most difficulties in using porphyrins is the solvent used for solubilizing them, through its physical and chemical characteristics [71]. To have an optimal photostability, the solvents used to dissolve the porphyrins should fulfill the following criteria: a high polarity (solvents with nitrogen), an aromatic character, and an alcoholic structure by forming hydrogen bonds between the alcoholic OH group and the N (NH) group from the porphyrin (**Figure 30**) [72].

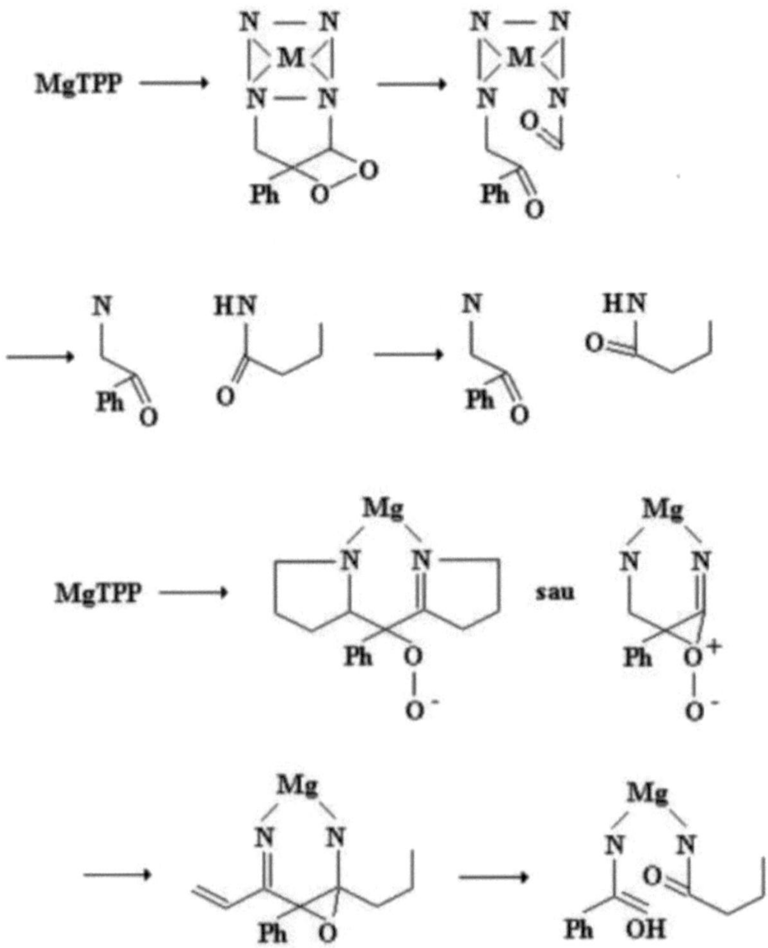

Figure 29. The scheme of the photodegradation process.

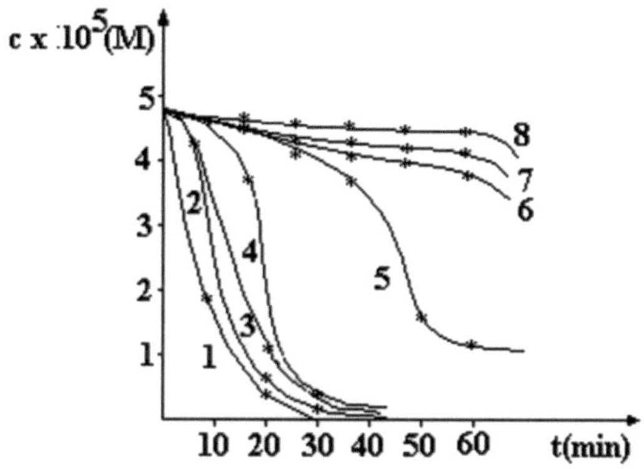

Figure 30. The photodegradation rate variation of porphyrins in various solvents: 1, benzene; 2, nitromethane; 3, piperidine; 4, pyridine; 5, nitrobenzene; 6, pentanol; 7, hexanol; 8, dodecanol.

The use of chlorinated solvents, which causes the formation of free-base porphyrin in diacid form and metallo-porphyrins, causes demetalation generated by the action of HCl. The demetalation constant varies with increasing their ability to be more easily photodegradable. The photodegradation rate is higher for the metallo-porphyrins with the d orbitals being full or completely free. The use of binary mixtures of solvents (e.g., DMSO:water (0.05%:99.95%)) is essential for the porphyrins solubilization and application (**Table 11**).

In conclusion, for a more stable porphyrin, the necessary criteria are porphyrins with meso-substituent having positive Hammett constant, metallo-porphyrins d0 or d10, regardless of the valence central, most unstable being the porphyrin with Mg, Zn, Cd, Pb, Al, aromatic solvents, and with OH groups or NO_2 [73, 74].

DMSO:water (%)	Non-sulfonated porphyrins	Sulfonated porphyrins
100:0–80:20	Neutral forms	Neutral + anions
80:20–50:50	Dimers	Dimers
50:50–25:75	Cations + aggregates	Neutral
50:50–37:73	–	Monocations
25:75–0.5:99.5	Aggregates (J) + monomers	Dications + aggregates

Table 11. Porphyrin forms in DMSO:water mixture.

6. Conclusions

Porphyrins and metallo-porphyrins are very versatile and efficient sensitizers and chemo-responsive materials, with a very extensive list of applications. The structures, chemical characteristics, optical properties, and photo-physical/photochemical properties have been detailed in this chapter. Specific applications of porphyrins and phthalocyanine, either as free base or metallic complexes, are exemplified.

Acknowledgements

This work received the financial support from the projects PN 16.31.02.04.03, PN 16.31.02.02.03, PNII 185/2014, and 120BG/2016.

Author details

Rodica-Mariana Ion

Address all correspondence to: rodica_ion2000@yahoo.co.uk

1 ICECHIM, Nanomedicine Research Group, Splaiul Independentei, Bucharest, Romania

2 Materials Engineering Department, Valahia University, Aleea Sinaia, Targoviste, Romania

References

[1] Baldea I, Ion RM, Olteanu DE, Nenu I, Tudor D, Filip AG. Photodynamic therapy of melanoma using new, synthetic porphyrins and phthalocyanines as photosensitisers—A comparative study. Clujul Medical. 2015;**88**(2):175–180

[2] Leznof CC, Lever ABP. Phthalocyanines, Properties and Applications. New York: VCH Publishers; 1989

[3] Wrobel D, Ion RM, Goc J. Photovoltaic and photoelectrical properties of porphyrin dyes. Journal of Molecular Structure. 1998;**450**:239. DOI: http://dx.doi.org/10.1016/S0022-2860(98)00434-7

[4] Frackowiak D, Planner A, Waszkowiak A, Boguta A, Manikowski H, Ion RM, Wiktorowicz K. Yield of ISC of Pcs evaluated on the basis of a time-resolved photothermal method. Journal of Photochemistry and Photobiology A: Chemistry. 2001;**141**:101–108. DOI: http://dx.doi.org/10.1016/S1010-6030(01)00438-5

[5] Leznoff CC, Lever ABP, editors. Phthalocyanines: Properties and Applications. Vols. 1–4. Cambridge: VCH Publishers (LSK) Ltd.; 1996

[6] Agirtas S, Ion RM, Bekaroglu O. Spectral study of the supramolecular assemblies porphyrins–phtalocyanines for PDT. Science Engineering, C: Bioimetic Materials Sensors and systems C. 2000;**7**:105–110. DOI: http://dx.doi.org/10.1016/S0928-4931(99)00125-3

[7] Ion RM, Yilmaz I, Bekaroglu O. Supramolecular assemblies of pyrydil-porphyrin and diazadithia phtalocyanine. Journal of the Serbian Chemical Society. 1998;**64**(7-8): 453–462

[8] Yeber MC, García G. Photocatalytic degradation of Kraft Lignin using the $S_2O_8^{2-}/Fe^0$/UV process: Optimization with multivariate analysis. Desalination and Water Treatment. 2014;**56**(7):1793–1801. DOI: http://dx.doi.org/10.1080/19443994.2014.952668

[9] (a) Ion RM, Gazdaru V. The metallo-porphyrins used in the degradation of Kraft-lignin. Progress in Catalysis. 1992;**2**:2

[10] Teodorescu L, Ion RM, Mocanu E, Culetu H, Badica D, Belsadski M. Porphyrinic photosensitizers used in photooxidation reaction. Revista de Chimie. 1988;**39**(2):132

[11] Fernández L, Esteves VI, Cunha Â, Schneider RJ, Tomé JPC. Photodegradation of organic pollutants in water by immobilized porphyrins and phthalocyanines. Journal of Porphyrins and Phthalocyanines. 2016;**20:**150. DOI: http://dx.doi.org/10.1142/S108842461630007X

[12] Ion RM, Teodorescu L, Mandravel C, Volanski E, Hillebrand M. The photochemical degradation of the porphyrinic photosensitizers used in the solar energy conversion. Revista de Chimie. 1990;**41**(2):129

[13] Teodorescu L, Ion RM. New aspects on the photodegradation of the porphyrinic photosensitizers. Revista de Chimie 1990;**41**(4):312

[14] Neagu M, Constantin C, Tampa M, Matei C, Lupu A, Manole E, Ion RM, Fenga C, Tsatsakis AM. Toxicological and efficacy assessment of post-transition metal (Indium) phthalocyanine for photodynamic therapy in neuroblastoma. Oncotarget. 2016;**7**:69718-69732. DOI: http://dx.doi.org/10.18632/oncotarget.11942

[15] Boucher LJ. Metal complexes of phthalocyanines. In: GA Melson Editor. Coordination Chemistry of Macrocyclic Compounds. 1979. pp. 461–516

[16] Mackintosh HJ, Budd PM, McKeown NB. Catalysis by microporous phthalocyanine and porphyrin network polymers. Journal of Materials Chemistry. 2008;**18**:573–578. DOI: http://dx.doi.org/10.1039/B715660J

[17] Link RG. In: Schranzer GN, editor. Transition Metals in Homogeneous Catalysis. New York: Marcel Dekker; 1971. p. 297

[18] Calvo-Flores FG, Dobado JA, Isac-Garcia J, Martin-Martinez FJ. Lignin and Lignans as Renewable Raw Materials, Chemistry Technology and Applications. Chichester, UK: John Wiley and Sons; 2015

[19] Kansal SK, Singh M, Sud D. Studies on TiO_2/ZnO photocatalysed degradation of lignin. Journal of Hazardous Materials. 2008;**153**(1-2):412–417. DOI: http://dx.doi.org/10.1016/j.jhazmat.2007.08.091

[20] Bard AJ. Photoelectrochemistry. Science. 1980;**207**:139–144. DOI: http://dx.doi.org/10.1126/science.207.4427.139

[21] Ziyadova TM, Burmistrov VA, Semeikin AS, Koifman OI. Oxidation of cobalt(II) tetraphenylporphyrinate with molecular oxygen. Macrocycles. 2015;**8**(3):274–278. DOI: http://dx.doi.org/10.6060/mhc140926z

[22] Barona-Castaño JC, Carmona-Vargas CC, Brocksom TJ, de Oliveira KT. Porphyrins as catalysts in scalable organic reactions. Molecules. 2016;**21**:310. DOI: http://dx.doi.org/10.3390/molecules21030310

[23] Groves JT, Nemo TE, Myers RS. Hydroxylation and epoxidation catalysed by iron-porphine complexes. Oxygen transfer from iodosylbenzene. Journal of the American Chemical Society. 1979;**101**:1032–1033. DOI: http://dx.doi.org/10.1021/ja00498a040

[24] Costas M. Selective C–H oxidation catalysed by metalloporphyrins. Coordination Chemistry Reviews. 2011;**255**:2912–2932. DOI: http://dx.doi.org/10.1016/j.ccr.2011.06.026

[25] Ribeiro S, Serra AC, Gonsalves AR. Efficient solar photooxygenation with supported porphyrins as catalysts. ChemCatChem. 2013;**5**:134–137. DOI: http://dx.doi.org/10.1002/cctc.201200532

[26] Srour H, le Maux P, Chevance S, Simonneaux G. Metal-catalysed asymmetric sulfoxidation, epoxidation and hydroxylation by hydrogen peroxide. Coordination Chemistry Reviews. 2013;**257**:3030–3050. DOI: http://dx.doi.org/10.1016/j.ccr.2013.05.010

[27] Ion RM, Brezoi DV. Nanocomposite based on PPy with porphyrin-m-oxo-dimers. Nanotrends. 2006;**2**(1):1–11

[28] Xu W, Dziedzic-Kocurek K, Yu M, Wu Z, Marcelli A. Spectroscopic study and electronic structure of prototypical iron porphyrins and their μ-oxo-dimer derivatives with different functional configurations, RSC Advances, 2014;**4**:46399–46406. DOI: http://dx.doi.org/10.1039/C4RA04685D

[29] Ion RM, Grigorescu M, Ureche A, Socoteanu RP, Licsandru D. Studies of some porphyrine m-oxo-dimers. The synthese, spectral characterization and catalytic applications of Me-P m-oxo-dimers. Progress in Catalysis . 2000;**9**(1-2):57–65

[30] Ion RM, Ureche-Fotea A, Socoteanu R, Licsandru D. Porphyrins and metalloporphyrins in chemical and photochemical catalysts. Progress in Catalysis. 1995;**4**(2):47

[31] Ion RM, Moise F. The correlation between the electronic structure and the catalytic structure of TNP-Co supported on different metallic oxides. Progress in Catalysis. 1992;**2**:13

[32] Ion RM, Bercu C. NMR parameters-photochemical reactivity correlation at TPP supported on metallic oxides. Revista de Chimie. 1995;46(7):656–658

[33] Ion RM. The photosensitizer effect of Me-TNP. Revista de Chimie. 1994;**45**(9):321–324

[34] Helesbeux JJ, Duval O, Guilet D, Séraphin D, Rondeau D, Richomme P. Regioselectivity in the ene reaction of singlet oxygen with ortho-Prenylphenol derivatives. Tetrahedron. 2003;**59**(27):5091–5104. DOI: http://dx.doi.org/10.1016/S0040-4020(03)00733-6

[35] Foote CS. Definition of type I and type II photosensitized oxidation. Journal of Photochemistry and Photobiology. 1991;**54**:659–659. DOI: http://dx.doi.org/10.1111/j.1751-1097.1991.tb02071x

[36] Bolton JR. Ultraviolet Applications Handbook. 3rd ed. Bolton Photosciences Inc. Edmonton, AB, Canada: ICC Lifelong Learn Inc; 2010

[37] Xiong Z, Xu Y, Zhu L, Zhao J. Photosensitized oxidation of substituted phenols on aluminum phthalocyanine-intercalated organoclay. International Journal of Environmental Science and Technology. 2005;**39**(2):651–657 .DOI: http://dx.doi.org/10.1021/es0487630

[38] Drozd D, Szczubiałka K, Skiba M, Kepczynski M, Nowakowska M. Porphyrin–nanoclay photosensitizers for visible light induced oxidation of phenol in aqueous media. The Journal of Physical Chemistry C. 2014;**118**(17):9196–9202. DOI: http://dx.doi.org/10.1021/jp500024h

[39] Darwent JR, Douglas P, Harriman A, Porter G, Richoux MC. Metal phthalocyanines for reduction of water to hydrogen. Coordination Chemistry Reviews. 1982;**44**:83. DOI: 10.1039/c3cc45025b

[40] Daraio ME, Aramendia PF, San Roman EA, Braslavsky SE. Journal of Photochemistry and Photobiology. 1991;**54**:367

[41] Khezer-Sobbi A, Wöhrle D, Schlettwein D. Stability of various porphyrins in solution and as thin film electrodes. Journal of the Chemical Society, Perkin Transactions. 1993;**2**:481–488. DOI: http://dx.doi.org/10.1039/P29930000399

[42] Wöhrle D, Schnurpfeil G, Knothe G. Efficient synthesis of phthalocyanines and related macrocyclic compounds in the presence of organic bases. Dyes and Pigments. 1992;**18**:91–102. DOI: http://dx.doi.org/10.1039/P29930000399

[43] Weber JH, Buch DH. Complexes derived from strong field ligands. Inorganic Chemistry. 1965;**4469**. DOI: http://dx.doi.org/10.1021/ic50026a006

[44] Achar BN, Fohlen GM, Parker JA, Kesha Vayya J. Oxidation and photooxidation of sulfur-containing compounds in the presence of water soluble phthalocyanine complexes. Indian Journal of Chemistry. 1988;**27A**:411.

[45] Linstead RP, Weiss FT. Phthalocyanines and related compounds. Journal of the Chemical Society. 1950;2975. DOI: http://dx.doi.org/10.1039/JR9500002975

[46] Ruppert G, Banner R, Heisler G, Novalic S. The photo-Fenton oxidation—A cheap and efficient wastewater treatment method. Chemoshere. 1993;**27**(8):1339. DOI: http://dx.doi.org/10.1016/0045-6535(93)90228-W

[47] Gernjak W, Krutzler T, Glaser A, Malato S, Caceres J, Bauer R, Fernández-Alba AR. Photo-Fenton treatment of water containing natural phenolic pollutants. Chemosphere. 2003;**50**(1):71–78 http://dx.doi.org/10.1016/S0045-6535(02)00403-4

[48] Schulte P, Bayer A, Kuhu K, Luy Th, Voljmer M. H_2O_2/O_3, H_2O_2/UV And H_2O_2/Fe^{2+} processes for the oxidation of hazardous wastes. Ozone Science Engineering. 1995;**17** (2):119. DOI: http://dx.doi.org/10.1080/01919519508547541

[49] Ion RM, Moise F, Gazdaru V, Bercu C, Badescu V. Spectral studies of the photodegradation reaction of Zn-TPP supported on celullosic material. Progress in Catalysis. 1994;**3**(1):9

[50] Patterson J. editor. Instrumental Waste Water Treatment Technology. Boston: Butterworth; 1985. p. 3

[51] Ion RM, Oprea F, Ion N, Bacinschi Z. As oliveira, Lf Vieira-Ferreira, new nanocrystalline materials based on ZnP and ZnPc adsorbed onto MC. Tehnomus XII. 2003;449–452

[52] Coca S, Dimonie M, Ion RM, Popescu L, Moise F, Vasilescu A. The open-ring polymerization of cycloolefins study with TPP-W. Journal of Molecular Catalysis. 1994;**90**:101–110. DOI: http://dx.doi.org/10.1016/0304-5102(93)E0309-5

[53] Sugimoto H, Aida T, Inoue S. Ring-opening polymerizations of lactone and epoxide initiated with aluminum complexes of substituted tetraphenylporphyrins. Molecular design of highly active initiators. Macromolecules. 1990;**23**(11). DOI: http://dx.doi.org/10.1021/ma00213a018

[54] Ion RM. The use of phthalocyanines and related complexes in photodynamic therapy. In: Tebello N, Vefa A, editors. Photosensitizers in Medicine, Environment and Security. 1st ed. Dordrecht//Heidelberg: Springer; 2012. pp. 315–349

[55] Constantin C, Matei CN, Tampa MS, Lupu A-R, Ion R-M, Neagu M. Gallium (III) phtalocyanine for photodynamic therapy of epithelial type transformed cells-approaches on SHSY5Y cell line. European Journal of Immunology. 2016;**46**:1168–1169. DOI: http://dx.doi.org/10.1002/eji.201670200

[56] Baldea I, Olteanu DE, Bolfa P, Tabaran F, Ion RM, Filip GA. Melanogenesis and DNA damage following photodynamic therapy in melanoma with two meso-substituted porphyrins. Journal of Photochemistry and Photobiology B: Biology. 2016;**161**:402–410. DOI: http://dx.doi.org/10.1016/j.jphotobiol.2016.06.012

[57] Galateanu B, Hudita A, Negrei C, Ion RM, Costache M, Stan M, Nikitovic D, Hayes AW, Spandidos DA, Tsatsakis AM, Ginghina O. Impact of multicellular tumor spheroids as an in vivo like tumor model on anticancer drug response. International Journal of Oncology. 2016;**48**(6):2295–2302. DOI: http://dx.doi.org/10.18632/oncotarget.11942

[58] Frackowiak D, Planner A, Ion RM, Wiktorowicz K. Incorporation of dye in resting and stimulated leukocytes. In: Daehne S, Resch-Gennger U, Wolfbais OA, editors. Near-infrared Dyes for High Technology Applications. Boston: Kluwer Academic Press; 1998. pp. 87–114

[59] Ion RM. Photodynamic nanomedicine strategies in cancer therapy and drug delivery. In: Serra PA, editor. Advances in Bioengineering, Edition: 1, Chapter: Photodynamic Nanomedicine Strategies in Cancer Therapy and Drug Delivery. Zagreb: InTech; 2015. pp. 253–287. DOI: 10.5772/59624

[60] Ion RM, Suica-Bunghez IR. Oxidative stress-based photodynamic therapy with synthetic sensitizers and/or natural antioxidants. In Gowder JST, editor. Basic Principles and Clinical Significance of Oxidative Stress. Zagreb: InTech; 2015. pp. 283–318. DOI: 10.5772/59293

[61] Ion RM, Brezoi DV. New nano-sized sensing drug and its clinical application. Solid State Phenomena. 2005;**106**:79–82. DOI: http://dx.doi.org/10.4028/www.scientific.net/SSP.106.79

[62] Baldea I, Olteanu DE, Bolfa P, Ion RM, Decea N, Cenariu M, Banciu M, Sesarman AV, Filip AG. Efficiency of photodynamic therapy on WM35 melanoma with synthetic porphyrins: Role of chemical structure, intracellular targeting and antioxidant defense. Journal of Photochemistry and Photobiology B Biology. 2015;**151**:142–152. DOI: http://dx.doi.org/10.1016/j.jphotobiol.2015.07.019

[63] Simionescu S, Teodorescu S, Ion RM, Nechifor GH. Polymer membranes for selective separation of ionizing forms of TPPS4 as drug in photodynamic therapy. Materiale Plastice. 2016;**53**(2):194–197

[64] Ion RM. Photodynamic therapy of cancer. A photocatalytic or a photosensitization process? Progress in Catalysis. 1997;**6**(1):55–64

[65] Ion RM. Photochemical production and quenching of singlet oxygen by the porphyrins used in photodynamic therapy of cancer. Romanian Journal of Biophysics. 1996;**6**(3-4):207

[66] Ion RM, Planner A, Wicktowicz K, Frakowiak D. The incorporation of various porphyrins into blood cells measured via flow cytometry, absorption and emission spectroscopy. Acta Biochimica Polonica. 1998;**45**(3):833–845

[67] Ion RM. Spectral analysis of the porphyrins incorporation into human blood. Journal of Biomedical Optics. 1999;**4**(3):319

[68] Ion RM. Porphyrins for tumor destruction in photodynamic therapy. Current Topics on Biophysics. 2000;**24**(1):30–42

[69] Ion RM, Sorescu AA, Nuta A. The photodegradation of Cl_2TiTPP. Proceeding of the GV-Conference, Zilina: Publishing Society; 2015: pp. 217–230

[70] Ion RM, Mandravel C. Spectroscopic studies of aluminium porphyrins. Bulgarian Chemical Communications. **29**:2217

[71] Ion RM. Porphyrins for tumor destruction in photodynamic therapy. Current Topics on Biophysics. 2000;**24**(1):30–42

[72] Wrobel D, Hanyz I, Bartowiak R, Ion RM. Prompt Fluorescence and time-resolved delayed luminescence of porphyrin in organic solvents and polymer matrices. Journal of Fluorescence. 1998;**8**(3):191. DOI: http://dx.doi.org/10.1023/A:1022524814908

[73] Olejarz B, Bursa B, Szyperska I, Ion RM, Dudkowiak A. Spectral properties and deactivation processes of anionic porphyrin coupled with TiO_2 nanostructure. International Journal of Thermophysics. 2010;**31**(1):163–171

[74] Ion RM, Coca S. Spectrophotometric study of the photodegradation reaction of TPP-AliBu. Progress in Catalysis. 1993;**1**:54, http://dx.doi.org/10.1016/0304-5102(93)E0309-5

Chapter 10

Phthalocyanines: Alternative Sensitizers of TiO_2 to be Used in Photocatalysis

William A. Vallejo Lozada, Carlos Diaz-Uribe,
Cesar Quiñones, Mileidy Lerma,
Catalina Fajardo and Karen Navarro

Additional information is available at the end of the chapter

http://dx.doi.org/10.5772/67880

Abstract

Currently, titanium dioxide is a most researched semiconductor in photocatalysis field; however, practical applications of TiO_2 are limited due to high band gap (3.2 eV). In last decades, researchers implemented several strategies to improve photoactivity of TiO_2 in visible electromagnetic spectrum. Titanium dioxide (TiO_2) sensitization for absorption of naturals and/or synthetics organic dyes is an important research subject in the field, and it is an efficient method to develop practical application in waste treatment. In this chapter, we review main theoretical aspects of sensitization process of TiO_2 by phthalocyanines and its effect in photocatalytic properties. In the last section, we review reports of photocatalytic systems.

Keywords: phthalocyanines, sensitization, TiO_2, photocatalysis, reactive oxygen species.

1. Introduction

Photocatalytic process has received great attention as it is one of the most promising technologies within renewable energy technology projects [1]. Treatment of recalcitrant compound and solar recovery of polluted water is an intensive field of research around world laboratories. Currently, titanium dioxide (TiO_2) is one of the most important photocatalytic materials for environmental purification; this material is inexpensive, it is available in abundance on earth surface, it is environmental safe, its energy band limits are identical to redox level of water. However, despite so many features, TiO_2 has drawbacks such as fast recombination rate of photogenerated electron-hole pair, low quantum yield in the photocatalytic reactions in aqueous solutions, and it has a high band gap value (3.2 eV, only photocatalytic active under UV irradiation), these drawbacks limit its practical applications [2, 3]. Different methodologies improve TiO_2 properties in visible range of electromagnetic spectrum: (a) metal and non-metal doping [4, 5], (b) ion implantation [6], (c) composites with others semiconductors (e.g., CdS, ZnO ZrO_2, V_2O_5, and WO_3) [7–10], quantum dots sensitization [11], and (d) TiO_2 sensitization by physical and/or chemical adsorption of synthetics and/or naturals organic dyes, this strategy improves the TiO_2 photophysical properties in the visible range of the electromagnetic spectrum, and it is considered to be one of the most efficient methods to develop practical applications [12–14].

Through this chapter, we review theoretical aspects of photocatalytic process, information about TiO_2, sensitization process, physical and chemical properties of phthalocyanines, and last section presents reports of phthalocyanines as sensitizers in photocatalytic systems.

2. Photocatalytic process

Since pioneer work of Fujishima on photochemical water splitting [15], several fields on renewables energy application have been greatly developed and photocatalysis has emerged as one of the most promising approach for the sustainable organic pollutants decomposition in both water and air [16]. The basic photocatalytic principle relies on five steps: (a) reactants diffusion to the surface of semiconductor, (b) adsorption of reactants on semiconductor surface, (c) reaction on the surface of semiconductor (electromagnetic radiation absorption with electrical excitation, charge transfers of photogenerated charge carriers and both direct electronic transfer to pollutant, and reactive oxygen species), (d) desorption of products from semiconductor surface, and (e) diffusion of products from the surface of the semiconductor [17]. **Figure 1** shows general photocatalytic process, the schematic reactions are as follows:

$$TiO_2 + h\nu_{(E>E_g)} \rightarrow TiO_2(h_{vb}{}^{+}/e_{cb}{}^{-}) \tag{1}$$

$$TiO_2(h_{vb}{}^{+}/e_{cb}{}^{-}) \rightarrow TiO_2 + \text{Heat; bulk or surface recombination} \tag{2}$$

$$TiO_2(h_{vb}{}^{+}/e_{cb}{}^{-}) \rightarrow TiO_2 + h\nu\text{; bulk or surface recombination} \tag{3}$$

where (vb) and (cb) are valence and conduction bands of semiconductor, respectively, and P is the pollutant. Under correct electromagnetic irradiation (.), semiconductor absorbs this radiation and one electron can be excited from (vb) to (cb), in this stage one electron-hole pair (charge carriers) is generated (Eq. (1)), after that, charge carriers may recombine through both bulk and surface recombination. In the absence of electric field, recombination process is very fast (nanoseconds) and the extra energy is released as phonon (heat) or some times photons are emitted (Eqs. (2) and (3)). Furthermore, after the charge transfer, inverse process charge transfer from the adsorbed species on TiO_2 surface can occur.

$$(e_{cb}{}^{-}) + O_{2(ad)} \rightarrow O_2{}^{\bullet -}{}_{(ad)} \tag{4}$$

$$O_2{}^{\bullet -}{}_{(ad)} + H_2O \rightarrow OH^{\bullet}{}_{(ad)} + OH^{-}{}_{(ad)} + O_{2(ad)} \tag{5}$$

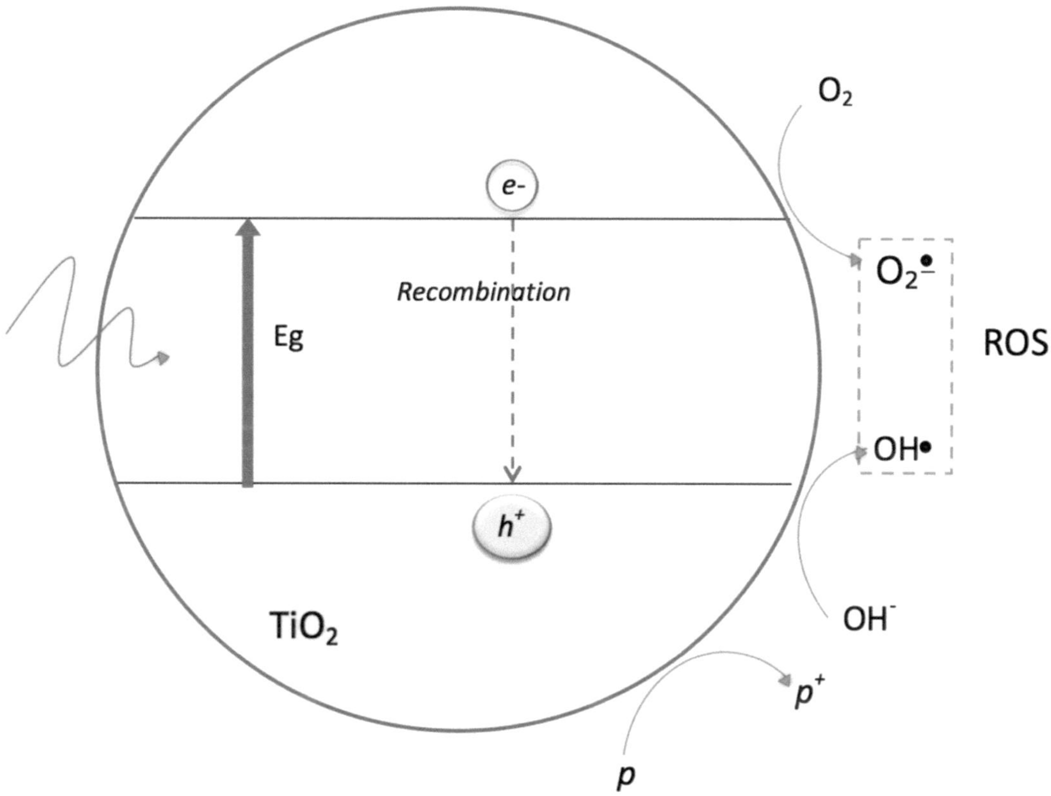

Figure 1. General photocatalityc process: absorption of electromagnetic radiation, charge pair genetarion, bulk and surface recombination, ROS generation.

Electron-hole pair separation follows different ways, oxygen prevents recombination by trapping electrons, and it generates superoxide radical anions, after that hydroxyl radicals are produced according to Eqs. (4) and (5). If the oxygen molecules are previously adsorbed on TiO_2 surface, it supports the electron transfer process.

$$(h_{vb}{}^{+}) + H_2O_{ad} \rightarrow OH^{\bullet}{}_{(ad)} + H^{+} \tag{6}$$

$$(h^{+}) + OH^{-}{}_{(ad)} \rightarrow OH^{\bullet}{}_{(ad)} \tag{7}$$

Furthermore, oxidation by holes yields more hydroxyl radicals molecules (Eqs. (6) and (7)). Hydroxyl radicals are powerful oxidizing species, they are considered as important species in the photocatalytic processes.

$$(h^{+}) + P_{(ad)} \rightarrow P^{\bullet+}{}_{(ad)} \tag{8}$$

$$OH^{\bullet}{}_{(ad)} + P_{(ad)} \rightarrow (P - OH^{\bullet})_{(ad)} \leftrightharpoons P^{\bullet+}{}_{(ad)} + OH^{-}{}_{(ad)} \tag{9}$$

The photodecomposition of pollutant proceeds through both parallel and consecutive reactions. Hydroxyl radicals can react directly with all organic pollutant (Eq. (9)). Some authors

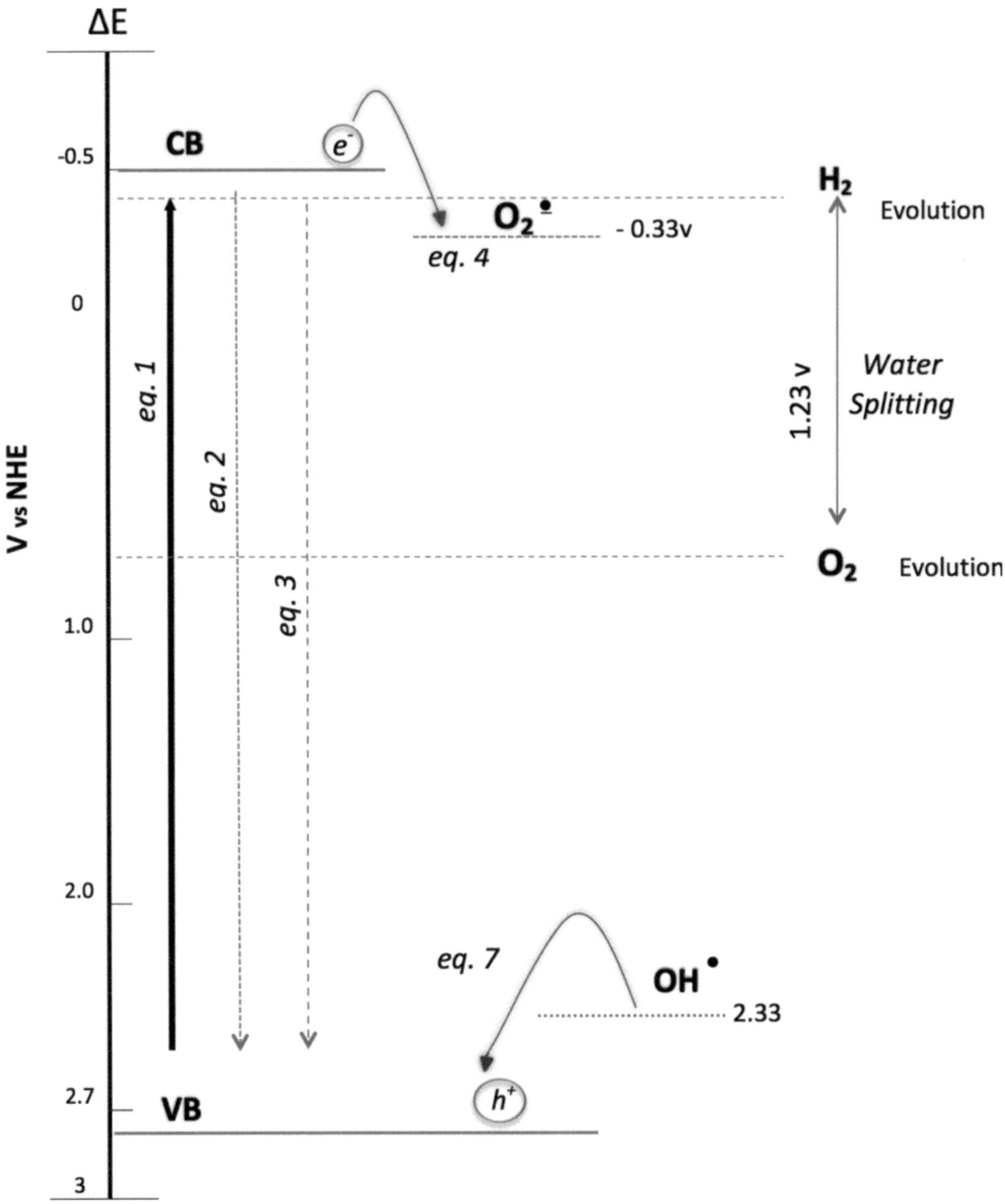

Figure 2. TiO_2 potential energy and ROS reaction generation to normal hydrogen electrode at pH = 7.0 (NHE) [20–22].

reported that photogenerated holes play a major role in TiO_2 photocatalysis. The redox potential of holes is thermodynamically suitable to oxidize almost any organic molecule [18]. Charge carrier trapping would suppress recombination and increase the lifetime of the separated electron and hole, other factors like e.g., surface area, crystallinity, trap density, etc. can affect photocatalytic performance of semiconductor [19].

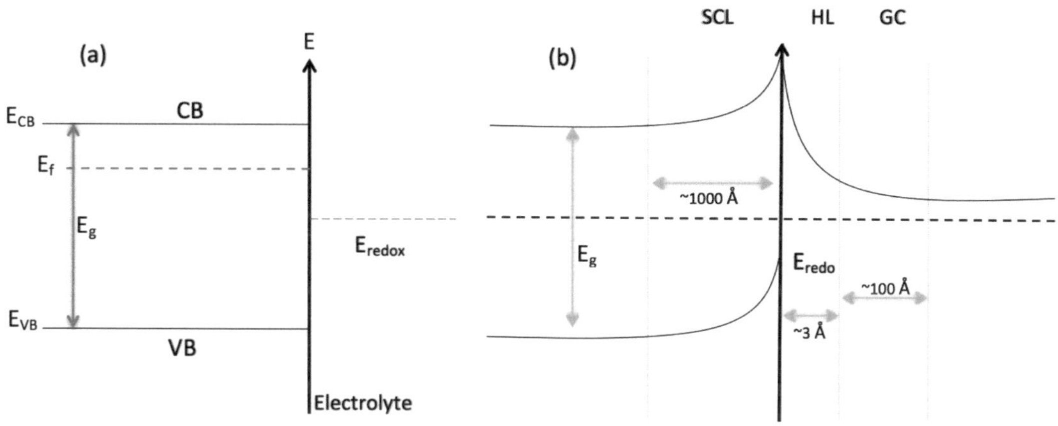

Figure 3. Energy band diagram for an n-type semiconductor before (a) and after (b) the equilibration of Fermi levels at the interface of semiconductor/electrolyte (inside, SCL: spatial charge region, HL: Helmholtz layer, GC: Gouy Chapman space). Adapted from Ref. [23].

Driving force of the photocatalytic process is the energy difference between position of band edges for the conduction/valence bands and the redox potential levels of solution. It determines the probability of the charge transfer processes. **Figure 2** shows band edge positions of TiO_2 and redox potential of different reactions of ROS generation. Furthermore, this figure shows TiO_2 potential energy related with normal hydrogen electrode at pH = 7.0 (NHE), in this scale, band gap of anatase-TiO_2 (3.2 eV) is located in a range −0.5 to 2.7 eV. Within this range, we can assign reactions (Eqs. (1)–(8)) in terms of energetic levels (redox potentials). Energy valence band of semiconductor must be lower than oxidant species (donor) and energy conduction band of semiconductor must be higher than reductant species (acceptor) [20–22].

Band bending in semiconductor occurs near to the junction due to the difference between potential redox of electrolyte and Fermi level (chemical potential) of semiconductor, **Figure 3** shows semiconductor/electrolyte junction. Helmholtz layer (Armstrong thickness) is generated on the semiconductor surface, this layer is generated by ions and/or molecules adsorbed on semiconductor surface, inside semiconductor as such as heterojuntion in solar cells, depletion zone is generated. Equilibrium between semiconductor surface and electrolyte is reached through flow of charge from one phase to another, and a band bending is formed within the semiconductor side; differences between Fermi levels of semiconductor and redox potential electrolyte determine band-bending in semiconductor. Space charge layer (SCL) is the region where bending band occurs and is characterized by the accumulation of charge at the surface. The extension of the SCL depends on the dielectric constant of the material and the concentration of donor impurities [23].

3. Titanium dioxide

Advantages of TiO_2 (e.g., it is a photocatalytic activity under UV irradiation, resistant to photocorrosion, innocuous to the nature, and inexpensive) have permitted to implement photocatalytic applications. Nowadays, TiO_2 is the most investigated material as a photocatalyst and a better understanding and improvement of catalytic reactions are main driving forces for surface investigations on TiO_2. **Figure 4** shows a comparison between TiO_2 and some photocatalysts with respect to NHE. Several semiconductors have reported photocatalytic activity or water splitting (e.g., Fe_2O_3 [24], CdS [25], Cu_2O [26], WO_3 [27], SnO_2 [28], ZnO [29]). Currently, more than 190 semiconductors have been assayed as suitable photocatalysts [30].

Titanium dioxide (TiO_2) has three polymorphs: (a) anatase, (b) rutile, and (c) brookite structures. In **Table 1**, some physical and chemical properties for these three polymorphs are

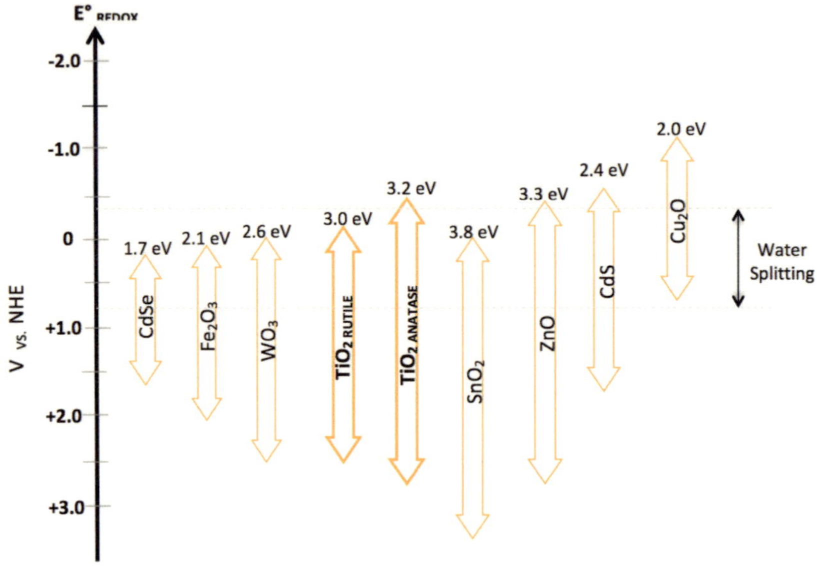

Figure 4. Band gap of some photocatalysts compared to the redox potential of different chemical species measured at 1 M, pH 7. Adapted from Ref. [22].

Crystal structure	System	Space group	Lattice constant (nm)			
			A	*b*	*c*	*c*/*a*
Rutile	Tetragonal	$D_{4h}^{14} - P4_2/mnm$	0.4584	–	0.2953	0.644
Anatase	Tetragonal	$D_{4h}^{19} - I4_1/amd$	0.3733	–	0.937	2.51
Brookite	Rhombohedral	$D_{2h}^{15} - Pbca$	0.5436	0.9166	0.5135	0.944
Density (kg/m^3)		Band gap energy (eV)		Standard heat capacity (J/mol °C)		
Rutile	4240	3.0 indirect		55.06		
Anatase	3830	3.2 indirect		55.52		
Brookite	4170	–		298.15		

Adapted from Ref. [31].

Table 1. TiO_2 bulk properties.

summarized. The thermodynamic most stable structure is brookite, while anatase is a metastable phase, it becomes rutile at high temperatures. All three polymorphs have octahedral structures that are differ in distortion of octahedron. In both anatase and brookite structures, octahedron is more distorted as oxygen atoms are very close. In rutile, the unit cell contains two TiO_2 units and the Ti and O coordination numbers are six and three, respectively. **Figure 5** shows the unit cells anatase and rutile [31, 32].

Titanium dioxide (TiO_2) is an n-type semiconductor, oxygen vacancies and Ti interstitials have been studied in rutile and anatase polymorphs, rutile-TiO_2 has an indirect band gap at 3.0 eV and anatase-TiO_2 has an indirect band gap at 3.2 eV [31]. The optical properties of TiO_2 can be determined by diffuse reflectance measurements that can be analyzed by Kubelka-Munk remission function [33]:

$$F(R_\alpha) = \frac{(1 - R_\alpha)^2}{2R_\alpha} \tag{10}$$

where R_α is the reflectance; $F(R_\alpha)$ is an indicative of the absorbance of the sample at particular wavelength value, it is proportional to the absorption constant of the material. The optical band gap by extrapolating the linear portion of the $(F(R_\alpha)hv)^2$ versus hv plot on the x-axis according to:

$$(F(R_\alpha)hv)^2 = A(hv - E_g) \tag{11}$$

where E_g is the band gap energy and A is a constant depending on the transition probability. Anatase polymorph has high photocatalytic activity; however due to high band gap value, TiO_2 is photoactive under UV irradiation. Currently, several strategies are used to improve the photoactivity of TiO_2 in the visible electromagnetic spectrum: metal doping, non-metal

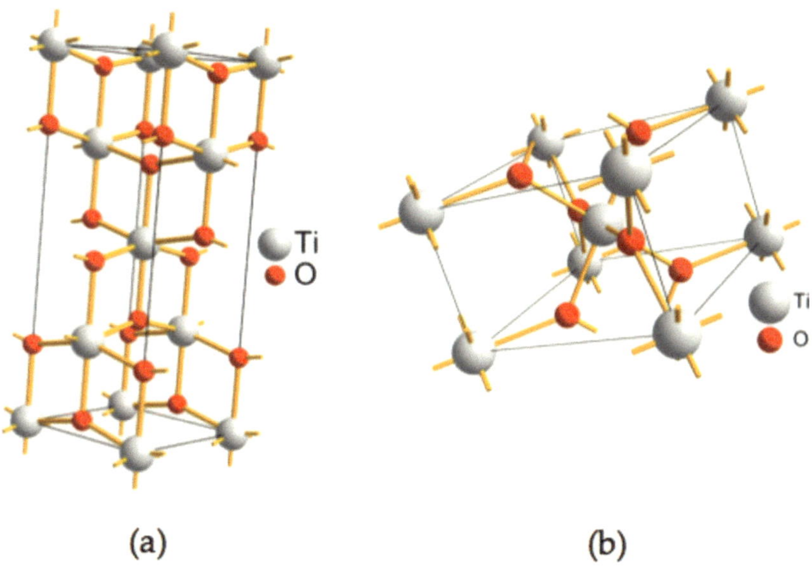

Figure 5. Crystal structures of TiO_2 polymorphs: (a) anatase and (b) rutile.

doping, composites with semiconductors having lower band gap energy, quantum dots, dye sensitization. In the next section, we shall delve into sensitization method [34].

4. Sensitization process

Currently, one of the main topic research on photocatalytic technologies is directed to use visible region of electromagnetic spectrum as source radiation its implementation allows to clean and cheap photochemical technologies are employed in wastewater treatment [35]. Titanium dioxide is photocatalytic active under UV radiation and sensitization improves the TiO_2 photoactivity in visible range of electromagnetic spectrum. In sensitization process, a molecular entity (sensitizer) alters other molecule (semiconductor) by energy transfer or electronic transfer from sensitizer to semiconductor after radiation absorption [36]. In this process, phthalocyanine absorbs the visible light, after that, it is excited to a state of greater energy leaving an electron in the lowest energy unoccupied molecular orbital (LUMO) orbital. This electron can be transferred to the conduction band of TiO_2, from where it is transferred to an oxygen molecule to produce superoxide anion (Eq. (4)). It begins degradation processes of contaminants such as dyes or even treatments of bacteria in water samples. **Figure 6** shows TiO_2 photosensitization for phthalocyanine through an ester bond. Several experimental works have confirmed that these phthalocyanines can be adhered on the TiO_2 surface if they are substituted by groups such as –COOH or $–SO_3$ which are capable of forming strong bonds on the surface of the semiconductor [37–39].

A photosensitizer (FS) is a molecule (e.g., organic or inorganic dyes), which induces photochemical and photophysical reactions after its excitation under electromagnetic radiation. Some requirements for FS are as follows [40–43]:

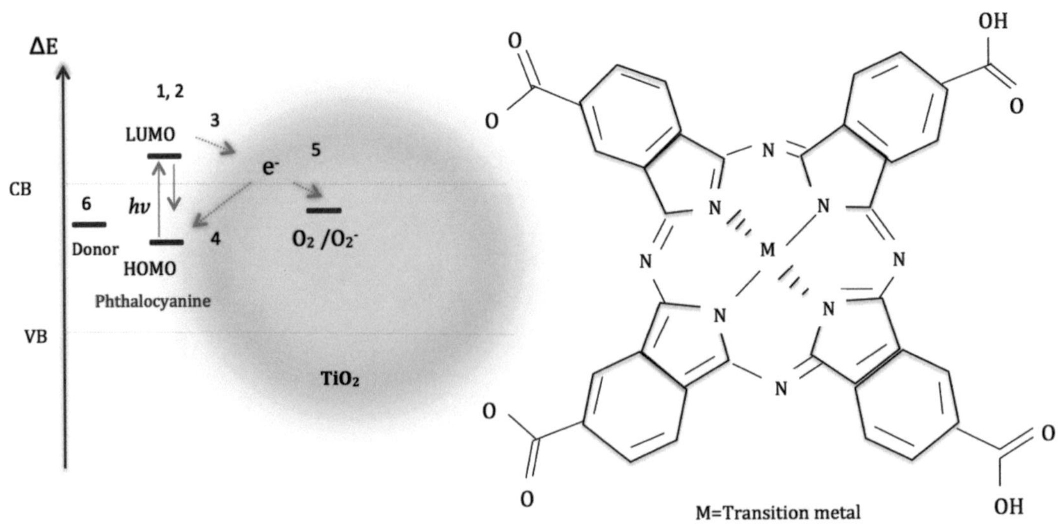

Figure 6. General scheme of TiO_2 photosensitization for phthalocyanines: metallic tetra carboxy phthalocyanines bonds to TiO_2 through carboxylate group ($—CO^{2-}$). Sensitization mechanism: (1) sensitizer excitation, (2) decay to basal state, (3) electron transfer, (4) recombination, (5) ROS generation, and (6) sensitizer is regenerated by electron donors (e.g., pollutant) [38].

http://dx.doi.org/10.5772/67880

- Intense visible light (400–750 nm) absorption, molar absorption coefficients greater than $10^4\ M^{-1}\ cm^{-1}$.
- High efficiency intersystem crossing singlet excited state (S^1) to the triplet excited state (T^1) and energy $T^1 > 95\ kJ\ mol^{-1}$ (energy singlet oxygen).
- Long life of excited states.
- High quantum yield of singlet oxygen production.
- High photostability.

Different organic dyes have been less investigated for sensitization of wide TiO_2 such as porphyrins, several transition-metal complexes and organic dyes (e.g., porphyrins and phthalocyanines) have been successfully employed as sensitizers in photocatalysis. After irradiation, photo-sensitizer in its basal state (^0FS) absorbs a photon of visible light ($10^{15}\ s^{-1}$), immediately it excites to singlet state (^1FS *) and returns to ground state by emitting fluorescence (10^6–$10^9\ s^{-1}$) or phosphorescence (10^{-2}–$10^4\ s^{-1}$). However, it could decay through intersystem crossing triplet state (^3FS*) that can react with a fundamental oxygen molecule through two types of reactions that lead to the generation of reactive oxygen species (ROS) that are highly cytotoxic for bacteria, fungi, organic pollutants, etc., such as (a) reaction type I yields superoxide and hydroxyl radical and (b) reaction type II yields singlet oxygen [44–46]. **Figure 7** shows general scheme of excitation and decay to sensitizer after irradiation absorption.

Phenalenone, bengal rose, methylene blue, ruthenium complexes, porphyrin derivatives, and phthalocyanines are the most used organic dyes as FS. **Table 2** shows the quantum yields to singlet oxygen production. Furthermore, **Figure 8** shows the typical structures of these sensitizers.

Reactive oxygen species (ROS) is the term applied to molecules more reactive than molecular O_2, oxygen disruption and excitation or/and sequential reduction of oxygen can generate these

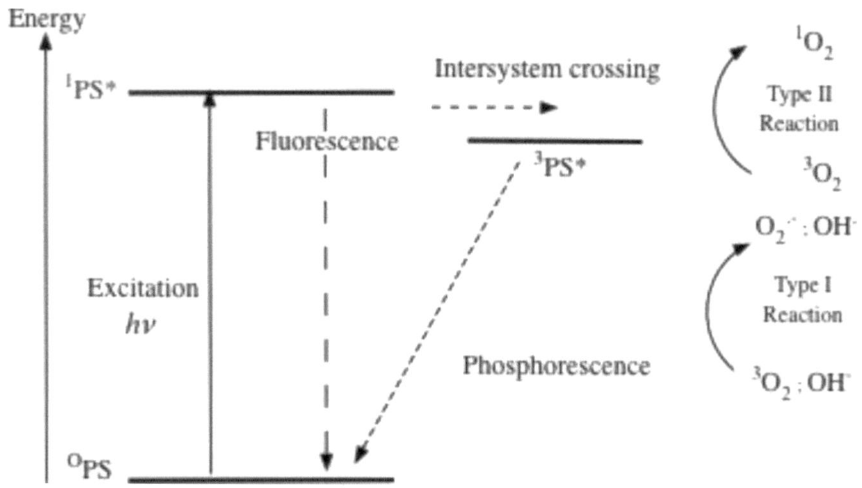

Figure 7. General scheme of excitation and decay of sensitizer after irradiation absorption. Adapted from Ref. [45].

Chemical compound	Quantum yield $^1O_2$ (ϕ_Δ)
Phenalenone	1.0
Bengal rose.	0.75
Methylene blue	0.50
Ruthenium complexes*	0.20–1.00
Phthalocyanines	0.60
Porphyrins	0.70

* For 2,2′-bipyridine chelating polyaceticheterocyclic ligands or 1,10-phenanthroline and their derivatives.

Table 2. List of several organic compounds and oxygen singlet quantum yield each one [42].

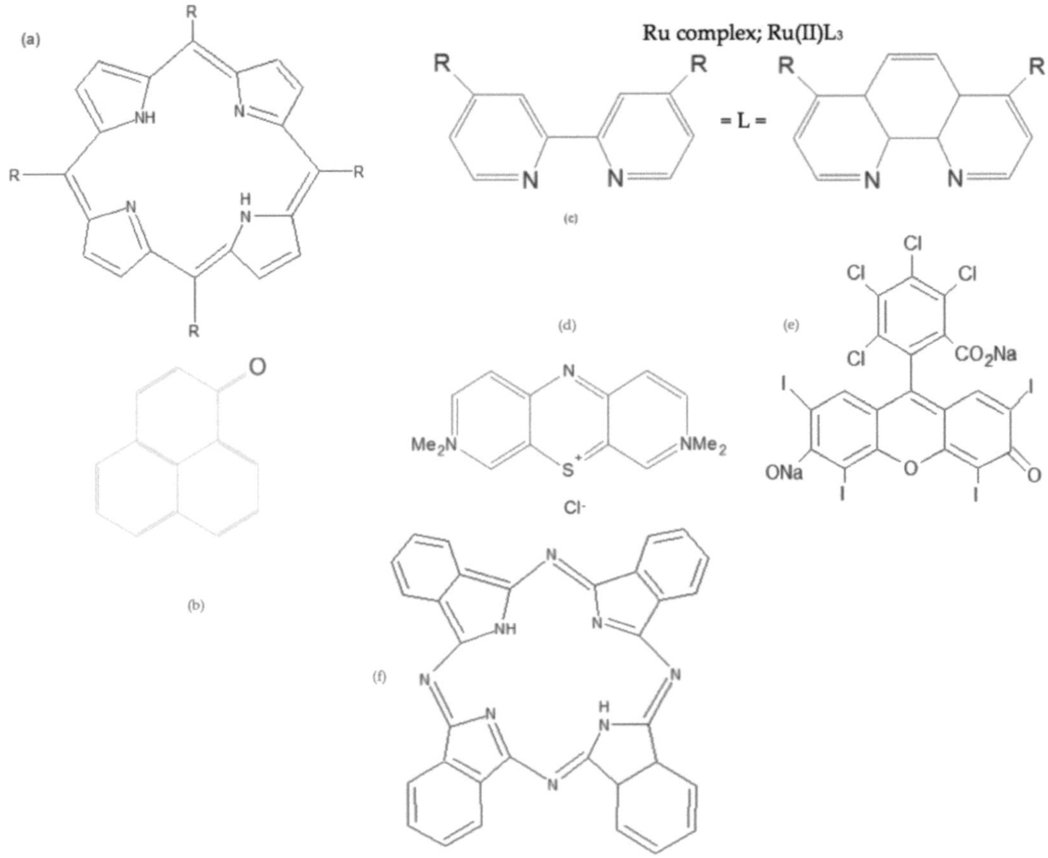

Figure 8. Chemical structures of typical sensitizers: (a) porphyrin, (b) phenalenone, (c) ruthenium complex, (d) methylene blue, (e) bengal rose, and (f) phthalocyanines [40].

species. **Figure 9** shows the sequence of reaction to ROS generation such as (a) O_2 can transforms into singlet oxygen ($^1O_2$) after electromagnetic radiation, (b) O_2 also can produce superoxide anion electron transfer and (c) in presence of ion hydronium anion, superoxide generates

http://dx.doi.org/10.5772/67880

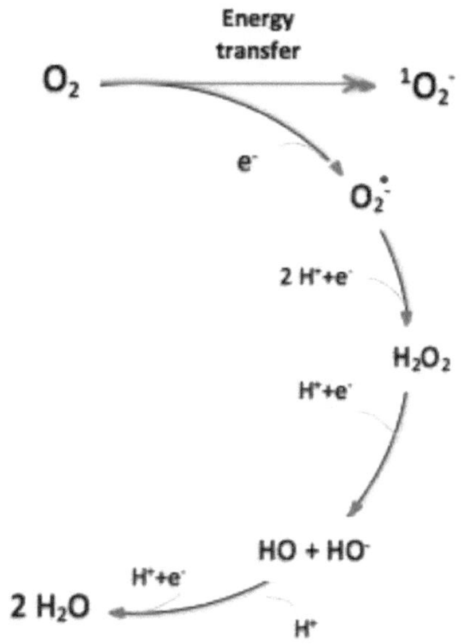

Figure 9. Scheme of cycle of main ROS generation. Adapted from Ref. [45].

peroxide hydrogen which produces hydroxyl radical and the hydroxyl ion, the latter can be protonated to form water [46, 47]. Reactive oxygen species (ROS) are highly reactive and they react to recalcitrant organic compounds and even bacteria [48].

5. Phthalocyanines: physical and chemical properties

Braun and Tcherniak synthesized phthalocyanines in 1907 (dark and insoluble solid known as acid phthalocyanines). **Figure 8** shows typical structure of phthalocyanines. Phthalocyanines are macrocycles composed by four isoindole groups; benzene and a pyrrole each form an isoindole group. The isoindole groups are linked together by four nitrogen atoms, thus phthalocyanines have eight nitrogen atoms and eight alternating carbon atoms with conjugated double bonds. Furthermore, four nitrogen atoms can act as ligands and coordinate an atom of some element (e.g., Zn, Cu, Ni, Fe, Co, etc.) to obtain metal complex phthalocyanines. Also, the benzene rings located at the periphery of the structure can be substituted with several functional groups to obtain several substituted phthalocyanines [49–55].

Metal phthalocyanines with carboxylic substituents (called metal tetra carboxy phthalocyanines), which are attractive for the sensitization process, exhibit high absorption coefficients in the visible region of the electromagnetic spectrum, high photo-stability to minimize photo-bleaching effects, and bound to TiO_2 surface through the carboxylate groups [56, 57]. Synthesis of metal phthalocyanines is usually carried out by transmetalation of acid phthalocyanines for using quinoline and 1-chloronaphthalene. Furthermore, Achar method is used in the synthesis of metal tetra carboxy phthalocyanines; it uses urea, anhydride trimethyl ether, and ammonium tetrathiomolybdate as catalyst. **Figure 10** shows Achar synthesis to Zn and Cu tetra carboxy phthalocyanines (TcPcZn and TcPcCu) [58].

Phthalocyanines properties have been extensively explored in several research fields, due to their 18 aromatic π electrons macrocycle phthalocyanines, similar to natural prophyrin, have high thermal and chemical stability and prominent electronic properties, and poor solubility in alcohols, ketones or ethers. Suitable solvents are those having high boiling points such as quinoline, trichlorobenzene, and some strong acids, however, their solubility is limited to a maximum amount 1 mg L^{-1} of solvent. Furthermore, for the metal-phthalocyanines, the solubility varies depending on the central atom; the TcPcCu and TcPcZn presented appreciable solubility in concentrated sulfuric acid and dimethylsulfoxide (DMSO) [59, 60]. Several phthalocyanines and their derivatives produce large quantities of singlet oxygen ($^1O_2$) and other reactive oxygen species (ROS) in the presence of light and molecular oxygen [57, 58]. They have been used to implement TFA and potential applications in photocatalysis in conjunction with TiO_2.

Optical properties are one of the most important physical chemistry characterizations of tetra carboxy phthalocyanines (TcPc). Typical UV-Vis absorption spectrum of TcPc in solution shows two bands: (a) Q bands located at near IR (600–800 nm), free-metal phthalocyanines shows four Q bands, and (b) soret band located at near UV (300–400 nm), both Q and soret bands correspond to $\pi \rightarrow \pi^*$. Phthalocyanines can be derived form dimmers even in dilute solutions, the self-associations are detected by the appearance of new absorption bands at higher energies. **Figure 11** shows absorption of UV-Vis spectrum of zn-tetra carboxy phthalocyanine dissolved in concentrated H_2SO_4. The Q-band of M-PCs has bathochromic effects through an extension of the

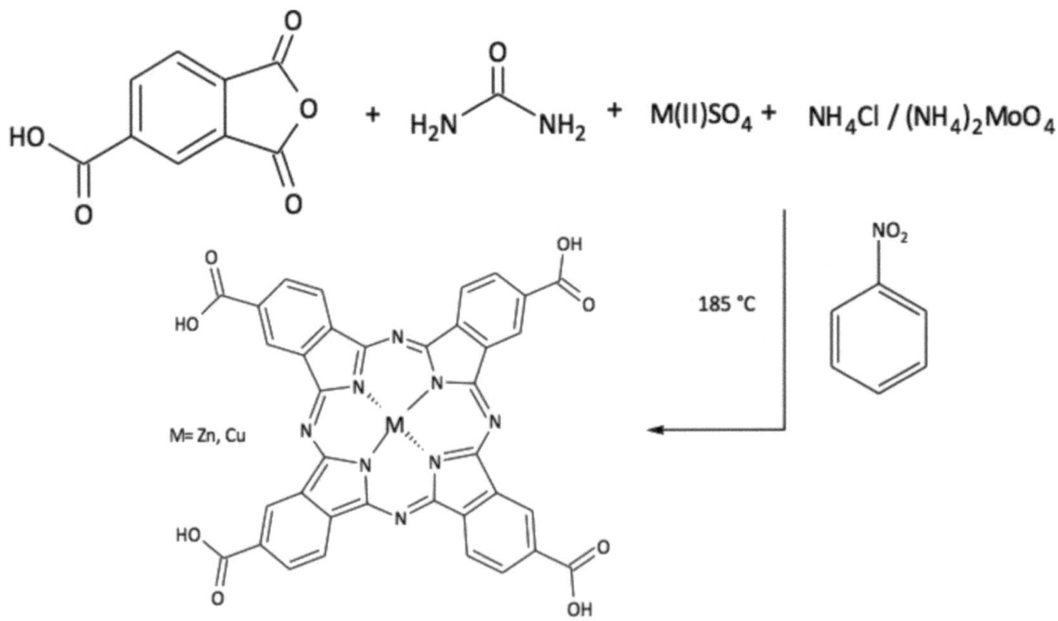

Figure 10. Synthesis of Zn and Cu tetra carboxy phthalocyanines (Method of Achar) [58].

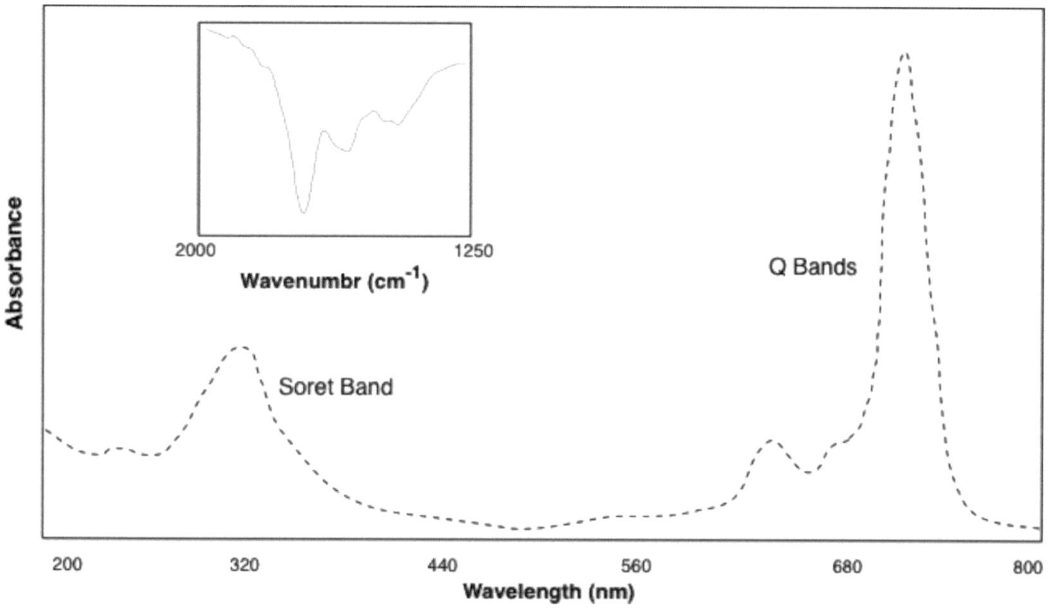

Figure 11. Absorption UV-Vis spectrum of zn-tetra carboxy phthalocyanine dissolved in concentrated H_2SO_4, (inside: shows carboxylates IR-signals to sensitization on TiO_2 thin films).

π conjugation system. Furthermore, shift of absorption maxima depends on change in electron distribution in the phthalocyanine macrocycle by the substituents [61, 62].

IR-spectra of tetracarboxyl phthalocyanines shows typical signals associated to —OH, —CH2, C—O, and beside specific signals located near 1650 cm^{-1} (C—N) and 3023 cm^{-1} (N—H), furthermore symmetric and asymmetric stretching furthermore signals associates to (—CO^{2-}) are reported to tetracarboxyl phthalocyanines [63]. The tetracarboxyl phthalocyanines can interact on TiO_2 surface by two ways, first through a very strong physical adsorption and second through chemical adsorption of reaction of carboxylic acids with group Ti—OH on TiO_2 surface; phthalocyanines could be absorbed as carboxylates on the semiconductor surface (**Figure 11**, shows carboxylates IR-signals) [64, 65].

6. Photocatalytic efficiency and perspectives

In typical photocatalytic test, visible and/or UV lamps are used as radiation source, catalyst is used in suspension and/or thin film form, temperature and pH maintain constant values. Sometimes O_2 is bubbling into solution and some authors incorporate ultrasonic bath. Prior to irradiation, the suspension is magnetically stirred in the dark to reach dye adsorption–desorption equilibrium on TiO_2 surface, photodegradation progress is followed by spectro-photometric methodology.

First-order decay in kinetic describes profile of photocatalytic degradation and others dyes [66]. Previous studies have showed that photocatalytic degradation rate of textile dyes in heterogeneous photocatalytic oxidation systems under UV-light illumination followed the Langmuir-Hinshelwood (L-H) kinetics model [67–69]:

$$v = -d[C]/dt = \frac{kK[C]}{1+K[C]} \tag{12}$$

where v is the rate of dye mineralization, k is the rate constant, C is the pollutant concentration, and K is the adsorption coefficient. Eq. (12) can be solved explicitly for t by using discrete changes in pollutant from the initial concentration to a zero reference point. However, apparent first-order model is used when the concentration of substrate is in the scale of millimoles $1 \gg K\,[C]$:

$$v - d[C]/dt = k_{app}[C] \tag{13}$$

$$[C]_t = [C_o]e^{k_{app}t} \tag{14}$$

where time (t) at minutes and k_{app} is the apparent reaction rate constant (k_{app} =k*K; min^{-1}); the assumption of a pseudo first-order model was used in several studies to characterize the effect of different experimental conditions on the degradation rate [70–72]. **Table 3** lists k_{app} values to several photocatalytic studies. Langmuir-Hinshelwood kinetics permits to compare different photocatalytic yield, and it is a reference methodology to analyze several conditions of photocatalytic tests to develop pilot and industrial applications.

Basic applications in photocatalysis comprise: (a) hydrogen production, (b) cleaning (e.g., filter to gas phase), (c) water waste treatment, (d) self-cleaning systems, and (e) emerging application (e.g., medical, mechanical, self-cooling, and others). Reports indicate that more than 1000 companies are interested in photocatalytic applications. Nowadays, the global market for photocatalyst material and products reach nearly to $1.5 billion in 2014 and forecast to reach nearly $2.9 billion in 2020. **Figure 12** shows photocatalytic products and their distribution in market. Main application includes exterior building materials, since 1990s several industries have installed photocatalytic filters with UV lamps inside air purifiers; applications in deodorization are implemented in hospitals and officers. Last year, both exterior and road materials have been increased [76–81].

Photocatalyst	Pollutant	k_{app} (min^{-1}) ×10^{-3}	Efficiency (%)
Degussa P-25-TiO_2 [72]	Styrene	44.5	95
PEG-2000 TiO_2 [73]	Rhodamine	0.93	–
Nanotubes arrays-TiO_2 /Ag nanoparticles [74]	Ethylene	0.32	–
Natural rubber-TiO_2 [75]	Methylene Blue	38.3	90
Iron (III) Tetracarboxyphthalocyanine-TiO_2 [58]	Methylene blue	27.9	30

Table 3. Langmuir-Hinshelwood (L-H) kinetics model to photocatalytic treatments.

http://dx.doi.org/10.5772/67880

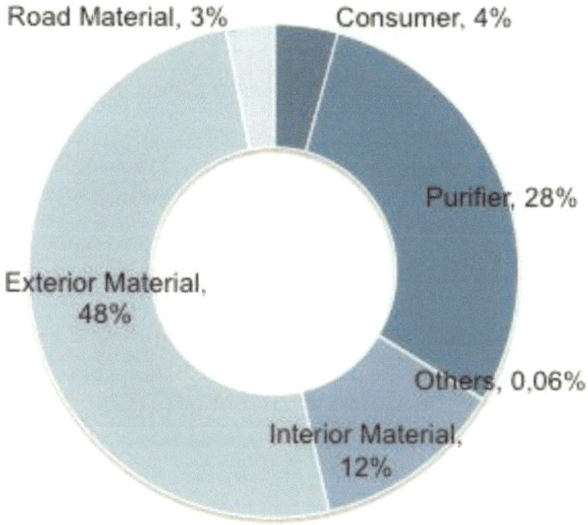

Figure 12. Photocatalytic products and their distribution in market [80].

Photocatalytic applications report suitable photodegradation efficiency in several systems (e.g., TiO_2, sol-gel, thin films, TiO_2/active carbon, TiO_2/sensitizer and others), intense research around world impulse this growing and currently several application in both industrial and pilot-scale have been developed to treatment different pollutants (e.g., hard metals, herbicides, dyes, drugs and others); several reports indicate photocatalytic system continue to grow as part of environmental technology developments based on renewable energies.

7. Conclusions

Titanium dioxide (TiO_2) is most used semiconductor for wastewater treatment and several strategies (e.g., quantum dots, noble metal deposition, coupled semiconductor, ion modification, doping sensitization, etc.) and allows improving photocatalytic activity in a visible range. In this chapter, we review some physical and chemical properties of phthalocyanines as sensitizers in photocatalytic systems. We present different aspects of basic principle and development of these systems. Sensitization is an economic and technological option to improve photo-activity of TiO_2 in visible range of electromagnetic spectrum. Photocatalytic technology has several advantages over its homologues in wastewater treatment, and nowadays, high efficiency of different systems has been developed in both laboratory and pilot-scale. Furthermore, industrial applications can be address to develop photocatalytic system treats several pollutants (two or more pollutants simultaneously). Finally, photocatalytic applications are a real option in the field of renewable energy technologies, and today, these are one of the most important environmental technologies around the world.

Author details

William A. Vallejo Lozada[1]*, Carlos Diaz-Uribe[1], Cesar Quiñones[2], Mileidy Lerma[1], Catalina Fajardo[1] and Karen Navarro[3]

*Address all correspondence to: williamvallejo@mail.uniatlantico.edu.co

1 Universidad del Atlántico, Barranquilla, Colombia

2 Instituto Politécnico Gran Colombiano, Bogotá, Colombia

3 Universidad Nacional de Córdoba, Cordoba, Argentina

References

[1] Zhanga W, Liu G, Shi L, Liu H, Wang T, Ye J. Engineering coordination polymers for photocatalysis. Nano Energy. 22 (2016) 149–168. DOI: 10.1016/j.nanoen.2016.01.029

[2] Mendonc V, Lopes OF, Fregonesi RP, Giraldic TR, Ribeirod C. TiO_2-SnO_2 heterostructures applied to dye photodegradation: the relationship between variables of synthesis and photocatalytic. Applied Surface Science. 298 (2014) 182–191. DOI: 10.1016/j.apsusc.2014.01.157

[3] Habib U, Asif T, Tapas Ma. Polypyrrole/TiO_2 composites for the application of photocatalysis. Sensors and Actuators B. 241 (2017) 1161–1169. DOI: 10.1016/j.snb.2016.10.019

[4] Murashkina A, Murzin P, Rudakova A, Ryabchuk V, Emeline A, Bahnemann D. Influence of the dopant concentration on the photocatalytic activity: Al-doped TiO_2. The Journal of Physical Chemistry. C. 119(44) (2015) 24695. DOI: 10.1021/acs.jpcc.5b06252.

[5] Valentin C, Pacchioni G. Trends in non-metal doping of anatase TiO_2: B, C, N and F. Catalysis Today. 206 (2013) 12–18. DOI: 10.1016/j.cattod.2011.11.030.

[6] Hou X, Hao F, Fan B, Gu X, Wu X, Liu A. Modification of TiO_2 photocatalytic films by V+ ion implantation. Nuclear Instruments and Methods in Physics Research Section B: Beam Interactions with Materials and Atoms. 243(1) January (2006) 99–102. DOI: 10.1016/j.nimb.2005.07.195.

[7] Kim J, Choi J, Lee Y, Hong J, Lee J, Yang J, Lee W, Hur H. Enhanced photocatalytic activity in composites of TiO_2 nanotubes and CdS nanoparticles. Chemical Communications. 48 (2006) 5024–5026. DOI: 10.1039/B612572G.

[8] Dulamita N, Maicaneanu A, Sayle D, Stanca M, Craciun R, Olea M, Afloroaei C, Fodor A. Ethylbenzene dehydrogenation on Fe_2O_3-Cr_2O_3-K_2CO_3 catalysts promoted with transitional metal oxides. Applied Catalysis A. 287 (2005) 9. DOI: 10.1016/j.apcata.2005.02.037.

[9] Salinas D, Guerrero S, Araya P. Transesterification of canola oil on potassium-supported TiO_2 catalysts. Catalysis Communications. 11 (2010) 773. DOI: 10.1016/j.catcom.2010.02.013.

[10] Cheng C, Amini A, Zhu C, Xu Z, Song H, Wang N. Enhanced photocatalytic performance of TiO_2-ZnO hybrid nanostructures. Scientific Reports 4, Article number: 4181 (2014). DOI: 10.1038/srep04181

[11] Pernik D, Tvrdy K, Radich J, Kamat P. Tracking the adsorption and electron injection rates of cdse quantum dots on TiO_2: linked versus direct attachment. The Journal of Physical Chemistry C. 115 (27) (2011) 13511–13519. DOI: 10.1021/jp203055d

[12] Ludin N, Mahmoud A, Mohamad A, Kadhum A, Sopian K, Karim N. Review on the development of natural dye photosensitizer for dye-sensitized solar cells. Renewable and Sustainable Energy Reviews. 31 (2014) 386–396. DOI: 10.1016/j.rser.2013.12.001

[13] Chiu-Hsuan L, Shie J, Tsai C, Yang Y, Chang C. Photocatalytic Decomposition of Indoor Air Pollution Using Dye-Sensitized TiO2 Induced by Anthocyanin and Ru Complexes. Journal of Clean Energy Technologies. 1 (2), April 2013. 10.7763/JOCET.2013.V1.27

[14] Huang C, Lv Y, Zhou Q, Kang S, Li X, Mu J. Visible photocatalytic activity and photoelectrochemical behavior of TiO_2 nanoparticles modified with metal porphyrins containing hydroxyl group. Ceramics International. 40 (2014) 7093–7098. DOI: 10.1016/j.ceramint.2013.12.042

[15] Fujishima A, Honda K. Electrochemical photolysis of water at a semiconductor electrode. Nature. 238 (1972) 37. DOI: 10.1038/238037a0

[16] Zong X, Wang L. Ion-exchangeable semiconductor materials for visible light-induced photocatalysis. Journal of Photochemistry and Photobiology C: Photochemistry Reviews. 18 (2014) 32–49. DOI: 10.1016/j.jphotochemrev.2013.10.001

[17] Serpone N, Emeline A. Suggested terms and definitions in photocatalysis and radiocatalysis. International Journal of Photoenergy. 4 (2002) 91. DOI: 10.1155/S1110662X02000144

[18] Ishibashia K, Fujishima A, Watanabe T, Hashimoto K. Quantum yields of active oxidative species formed on TiO_2 photocatalyst. Journal of Photochemistry and Photobiology A: Chemistry. 134 (2000) 139–142. DOI: 0.1016/S1010-6030(00)00264-1

[19] Maeda Y, Okumura M, Tsubota S, Kohyama M, Haruta M. Local barrier height of Au nanoparticles on a TiO_2(1 1 0)-(1 × 2) surface. Applied Surface Science. 222 (2004) 409–414. DOI: 10.1016/j.apsusc.2003.09.007

[20] Ola O, Maroto-Valer M. Review of material design and reactor engineering on TiO_2 photocatalysis for CO_2 reduction. Journal of Photochemistry and Photobiology C: Photochemistry Reviews. 24 (2015) 16–42. DOI: 10.1016/j.jphotochemrev.2015.06.001

[21] Comninellis C, Chen G. Electrochemistry for Environment. Ed. Springer. New York, 2010, Ch 16, pp. 399–405.

[22] Linsebigler A, Lu G, Yates J. Photocatalysis on TiO_2 surfaces: principles, mechanisms, and selected results, Chemical Reviews. 95 (3) (1995) 735–758. DOI: 10.1021/cr00035a013

[23] Garcia G, Tasso T, Valnice M. Enhancement of Photoelectrocatalysis Efficiency by Using Nanostructured Electrodes, Modern Electrochemical Methods in Nano, Surface and Corrosion Science, Dr. M. Aliofkhazraei (Ed.), InTech, (2014), Rijeka, Croatia. DOI: 10.5772/58333.

[24] Mishra M, Chun D. α-Fe_2O_3 as a photocatalytic material: a review. Applied Catalysis. A: General. 498 (2015) 126–141. DOI: 10.1016/j.apcata.2015.03.023

[25] Zhao D, Yang C. Recent advances in the TiO_2/CdS nanocomposite used for photocatalytic hydrogen production and quantum-dot-sensitized solar cells. Renewable and Sustainable Energy Reviews 54 (2016) 1048–1059. DOI: 10.1016/j.rser.2015.10.100

[26] Al-Jawhari H. A review of recent advances in transparent p-type Cu_2O-based thin film transistors. Materials Science in Semiconductor Processing. 40 (2015) 241–252. 10.1016/j.mssp.2015.06.063

[27] Georgiev J, Valov E, Armyanov S, Philippidis N, Poulios I, Sotiropoulos S. Bi-component semiconductor oxide photoanodes for the photoelectrocatalytic oxidation of organic solutes and vapours: a short review with emphasis to TiO_2–WO_3 photoanodes. Journal of Hazardous Materials. 211–212 (2012) 30–46. DOI: 10.1016/j.jhazmat.2011.11.069

[28] Gupta S, Tripathi M. A review of TiO_2 nanoparticles. The Chinese Science Bulletin. 56 (16) (2011) 1639–1657. DOI: 10.1007/s11434-011-4476-1

[29] Morasae Samadi, Mohammad Zirak, Amene Naseri, Elham Khorashadizade, Moshfegh Alirez Z. Recent progress on doped ZnO nanostructures for visible-light photocatalysis. Thin Solid Films 605 (2016) 2–19. DOI: 10.1016/j.tsf.2015.12.064

[30] Hassan M, Zhao Y, Xie B. Employing TiO2 photocatalysis to deal with landfill leachate: current status and development. Chemical Engineering Journal. 285 (2016) 264–275. 10.1016/j.cej.2015.09.093

[31] Diebold U. The surface science of titanium dioxide. Surface Science Reports. 48 (2003) 53–229. DOI: 0.1016/S0167-5729(02)00100-0

[32] Rahimi N, Pax R, MacA A. Review of functional titanium oxides. I: TiO_2 and its modifications. Progress in Solid State Chemistry 44 (2016) 86–105.

[33] Murphy A. Band-gap determination from diffuse reflectance measurements of semiconductor films, and application to photoelectrochemical water-splitting. Solar Energy Mater. Solar Cells. 91 (2007) 1326–1337. 10.1016/j.solmat.2007.05.005

[34] Varshney G, Kanel S, Kempisty D, Varshney V, Agrawal A, Sahle-Demessie E, Varma R, Nadagoud M. Nanoscale TiO_2 films and their application in remediation of organic pollutants. Coordination Chemistry Reviews. 306 (2016) 43–64. DOI: 10.1016/j.ccr.2015.06.011

[35] Ferro S, Ricchelli F, Monti D, Mancini G, Jori G. Efficient photoinactivation of methicillin-resistant Staphylococcus aureus by a novel porphyrin incorporated into a poly-cationic liposome. The International Journal of Biochemistry & Cell Biology. 39 (2007) 1026–1034. DOI: 10.1016/j.biocel.2007.02.001

[36] Isci Ü, Beyreis M, Tortik N,Topal SZ, Glueck M. Methylsulfonyl Zn phthalocyanine: a polyvalent and powerful hydrophobic photosensitizer with a wide spectrum of photodynamic applications. Photodiagnosis and Photodynamic Therapy. 13 (2016) 40–47. DOI: 10.1016/j.pdpdt.2015.10.010

[37] Wilkinson F, Helman WP, Ross AB, et al. Quantum yields for the photosensitized formation of the lowest electronically excited. Journal of Physical and Chemical Reference Data. 22 (1993) 113–262. DOI: 10.1063/1.555934

[38] Granados G, Páez CA, Martínez F, Páez-Mozo E Photocatalytic degradation of phenol on TiO_2 and TiO_2/Pt sensitized with metallophthalocyanines. Catalysis Today. 107–108 (2005) 589–594. DOI: 10.1016/j.cattod.2005.07.021.

[39] Vargas E, Vargas R, Núñez O. A TiO2 surface modified with copper (II) phthalocyanine-tetrasulfonicacid tetrasodium salt as a catalyst during photoinduced dichlorvos mineralization by visible solar light. Applied Catalysis B: Environmental. 156–157 (2014) 8–14. DOI: 10.1016/j.apcatb.2014.02.045

[40] Islam A, Sugihara H, Hara K, Singh LP, Katoh P, Yanagida M, et al. Dye sensitization of nanocrystalline titanium dioxide with square planar platinum (II). Inorganic chemistry. 40 (2001) 5371–5380. DOI: 10.1021/ic010391

[41] Braun AM, Maurette MT, Oliveros E. Photochemical Technology. John Wiley & Sons. Chichester, (1991).

[42] García Fresnadillo D. Fotosensibilización y el sensibilizador:síntesis, propiedades y limitaciones. in; J. Blanco, M. Blesa (Eds.), Solar Safe Water. Puerto Iguazú, (2005), pp. 227–242.

[43] Paczkowskit J, Neckers DC. Photochemical properties of rose bengal. 11. Fundamental studies in heterogeneous energy transfer. Macromolecules. 18 (12) (1985) 2412–2418. DOI: 10.1021/ma00154a013

[44] Thomas J. L., Allen N. S. The degradation of dyed cotton fabrics by the sensitised production of singlet oxygen via an aqueous soluble phthalocyanine dye. Dyes and Pigments. 53 (2002) 195–217. DOI: 10.1016/S0143-7208(02)00020-7

[45] Konan Y. N, Gurny R, Allémann E. State of the art in the delivery of photosensitizers for photodynamic therapy. Journal of Photochemistry and Photobiology B: Biology. 66 (2002) 89–106. DOI: 10.1016/S1011-1344(01)00267-6

[46] Krumova K, Cosa G. Chapter 1: overview of reactive oxygen species, in singlet oxygen: applications in Biosciences and Nanosciences. Vol. 1 (2016), Rsc Publising Washington DC, USA. pp. 1–21. DOI: 10.1039/9781782622208-00001

[47] Chen P, Wang F, Chen Z. F, Zhang Q, Su Y, Shen L, et al. Study on the photocatalytic mechanism and detoxicity of gemfibrozil by a sunlight-driven TiO_2/carbon dots photocatalyst: the significant roles of reactive oxygen species. Applied Catalysis B: Environmental. 204 (2017) 250–259. DOI: 10.1016/j.apcatb.2016.11.040

[48] Celli J. P, Spring B. Q, Rizvi I, Evans C. L, Samkoe K. S, Verma S, Pogue B. W, Hasan T, Imaging and photodynamic therapy: mechanisms, monitoring and optimization. Chemical Reviews. 110 (5) (2010) 2795–2838. DOI: 10.1021/cr900300

[49] C. C. Leznoff (Editor), A. B. P. Lever (Editor). Phthalocyanines, Properties and Applications, Volume 1, Wiley, Hoboken, New Jersey, 1989; ISBN: 978-0-471-18720-2

[50] Lopez T, Ortiz E, Alvarez M, Navarrete J, Odriozola JA, Martinez-Ortega F. Study of the stabilization of zinc phthalocyanine in sol-gel TiO_2 for photodynamic therapy applications. Nanomedicine: Nanotechnology, Biology, and Medicine. 6 (6) (2010) 777–785. DOI: 10.1016/j.nano.2010.04.007

[51] Banerjee A, Kundu B, Pal A. J. Introducing immobilized metal phthalocyanines as spin-injection and detection layers in organic spin-valves: spin-tunneling and spin-transport regimes. Organic Electronics. 41 (2017) 173–178. DOI: 10.1016/j.orgel.2016.10.045

[52] Chauke V. P, Arslanoglu Y, Nyokong T. Synthesis and photophysical behaviour of tantalum and titanium phthalocyanines in the presence of gold nanoparticles: photocatalysis towards the oxidation of cyclohexene. Journal of Photochemistry and Photobiology A: Chemistry. 221 (2011) 38–46. DOI: 10.1016/j.jphotochem.2011.04.020

[53] Gülmez A. D, Polyakov M. S, Volchek V. V, Kostakoğlu S. T, Esenpinar A. A, Basova T. V, et al. Tetrasubstituted copper phthalocyanines: correlation between liquid crystalline properties, films alignment and sensing properties. Sensors and Actuators B. 241 (2017) 364–375. DOI: 10.1016/j.snb.2016.10.073

[54] Tortelli S, Cavazzini M, Orlandi S, Pozzi G, Pecnikaj I, et al. Property tuning in unsymmetrical alkoxy zinc phthalocyanines by introduction of perfluoro-tert-butoxy end groups. Journal of Fluorine Chemistry. 188 (2016) 110–116. DOI: 10.1016/j.jfluchem.2016.06.012

[55] Köksoy M. A, Köksoy B, Durmuş M, Bulut M. Preparation, characterization and photophysicochemical properties of novel tetra7-(diethyl 2-methylmalonatoxy)-3-p-oxyphenyl)coumarin-substituted zinc(II) and indium(III)chloride phthalocyanines. Journal of Organometallic Chemistry. 822 (2016) 125–134 . DOI: 10.1016/j.jorganchem.2016.08.009

[56] Chen Z, Zhou S, Chen J, Li L, Hua P, et al. An effective zinc phthalocyanine derivative for photodynamic antimicrobial chemotherapy. Journal of Luminescence. 152 (2014) 103–107. DOI: 10.1016/j.jlumin.2013.10.067

[57] Vallejo W, Diaz C, Cantillo A. Methylene blue photocatalytic degradation under visible irradiation on TiO_2 thin films sensitized with Cu and Zn tetracarboxy-phthalocyanines. Journal of Photochemistry and Photobiology A: Chemistry 299 (2015) 80–86. DOI: 10.1016/j.jphotochem.2014.11.009

[58] Achar B, Parker G, Keshavaya J. Preparation and structural investigations of Cu (II), Co (II), Ni (II) and Zn (II) derivatives of 2, 9, 16, 23-phthalocyanine tetracarboxylic acids. Indian Journal of Chemistry. 27 (1986) 411.

[59] Mantareva V, Durmuş M, Aliosman M, Stoineva I, Angelov I. Lutetium(III) acetate phthalocyanines for photodynamic therapy applications: synthesis and photophysicochemical properties. Photodiagnosis and Photodynamic Therapy. 14 (2016) 98–103. DOI: http://dx.doi.org/doi:10.1016/j.pdpdt.2016.02.008

[60] Ebrahimian A, Zanjanchi M. A, Noei H, Arvand M, Wang Y. TiO_2 nanoparticles containing sulphonated cobalt phthalocyanine: preparation, characterization and photocatalytic

performance. Journal of Environmental Chemical Engineering. 2 (2014) 484–494. DOI: 10.1016/j.jece.2014.01.022

[61] Bovill A. J, McConnel A. A, Nimmo J. A, Smith W. E. Resonance Raman spectra of a-copper phthalocyanine. The Journal of Physical Chemistry. 90 (4) (1986) 569–575.

[62] Sakamoto K, Ohno-Okumura E. Syntheses and functional properties of phthalocyanines. Materials. 2 (2009) 1127–1179. DOI: 10.3390/ma2031127

[63] Konstantinou I. K, Albanis T. A. TiO_2-assisted photocatalytic degradation of azo dyes in aqueous solution: kinetic and mechanistic investigations: a review. Applied Catalysis B: Environmental. 49 (2004) 1–14. DOI: 10.1016/j.apcatb.2003.11.010

[64] Lakshmi S, Renganathan R, Fujita S. Study on TiO_2-mediated photocatalytic degradation of methylene blue. Journal of Photochemistry and Photobiology A: Chemistry. 88 (1995) 163–167.

[65] Seoudi R, El-Bahy G. S, El Sayed Z. A. FTIR, TGA and DC electrical conductivity studies of phthalocyanine and its complexes. Journal of Molecular Structure. 753 (2005) 119–126. DOI: 10.1016/j.molstruc.2005.06.003

[66] Ghasemi S, Rahimnejad S, Rahman Setayesh S, Hosseini M, Gholami M. R. Kinetics investigation of the photocatalytic degradation of acid blue 92 in aqueous solution using nanocrystalline TiO_2 prepared in an ionic liquid. Progress Reaction Kinetics and Mechanism. 34 (2009) 55–76. DOI: 10.3184/146867809X413247

[67] Arami M, Yousefi Limaee N, Mahmoodi N. M, Salman Tabrizi N. Equilibrium and kinetics studies for the adsorption of direct and acid dyes from aqueous solution by soy meal hull. The Journal of Hazardous Materials B. 135 (2006) 171–179. DOI: 10.1016/j.jhazmat.2005.11.044

[68] Konstantinou I, Albanis T. TiO_2-assisted photocatalytic degradation of azo dyes in aqueous solution: kinetic and mechanistic investigations: a review. Applied Catalysis B: Environmental. 49 (2004) 1–14. 10.1016/j.apcatb.2003.11.010

[69] Faraji M, Mohaghegh N. Ag/TiO_2-nanotube plates coated with reduced graphene oxide as photocatalysts. Surface and Coatings Technology. 288 (2016) 144–150.

[70] Lakshmi S, Renganathan R, Fujita S. Study on TiO_2-mediated photocatalytic degradation of methylene blue. Journal of Photochemistry Photobiology. A. Chemistry. 88 (1995) 163–167. DOI: /10.1016/1010-6030(94)04030-6

[71] Zhang T. Photooxidative N-demethylation of methylene blue in aqueous TiO_2 dispersions under UV irradiation. Journal of Photochemistry and Photobiology A: Chemistry. 140 (2001) 163–172. DOI: 10.1016/S1010-6030(01)00398-7

[72] Taffarel S, Lansarin M, Moro C. Styrene photocatalytic degradation reaction kinetics. Brazilian Chemical. Society. 22(10) (2011) 1872–1879. DOI: 10.1590/S0103-5053201100 1000007

[73] Hu X, Zhu Q, Gu Z, Zhang N, Liu N, Stanislaus M, Li D, Yang Y. Wastewater treatment by sonophotocatalysis using PEG modified TiO_2 film in a circular photocatalytic-ultrasonic system. Ultrasonics Sonochemistry. 36 (2017) 301–308. DOI: 10.1016/j.ultsonch.2016.12.008

[74] Zhang Q, Ye S, Chen X, Song X, Li L, Huang X. Photocatalytic degradation of ethylene using titanium dioxide nanotube arrays with Ag and reduced graphene oxide irradiated by γ-ray radiolysis. Applied Catalysis B: Environmental. 203 (2017) 673–683. DOI: 10.1016/j.apcatb.2016.10.034

[75] Jawad A, Mubarak N, Ishak M, Ismail K, Nawawi W. Kinetics of photocatalytic decolourization of cationic dye using porous TiO_2 film. Journal of Taibah University for Science. 10 (2016) 352–362. DOI: 10.1016/j.jtusci.2015.03.007

[76] J. Cho H, Eom Y, Jeon S, Lee T. A pilot-scale TiO_2 photocatalytic system for removing gas-phase elemental mercury at Hg-emitting facilities. Journal of Industrial and Engineering Chemistry. 19 (2013) 144–149. 10.1016/j.jiec.2012.07.016

[77] Parra S, Malato S, Pulgarin C. New integrated photocatalytic-biological flow system using supported TiO_2 and fixed bacteria for the mineralization of isoproturon. Applied Catalysis B: Environmental. 36 (2002) 131–144. DOI: 10.1016/S0926-3373(01)00293-4

[78] Chong M, Jin B, Chow C, Saint C. Recent developments in photocatalytic water treatment technology: a review. Water research. 44 (2010) 2997–3027. DOI: 10.1016/j.watres.2010.02.039

[79] Spasiano D, Marotta R, Malato S, Fernandez-Ibañnez P, Di Somma I. Solar photocatalysis: materials, reactors, some commercial, and pre-industrialized applications. A comprehensive approach. Applied Catalysis B: Environmental. 170 (2015) 90–123. DOI: 10.1016/j.apcatb.2014.12.050

[80] Gagliardi M. Photocatalysts: technologies and global markets. BBC Reports. Wellesley, USA, 2015

[81] Freiman S. Global Road For Ceramic and Glass Technology. Wiley, Hoboken, New Jersey, 2007. ISBN100-470-10491-0. pp. 673–675.